T0203094

Graduate Texts in Mathematics 77

Erich Hecke

Lectures on the Theory
of Algebraic Numbers

Translated by George U. Brauer and Jay R. Goldman
with the assistance of R. Kotzen

Springer-Verlag
New York   Heidelberg   Berlin

Erich Hecke

formerly of
Department of Mathematics
Universität Hamburg
Hamburg
Federal Republic of Germany

Translators:

George U. Brauer
Jay R. Goldman

School of Mathematics
University of Minnesota
Minneapolis, MN 55455
USA

AMS Classification (1980) 12-01

Library of Congress Cataloging in Publication Data

Hecke, Erich, 1887–1947.
   Lectures on the theory of algebraic numbers.
   (Graduate texts in mathematics; 77)
   Translation of: Vorlesung über die Theorie der algebraischen Zahlen.
   Bibliography: p.
   1. Algebraic number theory.   I. Title.   II. Series.
QA247.H3713   512′.74   81-894
                        AACR2

Title of the German Original Edition: Vorlesung über die Theorie der algebraischen
Zahlen. Akademische Verlagsgesellschaft, Leipzig, 1923.

Printed in the United States of America.

9 8 7 6 5 4 3 2 1

ISBN 0-387-90595-2   Springer-Verlag   New York   Heidelberg   Berlin
ISBN 3-540-90595-2   Springer-Verlag   Berlin   Heidelberg   New York

Translators' Preface

. . . if one wants to make progress in mathematics one should study the masters not the pupils.

N. H. Abel

Hecke was certainly one of the masters, and in fact, the study of Hecke L-series and Hecke operators has permanently embedded his name in the fabric of number theory. It is a rare occurrence when a master writes a basic book, and Hecke's *Lectures on the Theory of Algebraic Numbers* has become a classic. To quote another master, André Weil: "To improve upon Hecke, in a treatment along classical lines of the theory of algebraic numbers, would be a futile and impossible task."

We have tried to remain as close as possible to the original text in preserving Hecke's rich, informal style of exposition. In a very few instances we have substituted modern terminology for Hecke's, e.g., "torsion free group" for "pure group."

One problem for a student is the lack of exercises in the book. However, given the large number of texts available in algebraic number theory, this is not a serious drawback. In particular we recommend *Number Fields* by D. A. Marcus (Springer-Verlag) as a particularly rich source.

We would like to thank James M. Vaughn Jr. and the Vaughn Foundation Fund for their encouragement and generous support of Jay R. Goldman without which this translation would never have appeared.

Minneapolis
July 1981

George U. Brauer
Jay R. Goldman

v

Author's Preface to the German Original Edition

The present book, which arose from lectures which I have given on various occasions in Basel, Göttingen, and Hamburg, has as its goal to introduce the reader without any knowledge of number theory to an understanding of problems which currently form the summit of the theory of algebraic number fields. The first seven chapters contain essentially nothing new; as far as form is concerned, I have drawn conclusions from the development of mathematics, in particular from that of arithmetic, and have used the notation and methods of group theory to develop the necessary theorems about finite and infinite Abelian groups. This yields considerable formal and conceptual simplifications. Nonetheless there will perhaps be some items of interest for the person who is familar with the theory, such as the proof of the fundamental theorem on Abelian groups (§8), the theory of relative discriminants (§36, 38) which I deal with by the original construction of Dedekind, and the determination of the class number without the zeta-function (§50).

The last chapter, Chapter VIII, leads the reader to the summit of the modern theory. This chapter yields a new proof of the most general quadratic reciprocity law in arbitrary algebraic number fields, which by using the theta function, is substantially shorter than those proofs known until now. Even if this method is not capable of generalization it has the advantage of giving the beginner an overview of the new kinds of concepts which appear in algebraic number fields, and from this, of making the higher reciprocity theorems more easily accessible. The book closes with the proof of the existence of the class field of relative degree two, which is obtained here as a consequence of the reciprocity theorem.

As prerequisites only the elements of differential and integral calculus and of algebra, and for the last chapter the elements of complex function theory, will be assumed.

I am indebted for help with corrections and various suggestions to Messrs. Behnke, Hamburger, and Ostrowski. The publisher has held the plan of the book, conceived already before the war, with perserverance which is worthy of thanks, and despite the most unfavorable circumstances, has made possible the appearance of the book. My particular thanks are due to him for his pains.

Mathematical Seminar   Erich Hecke
Hamburg
March 1923

Contents

CHAPTER I

Elements of Rational Number Theory

§1 Divisibility, Greatest Common Divisors, Modules, Prime Numbers, and the Fundamental Theorem of Number Theory

For the time being the objects of arithmetic are the whole numbers, $0, \pm 1, \pm 2, \ldots$ which can be combined by addition, subtraction, multiplication and division (not always) to form integers. Higher arithmetic uses methods of investigation analogous to those of real or complex numbers. Moreover it also uses analytic methods which belong to other areas of mathematics, such as infinitesimal calculus and complex function theory, in the derivation of its theorems. Since these will also be discussed in the latter part of this book, we will assume as known the totality of complex numbers, a number domain, in which the four types of operations (except division by 0) can be carried out unrestrictedly. The complex domain is usually developed more precisely in the elements of algebra or of differential calculus. In this domain the number 1 is distinguished as the one which satisfies the equation

$$1 \cdot a = a$$

for each number a. All successive integers are obtained by the process of addition and subtraction from the number 1, and if the process of division is then carried out the set of *rational numbers* is obtained as the totality of quotients of integers. Later, from §21 on, the concept of "integer" will be subjected to an essential extension.

In this introductory part the basic facts of rational arithmetic will be presented, briefly, as far as they concern divisibility properties of integers.

1

While, from two rational integers a, b, integers are always obtained in the form $a + b, a - b$, and $a \cdot b$, a/b need not be an integer. If a/b is an integer, a special property of a and b is present, which we wish to express by the symbol $b\,|\,a$, in words: b *divides* a, or b goes evenly into a, or b is a *divisor* (factor) of a, or a is a multiple of b. Each integer $a\,(\neq 0)$ has the trivial divisors $\pm a$, ± 1; a and $-a$ have the same divisors; the only numbers which divide every number are the two "units" 1 and -1. An integer a, different from zero, always has only finitely many divisors, as these cannot be larger in absolute value than $|a|$; on the other hand every non-zero integer divides 0.

If $b \neq 0$ and integral, then, among the multiples of b which are not larger than a given integer a there is exactly one largest multiple, say qb, and therefore $a - qb = r$ is a non-negative integer which is less than $|b|$. This integer r, uniquely determined by a and b by the requirement

$$a = qb + r, \qquad q \text{ integral}, \ 0 \le r < |b|$$

is called the remainder of the division of a by b, or the *remainder of a modulo b*. The statement $b\,|\,a$ is thus equivalent to $r = 0$.

If we now direct our attention to the common divisors c of two integers a, b which satisfies $c\,|\,a$ and $c\,|\,b$, then there is, to begin with, a uniquely determined *greatest common divisor* (abbreviated GCD); we denote it by $(a, b) = d$. According to this definition we always have $d \ge 1$. In order to find properties of this number (a, b) we consider that we always have $d\,|\,ax + by$ for all integers x, y. If we now consider the set of all numbers $L(x, y) = ax + by$, where x, y runs through all the integers, then d is obviously also the GCD of all $L(x, y)$; for it divides all $L(x, y)$ and there is no larger number with this property, since there can be no larger number which divides both $a = L(1, 0)$ and $b = L(0, 1)$. Among the positive integers $L(x, y)$, let $d_0 = L(x_0, y_0)$ be the smallest; thus from

$$L(x, y) > 0 \quad \text{it immediately follows that} \quad L(x, y) \ge d_0. \qquad (1)$$

We now show that each $n = L(x, y)$ is a multiple of d_0 and that $d = d_0$. Let the remainder r of $n \bmod d_0$ be determined by

$$r = n - q \cdot d_0 = L(x - qx_0, y - qy_0).$$

Here we have $0 \le r < d_0$; however by (1) it would follow from $r > 0$ that $r \ge d_0$. Thus we can have only $r = 0$, i.e., $n = qd_0$. Accordingly the numbers $L(x, y)$ are identical with the multiples of d_0 for each multiple $qd_0 = L(qx_0, qy_0)$ also appears among the $L(x, y)$. Consequently d_0 is likewise the GCD of all $L(x, y)$, hence it is identical with d. In particular this yields:

Theorem 1. *If $(a, b) = d$, then the equation*

$$n = ax + by$$

is solvable with integers x, y if and only if $d\,|\,n$.

Moreover it follows from this that every common divisor of a and b divides the GCD of a, b.

To ascertain the GCD one uses, as is well-known, a process which goes back to Euclid, the so-called Euclidean algorithm. The main point of this algorithm consists of reducing the calculation of (a, b) to the calculation of the GCD of two smaller numbers. It follows from $a = qb + r$ that the common divisors of a and b are identical with those of b and r, hence we have $(a, b) = (b, r)$. Assume $a > 0$, $b > 0$ for the sake of convenience, set $a = a_1, b = a_2$ because of symmetry, and then let the remainder of a_1 mod a_2 be a_3. In general let

a_{i+2} be the remainder of a_i mod a_{i+1} for $i = 1, 2, \ldots$

as long as the remainder can be determined, that is, $a_{i+1} > 0$, and indeed let

$$a_i = q_i a_{i+1} + a_{i+2}, \qquad 0 \le a_{i+2} < a_{i+1}.$$

Since, according to this procedure, the a_i form a monotone decreasing sequence of integers for $i \ge 2$, the process must reach an end after finitely many steps, which will occur when the remainder becomes zero. Suppose $a_{i+2} = 0$. Since

$$(a_1, a_2) = (a_2, a_3) = \cdots (a_i, a_{i+1})$$
$$= (a_{i+1}, a_{i+2}) = (a_{k+1}, a_{k+2})$$
$$= (a_{k+1}, 0) = a_{k+1},$$

the last non-vanishing remainder a_{k+1} is the GCD sought.

In the proof of Theorem 1 we have used only one property of the set of numbers $L(x, y)$, namely the property that this set is a module. Here we define:

Definition. A system S of integers is a *module* if it contains at least one number different from 0 and if along with m and n, $m + n$ and $m - n$ also always belong to S.

Thus if m belongs to S, then $m + m = 2m$, $m + 2m = 3m \cdots$ belong to S; moreover $m - m = 0$, $m - 2m = -m$, $m - 3m = -2m \cdots$ belong to S. Hence, in general, mx belongs to S for each integer x provided m belongs to S, and consequently $mx + ny$ also belongs to S for integers x, y if this holds for m, n.

We can prove the following very general theorem about modules with the help of the proof of Theorem 1.

Theorem 2. *The numbers in a module S are identical with the multiples of certain number d. d is determined by S up to the factor ± 1.*

For the proof we consider that S contains positive numbers in any case. Let d be the smallest positive number occurring in S. If n belongs to S, then by what has gone before, $n - qd$ also belongs to S for each integer q, in particular so must the remainder of n mod d, which is $<d$ but ≥ 0, and thus must $=0$. Consequently each n from S is a multiple of d and since d belongs to S so do all multiples of d. Let d' be a second number which also has the property: the numbers of S are identical with the multiples of d'—then d must be a multiple of d' and conversely, that is, $d' = \pm d$.

If in an arbitrary linear form $a_1 x_1 + a_2 x_2 + \cdots + a_n x_n$ with integral coefficients one lets the $x_1, \ldots, x_n$ run through all integers, then the range of values defined in this way is obviously a module. Hence in particular we have

Theorem 3. *The range of values of an arbitrary linear form in n variables with integral coefficients, not all vanishing, is identical with the range of values of a certain form of one variable $d \cdot x$. Here d is the GCD of the coefficients of the original form.*

In order that the equation (a so-called Diophantine equation)

$$k = a_1 x_1 + a_2 x_2 + \cdots + a_n x_n$$

be solvable in integers $x_1, \ldots, x_n$, it is necessary and sufficient that the GCD of $a_1, \ldots, a_n$ divides k.

If $(a, b) = 1$, we call a and b *coprime* or *relatively prime*. By Theorem 1, in order that $(a, b) = 1$, the solvability of

$$ax + by = 1$$

in integers x, y is necessary and sufficient.

As the most important rule of calculation with the symbol (a, b) we state:

Theorem 4. *For every three integers a, b, c, where $c > 0$*

$$(a, b)c = (ac, bc). \tag{2}$$

In fact if $(a, b) = d$, then the equation $acx + bcy = cd$ follows by Theorem 1 from the known solvable equation $ax + by = d$; consequently cd is a multiple of (ac, bc), again by Theorem 1. On the other hand, however, cd is a common divisor of ac, bc and hence must be equal to (ac, bc).

In addition we note the concept of *least common multiple* of two numbers a and b. This is the smallest positive number v which is divisible by a as well as by b. For this number we have

$$v = \frac{|a \cdot b|}{d}, \quad \text{where } (a, b) = d. \tag{3}$$

For by (2),

$$\left(\frac{a}{d}, \frac{b}{d}\right) = 1, \qquad v = \left(\frac{a}{d}v, \frac{b}{d}v\right).$$

However ab/d is a common divisor of $(a/d)v$ and $(b/d)v$ and thus it divides v, that is, $v \geq |ab|/d$; on the other hand, ab/d is a number which is divisible by a as well as by b, and consequently it has absolute value $\geq v$. Hence ab/d can only be $= \pm v$.

Since the numbers divisible by a and by b form a module and v is the smallest positive number occurring in it, every number divisible by a and by b must be a multiple of v.

We now turn to the *multiplicative decomposition* of a number a. If, except for the trivial decomposition into integral factors, in which one factor is ± 1 and the other is $\pm a$, there is no other, we call a a *prime number* (or *prime*). Such numbers exist, e.g., $\pm 2, \pm 3, \pm 5, \ldots$. We do not wish to count the units ± 1 as prime numbers. If, for the sake of simplicity, we restrict ourselves to the decomposition of positive numbers a into positive factors we see first of all that every $a > 1$ is divisible by at least one positive prime number since the smallest positive factor of a, which is > 1, obviously can only be a prime. Now we split off a prime number p_1 from the number a by the decomposition $a = p_1 a_1$, if $a_1 > 1$ we again split off another prime p_2 from a_1 by $a_1 = p_2 a_2$, and so on. Since the $a_1, a_2, \ldots$ form a decreasing sequence of positive integers we must arrive at an end of the process after finitely many steps, that is, some a_k must be $= 1$. With this, a is represented as a product of primes $p_1 \cdot p_2 \cdots p_k$. Hence the primes are building blocks from which each integer can be built up by multiplication. We now have

Theorem 5. (Fundamental Theorem of Arithmetic). *Each positive number > 1 can be represented in one—and except for the order of the factors—in only one way as a product of primes.*

For this it is sufficient to show that a prime p can divide a product of two numbers $a \cdot b$ only if it divides at least one factor. But this follows from Theorem 4. Namely, if the prime number does not divide a, then as a prime it cannot have any factor at all in common with a, hence $(a, p) = 1$. Then for each positive integer b, we have by Theorem 4

$$(ab, pb) = b.$$

Now if $p \mid ab$, then we must also have $p \mid b$, i.e., the prime p divides the other factor b of the product ab. This theorem carries over at once to a product of several factors.

In order to prove Theorem 5 we consider two representations of a positive number a as a product of powers of distinct positive primes p_i, q_i,

$$p_1^{a_1} p_2^{a_2} \cdots p_r^{a_r} = q_1^{b_1} q_2^{b_2} \cdots q_k^{b_k}.$$

By what was just proved each prime q divides at least one prime factor of the left-hand side and is thus identical with some p_k. Thus the $q_1, \ldots, q_k$ agree with $p_1, \ldots, p_r$, except possibly for order; hence we also have $k = r$. We choose the numbering so that $p_i = q_i$. Now if corresponding exponents were not equal, say $a_1 > b_1$, then after division of the equation by $q_1^{b_1}$ it follows that the left-hand side still has the factor $p_1 = q_1$, but the right-hand side no longer has this factor. Hence $a_1 = b_1$ and in general $a_i = b_i$.

With this theorem about the *unique* decomposition of each number into prime factors we have a substantially different method of deciding the questions treated above, e.g., whether a given number b divides another number a, how (a, b) or the least common multiple of a and b is found, etc. Specifically, if we think of a and b as decomposed into their prime factors $p_1, \ldots, p_r$,

$$a = p_1^{a_1} p_2^{a_2} \cdots p_r^{a_r}$$
$$b = p_1^{b_1} p_2^{b_2} \cdots p_r^{b_r},$$

where zero is also allowed for the exponents a_i, b_i, then obviously $b \mid a$ holds if and only if we always have $a_i \geq b_i$. Moreover we have

$$(a, b) = p_1^{d_1} p_2^{d_2} \cdots p_r^{d_r}, \qquad d_i = \min(a_i, b_i), i = 1, 2, \ldots, r,$$
$$v = p_1^{c_1} p_2^{c_2} \cdots p_r^{c_r}, \qquad c_i = \max(a_i, b_i), i = 1, 2, \ldots, r.$$

The existence of infinitely many primes follows immediately from the fact that

$$z = p_1 \cdot p_2 \cdots p_n + 1$$

is a number which is not divisible by any of the primes $p_1, \ldots, p_n$. Hence z is divisible by at least one prime number distinct from $p_1, \ldots, p_n$ and consequently if there are n primes, then there are $n + 1$ primes.

§2 Congruences and Residue Classes

By the preceding section, an integer $n \neq 0$ immediately determines a distribution of all integers according to the remainder which they yield mod n. We assign two integers a and b which have the same remainder mod n to the same *residue class mod n* or more simply, the same class mod n, and write

$$a \equiv b \ (\text{mod } n), \qquad (a \text{ is congruent to } b \text{ modulo } n),$$

which is equivalent to $n \mid a - b$. If a is not congruent to b relative to the modulus n we write $a \not\equiv b \ (\text{mod } n)$. $a \equiv 0 \ (\text{mod } n)$ asserts that a is divisible by n. Each number is called a *representative* of its class. Since the different remainders mod n are the numbers $0, 1, 2, \ldots, |n| - 1$, the number of different residue classes mod n is $|n|$. The following easily verified rules hold for calculations with congruences: if a, b, c, d, n are integers, $n \neq 0$, then we have:

(i) $a \equiv a \ (\text{mod } n)$.
(ii) If $a \equiv b \ (\text{mod } n)$, then $b \equiv a \ (\text{mod } n)$.
(iii) If $a \equiv b \ (\text{mod } n)$ and $b \equiv c \ (\text{mod } n)$, then $a \equiv c \ (\text{mod } n)$.
(iv) If $a \equiv b \ (\text{mod } n)$ and $c \equiv d \ (\text{mod } n)$, then $a \pm c \equiv b \pm d \ (\text{mod } n)$.
(v) If $a \equiv b \ (\text{mod } n)$, then $ac \equiv bc \ (\text{mod } n)$.

In general from $a \equiv b \pmod n$ and $c \equiv d \pmod n$ it follows that $ac \equiv bd \pmod n$. In particular we have $a^k \equiv b^k \pmod n$ for each positive integer k whenever $a \equiv b \pmod n$. By repeated application of (iv) and (v) we obtain: if $a \equiv b \pmod n$, then $f(a) \equiv f(b) \pmod n$ when $f(x)$ is an integral rational function of x (polynomial in x) with integral coefficients.

Hence, to put it briefly, we can calculate with congruences of the same modulus in exactly the same way as with equations as far as the integral rational operations (addition, subtraction, multiplication) are concerned. With division it is different. If $ca \equiv cb \pmod n$, it does not follow that $a \equiv b \pmod n$, for the hypothesis means $n | c(a - b)$. Now if $(n, c) = d$, we further have

$$\left(\frac{n}{d}, \frac{c}{d} \right) = 1, \qquad \frac{n}{d} \Big| \frac{c}{d}(a - b);$$

hence by Theorem 4,

$$\frac{n}{d} \Big| a - b, \quad \text{i.e.,} \quad a \equiv b \left(\text{mod} \frac{n}{d} \right).$$

For example: It does not follow from $5 \cdot 4 \equiv 5 \cdot 1 \pmod{15}$ that $4 \equiv 1 \pmod{15}$, but rather only $\text{mod}(15/5) = 3$. Hence we have

Theorem 6. *If* $ca \equiv cb \pmod n$, *then*

$$a \equiv b \left(\text{mod} \frac{n}{d} \right), \quad \text{where } (c, n) = d,$$

and conversely.

In connection with this there is the fact:

A product of two integers may be congruent to zero mod n although neither of the factors has this property.

For example $2 \cdot 3 \equiv 0 \pmod 6$ although neither 2 nor 3 is $\equiv 0 \pmod 6$. Concerning the connection between congruences relative to *different moduli* we see directly from the definition: if a congruence holds mod n, then it also holds modulo each factor of n, in particular also modulo $-n$. Furthermore, if

$$a \equiv b \pmod{n_1} \quad \text{and} \quad a \equiv b \pmod{n_2},$$

then

$$a \equiv b \pmod v,$$

where v is the least common multiple of n_1 and n_2.

Since the residue classes modulo n and the residue classes modulo $-n$ coincide, it is sufficient to investigate the residue classes modulo a positive n.

A system of n integers which contains exactly one representative from each residue class mod n will be called *a complete system of residues mod n.*

Since a complete system of residues mod n consists of $|n|$ distinct numbers, $|n|$ incongruent numbers modulo n always form a complete system of residues mod n, e.g., the numbers $0, 1, 2, \ldots, |n| - 1$. More generally

Theorem 7. *If $x_1, x_2, \ldots, x_n$ forms a complete system of residues mod $n (n > 0)$, then $ax_1 + b, \ldots, ax_n + b$ is also such a system, as long as a and b are integers and $(a, n) = 1$.*

For by Theorem 6 the n numbers $ax_i + b$ $(i = 1, 2, \ldots, n)$ are likewise incongruent numbers modulo n.

A representation of a residue system with respect to a composite modulus, which is often useful, is given by the following:

Theorem 8. *If $a_1, a_2, \ldots, a_n$ are pairwise relatively prime integers, then a complete residue system mod A, where $A = a_1 a_2 \cdots a_n$, is obtained in the form*

$$L(x_1, \ldots, x_n) = \frac{A}{a_1} c_1 x_1 + \frac{A}{a_2} c_2 x_2 + \cdots + \frac{A}{a_n} c_n x_n$$

if the x_i independently run through a complete residue system mod a_i $(i = 1, 2, \ldots, n)$. Here the c_i may be arbitrary integers relatively prime to a_i.

The number of these L values is $|A|$ and they are incongruent mod A since from the congruence mod A

$$L(x_1, \ldots, x_n) = L(x'_1, \ldots, x'_n) \pmod{A}$$

the same congruence follows modulo each a_i. Since

$$\frac{A}{a_k} \equiv 0 \pmod{a_i} \quad \text{for } k \neq i,$$

we have for $i = 1, 2, \ldots, n$

$$c_i \frac{A}{a_i} x_i \equiv c_i \frac{A}{a_i} x'_i \pmod{a_i}.$$

Moreover by Theorem 6, since $(c_i, a_i) = 1$ and $(A/a_i, a_i) = 1$, we get $x_i \equiv x'_i$ $\pmod{a_i}$. Two numbers L, as they occur in Theorem 8, are thus always incongruent mod A.

In exactly the same way one can prove that one obtains a complete system of residues mod $a \cdot b$ if we let the quantity x in $x + by$ run through a complete system of residues mod b, and independently let the quantity y run through a complete system of residues mod a.

A characteristic of each residue class mod n is the greatest common divisor which an arbitrary number from the class has in common with n. This really depends only on the class, since if $a \equiv b \pmod{n}$, then $a = b + qn$ with integral q, and hence each common factor of a and n is also a common factor of b and n and conversely. Thus it makes sense to speak of the *GCD of a residue class* mod n *and* n.

In particular we ask for *the number of residue classes* mod n *which are relatively prime to* n. This number is the Euler function $\varphi(n)$. To begin with, $\varphi(n)$ is easily determined for the case $n = p^k$, a power of a positive prime p, as $\varphi(p^k)$ is the number of those numbers among $1, \ldots, p^k$ which are not divisible by p. Among these the number divisible by p is the number of multiples of p between 1 and p^k, hence p^{k-1}, and thus

$$\varphi(p^k) = p^k - p^{k-1} = p^k\left(1 - \frac{1}{p}\right).$$

In order to determine $\varphi(n)$ for composite n we now prove the

Lemma. $\varphi(ab) = \varphi(a)\varphi(b)$ if $(a, b) = 1$.

One obtains, by Theorem 8, a complete system of residues mod ab in the form $ax + by$, if x runs through a complete system of residues mod b, and y runs through a complete system of residues mod a. However, in order that such a number be relatively prime to ab, i.e., relatively prime to a as well as to b, it is necessary and sufficient that $(ax, b) = 1$ and $(by, a) = 1$, i.e., since $(a, b) = 1$: $(x, b) = 1$ and $(y, a) = 1$. Hence one obtains the numbers $ax + by$ relatively prime to ab if we let x run through the residue classes which are relatively prime to b mod b, and y run through those relatively prime to a mod a; hence the lemma is proved. By repeated application, if n is decomposed into its positive prime factors, we obtain:

for $n = p_1^{a_1} p_2^{a_2} \cdots p_r^{a_r}$,

$$\varphi(n) = \varphi(p_1^{a_1}) \cdot \varphi(p_2^{a_2}) \cdots \varphi(p_r^{a_r}) = n \prod_{p|n}\left(1 - \frac{1}{p}\right). \tag{4}$$

In the product p must run through all positive primes which divide n.

The complete system of residue classes mod n relatively prime to n is called a *reduced system of residues* mod n. It contains $\varphi(n)$ classes, and a system of one representative from each class is called a *complete reduced system of residues* mod n. As in Theorem 7 one proves:

If $x_1, x_2, \ldots, x_h$ *is a complete reduced system of residues* mod n, *then* $ax_1, ax_2, \ldots, ax_h$ *is also such a system, provided* $(a, n) = 1$.

From this we obtain a highly important fact about each number a relatively prime to n. Since each of the numbers $ax_1, \ldots, ax_h$ is congruent mod n to one of the numbers $x_1, \ldots, x_h$ by the above, then the product of the numbers $ax_1, \ldots, ax_h$ is congruent to the product $x_1 \cdots x_k$, that is,

$$a^h x_1 x_2 \cdots x_h \equiv x_1 x_2 \cdots x_h \pmod{n}$$

and since each x is relatively prime to n, we obtain

$$a^h \equiv 1 \ (\text{mod} \ n),$$

and with this, since $h = \varphi(n)$,

Theorem 9. (Fermat's Theorem). *For each number a relatively prime to n*

$$a^{\varphi(n)} \equiv 1 \ (\text{mod} \ n).$$

In particular if n is a prime $p \ (>0)$, then $\varphi(p) = p - 1$, and after multiplication by a we have, for each integer a, the congruence

$$a^p \equiv a \ (\text{mod} \ p). \tag{5}$$

The significance of this theorem and the kernel of its proof really becomes understandable in Chapter II when we introduce the general group concept into these investigations. The theorem contains a statement about the solutions of the congruence $x^p - x \equiv 0 \ (\text{mod} \ p)$ and forms the basis for the theory of higher congruences.

§3 Integral Polynomials, Functional Congruences, and Divisibility mod p

If we let ourselves be guided in the further development of the ideas presented up to now by the analogies with algebra, then the next goal is the investigation of polynomials $f(x)$ with integral coefficients with regard to their behavior relative to a modulus n, and then the question of solvability of a congruence $f(x) \equiv 0 \ (\text{mod} \ n)$ in integers x.

By an *integral polynomial* $f(x) = c_0 + c_1 x + \cdots + c_k x^k$ we understand such a polynomial, where $c_0, c_1, \ldots, c_k$ are integers. Two integral polynomials $f(x)$ and $g(x)$, where $g(x) = a_0 + a_1 x + \cdots + a_k x^k$, are *said to be congruent modulo n* or

$$f(x) \equiv g(x) \ (\text{mod} \ n),$$

if

$$c_i \equiv a_i \ (\text{mod} \ n) \quad \text{for} \ i = 0, 1, 2, \ldots, k.$$

(For constants, i.e., polynomials of degree 0, this concept of congruence agrees with the one used up to now.) Thus this definition concerns the behavior of $f(x)$ and $g(x)$ identically in the variable x, not only for special values of x. For this reason even if for all integer values x_0 we have

$$f(x_0) \equiv g(x_0) \ (\text{mod} \ n),$$

the polynomials $f(x)$ and $g(x)$ need not be congruent as the example

$$x^p \equiv x \ (\text{mod} \ p)$$

(for p a prime) shows. By Fermat's theorem this is a correct *numerical congruence* for each integer x, but the polynomials x^p and x are not congruent to each other.

For these functional congruences exactly the same rules of calculation (i)–(v) in §2 hold as for numerical congruences, and the proof is likewise simple; for this reason we will not go into it.

Definition. For two integral polynomials $f(x)$ and $g(x)$, $f(x)$ is said to be *divisible* by $g(x)$ mod n if there is an integral polynomial $g_1(x)$ such that

$$f(x) \equiv g(x)g_1(x) \;(\text{mod } n).$$

If moreover a is an integer such that

$$f(a) \equiv 0 \;(\text{mod } n),$$

then a is called a *root of $f(x)$ mod n*.

If a is a root of $f(x)$ mod n and $a \equiv b \;(\text{mod } n)$, then obviously b is also a root of $f(x)$ mod n.

The connection between roots mod n and divisibility mod n is shown by the following fact:

Theorem 10. *If a is a root of the integral polynomial $f(x)$ mod n, then $f(x)$ is divisible by $x - a$ mod n and conversely.*

Since $f(a) \equiv 0 \;(\text{mod } n)$ we have

$$f(x) \equiv f(x) - f(a) \;(\text{mod } n).$$

However $(f(x) - f(a))/(x - a)$ is an integral polynomial, $g(x)$, since for each positive m

$$\frac{x^m - a^m}{x - a} = x^{m-1} + ax^{m-2} + a^2x^{m-3} + \cdots + a^{m-2}x + a^{m-1}$$

is an integral polynomial and $f(x) - f(a)$ is an integral combination of expressions $x^m - a^m$. Hence

$$f(x) \equiv (x - a)g(x) \;(\text{mod } n).$$

The converse is trivial.

However if f, g, g_1 are integral polynomials and

$$f(x) \equiv g(x)g_1(x) \;(\text{mod } n),$$

then a root a of $f(x)$ mod n need *not* be a root of $g(x)$ or $g_1(x)$ mod n, as one might conjecture by analogy with algebra. For example, we have

$$x^2 \equiv (x - 2)(x - 2) \;(\text{mod } 4).$$

4 is a root of x^2 mod 4 but not a root of $x - 2$ mod 4. Only for prime moduli do we have

Theorem 11. *If $f(x) \equiv g(x)g_1(x)$ (mod p), where p is a prime, then each root of $f(x)$ mod p is a root of at least one of the two polynomials $g(x), g_1(x)$ mod p.*

If for the integer a, $f(a) \equiv 0$ (mod p), then

$$g(a) \cdot g_1(a) \equiv f(a) \equiv 0 \text{ (mod } p).$$

If the prime p divides the product $g(a) \cdot g_1(a)$, then it divides one of the two factors.

Theorem 12. *An integral polynomial $f(x)$ of degree k has no more than k incongruent roots modulo a prime p, unless $f(x) \equiv 0$ (mod p), in which case all coefficients are divisible by p.*

The theorem is true for the polynomials of degree 0, the constants. For if $f(x) = c_0$ is independent of x, then $f(x) \equiv 0$ (mod p) has either 0 solutions— when p does not divide c_0—or it has more than 0 solutions—namely every integer if c_0 is divisible by p, that is, the polynomial $f(x) \equiv 0$ (mod p). Suppose now that our theorem has been proved for polynomials of degree $\leq k - 1$. Then we show it is correct for polynomials of degree k. If a is a root of $f(x)$ (mod p), then by the proof of Theorem 10 we may set

$$f(x) \equiv (x - a)f_1(x) \text{ (mod } p),$$

where $f_1(x)$ is of degree at most $k - 1$. By Theorem 11 each root of $f(x)$ mod p is either a root of $f_1(x)$ or a root of $x - a$ mod p (or both). However $x - a \equiv 0$ (mod p) has only one incongruent solution and $f_1(x) \equiv 0$ (mod p) has either at most $k - 1$ incongruent solutions, in which case $f(x)$ has at most $k - 1 + 1 = k$ solutions, or the polynomial $f_1(x) \equiv 0$ (mod p). In the latter case the polynomial $f(x)$ is $\equiv 0$ (mod p). Thus the theorem is proved by complete induction.

The theorem is not correct for composite moduli, as the example $x^2 - 1$ modulo 8 shows. This second-degree polynomial has four incongruent roots mod 8, namely $x = 1, 3, 5, 7$.

Theorem 13. *If for two integral polynomials $f(x)$ and $g(x)$*

$$f(x) \cdot g(x) \equiv 0 \text{ (mod } p), \quad p \text{ a prime,}$$

then either $f(x) \equiv 0$ (mod p) or $g(x) \equiv 0$ (mod p) or both.

Suppose the theorem is false, i.e., neither $f(x)$ nor $g(x)$ is $\equiv 0$ (mod p). Then let all terms of $f(x)$ and $g(x)$ which are divisible by p be omitted and two nonvanishing polynomials $f_1(x), g_1(x)$ are obtained, all of whose coefficients are not divisible by p, while at the same time

$$f(x) \equiv f_1(x) \,(\text{mod } p),$$
$$g(x) \equiv g_1(x) \,(\text{mod } p);$$

it follows that

$$f_1(x)g_1(x) \equiv 0 \,(\text{mod } p).$$

The highest-degree term in $f_1(x)g_1(x)$ must thus be $\equiv 0 \,(\text{mod } p)$ on the one hand, on the other hand however it is equal to the product of the highest terms of $f_1(x)$ and $g_1(x)$. Since p is a prime and all terms of $f_1(x)$ and $g_1(x)$ are not divisible by p, the product of such terms is also not divisible by p. Consequently the hypothesis is false, and the theorem is proved.

Definition. An integral polynomial is called *primitive* if its coefficients are relatively prime, i.e., if for each prime p, $f(x) \not\equiv 0 \,(\text{mod } p)$.

Then Theorem 13 obviously allows the following formulation:

Theorem 13a (Theorem of Gauss). *The product of two primitive polynomials is again a primitive polynomial.*

§4 Congruences of the First Degree

The polynomials of degree 1 and their roots mod n can be dealt with easily. This leads to the theory of congruences with one or several unknowns.

Let the integers a, b, n $(n > 0)$ be given. What statements may be made about the solutions x, in integers, of

$$ax + b \equiv 0 \,(\text{mod } n)? \qquad (6)$$

Since all the numbers of a residue class appear at once as solutions, if there are any, we ask only for the incongruent solutions mod n. The answer is

Theorem 14. *The congruence (6) has exactly one solution mod n if $(a, n) = 1$.*

For by Theorem 7, $ax + b$ falls exactly once into the residue class 0 if x runs through a complete system of residues mod n.

If, however, $(a, n) = d$ and (6) is solvable, then the congruence is also true mod d and for b it yields the condition

$$b \equiv 0 \,(\text{mod } d).$$

Then by Theorem 6, (6) is equivalent to

$$\frac{a}{d}x + \frac{b}{d} \equiv 0 \left(\text{mod } \frac{n}{d}\right)$$

and this equation has, by Theorem 14, exactly one solution x_0 mod(n/d). All solutions of (6) are thus the numbers

$$x = x_0 + \frac{n}{d} y$$

with integral y and among these there are exactly d different ones mod n. They are obtained if y is allowed to run through a complete residue system mod d.

In the case $(a, n) = d > 1$, (6) is thus solvable if and only if $d|b$. Then the number of distinct solutions mod n is equal to d.

The congruence (6) is equivalent to an equation $ax + b = nz$, with z integral, i.e., its solution is equivalent to the Diophantine equation $ax - nz = -b$. Of course an application of Theorem 1 to this equation also leads to the above result. In particular, if $(a, n) = 1$, the congruence

$$aa' \equiv 1 \ (\text{mod } n)$$

always has exactly one solution a' determined mod n, and the solution of the more general congruence $ax + b \equiv 0 \ (\text{mod } n)$ is obtained, by multiplying by a', in the form

$$x \equiv -a'b \ (\text{mod } n).$$

Moreover by Theorem 9 we can take the number $a^{\varphi(n)-1}$ for a'.

We can consider several linear congruences, with one unknown x but relative to different moduli brought into the form

$$x \equiv a_1 \ (\text{mod } n_1), \qquad x \equiv a_2 \ (\text{mod } n_2), \qquad \ldots, \qquad x \equiv a_k \ (\text{mod } n_k). \quad (7)$$

If x and y are two numbers which satisfy this system, then $x - y$ is divisible by each n_i, hence also by the least common multiple v of $n_1, \ldots, n_k$, that is, $x \equiv y \ (\text{mod } v)$; conversely, if x is a solution of (7), and $x \equiv y \ (\text{mod } v)$, then y is also a solution of (7). Thus the solutions of (7), in case such a solution exists, are uniquely determined mod v. We are interested only in the most important case:

Theorem 15. *The k congruences (7) have exactly one solution determined* mod $n_1 n_2 \cdots n_k$ *if the moduli are pairwise relatively prime.*

For with Theorem 8 in mind let us set

$$x = \frac{v}{n_1} x_1 + \frac{v}{n_2} x_2 + \cdots + \frac{v}{n_k} x_k \qquad (v = n_1 n_2 \cdots n_k)$$

and determine the x_i from the congruences

$$\frac{v}{n_i} x_i \equiv a_i \ (\text{mod } n_i) \qquad (i = 1, 2, \ldots, k)$$

which is always possible by Theorem 14 on account of the hypothesis. An x obtained in this way is a solution of (7).

The investigation of the roots of polynomials of higher degree mod n then leads to congruences of higher degree in one unknown. In order to be able to attack the elements of this much more complicated theory we must think through the calculations with residue classes more precisely. We will encounter the essential relationships which were presented here several times, in the following sections, in still different forms, so that it is useful to extract the concept which is capable of so many different kinds of realizations and to make it the object of the investigation. This is the *group concept*. The following chapter is devoted to it.

CHAPTER II

Abelian Groups

§5 The General Group Concept and Calculation with Elements of a Group

Definition of a Group. A system S of elements $A, B, C \ldots$ is called a *group* if the following conditions are satisfied:

(i) *There is a prescription (rule of composition) given according to which from an element A and an element B, a unique element of S, say C, is always obtained.*

We express this relation symbolically

$$C = AB \quad \text{or} \quad (AB) = C.$$

This composition need not be commutative with respect to the elements A and B, that is, AB and BA may be different.

(ii) *The associative law is true for this composition:* For every three elements A, B, C,

$$A(BC) = (AB)C.$$

(iii) *If A, A', B are any three elements of S, then the following are to hold:*

$$\text{If } AB = A'B, \text{ then } A = A'.$$
$$\text{If } BA = BA', \text{ then } A = A'.$$

(iv) *For every two elements A, B, in S, there is an element X in S such that $AX = B$ and an element Y in S such that $YA = B$.*

If the system S contains only finitely many different elements—let their number be h—then (iv) is automatically satisfied as a consequence of (i) and (iii). To prove this, let X in AX run through the h different elements $X_1, \ldots X_h$

of the group. Then, by (i), AX always represents an element of the group, and by (iii) the h elements so obtained differ from one another. Consequently in this way each element of the group appears exactly once, in particular this holds for the element B, thus there is an X such that $AX = B$. In an analogous fashion one can deduce the second part of (iv).

If the group contains infinitely many different elements it is called an *infinite group*; otherwise it is called a *finite group of order h*, where h is the number of its elements.

The group property does not automatically belong to a system S but only with respect to a definite type of composition. With one type of composition S may be a group, while the same elements need not form a group under a different kind of composition.

Examples of groups are the system of all integers with composition by addition and the system of all positive numbers (integers and fractions) with composition by multiplication.

On the other hand the system of positive integers alone with composition by multiplication does not form a group, because requirement (iv) is not satisfied.

Furthermore if we consider two integers as equal whenever they are congruent relative to a definite modulus n, then the system of residues mod n with composition by addition forms a finite group of order n.

In exactly the same way the system of residues mod n, which are relatively prime to n, with composition by multiplication forms a group of order $\varphi(n)$. In all these examples the rule for composition is commutative. An example of a noncommutative group is the system of all rotations of a regular body, e.g., a die, about its midpoint which brings the body back to cover itself. Here the composition of two such rotations A and B, which is called AB, is to be that rotation which is obtained if first B and then A is performed.

The set of all permutations of n digits forms a finite group. Composition of the permutation A with B means the permutation AB which results from the performance of B followed by the performance of A.

If two groups $\mathfrak{G}_1$ and $\mathfrak{G}_2$ are given whose elements are to be denoted by the indices 1 and 2 respectively and if a well-defined invertible correspondence (denoted by $\rightarrow$) can be exhibited such that if $A_1 \rightarrow A_2$ and $B_1 \rightarrow B_2$, then $A_1 B_1 \rightarrow A_2 B_2$, then we call the two groups $\mathfrak{G}_1$ and $\mathfrak{G}_2$ *isomorphic*. Two isomorphic groups are only distinguished by the way in which the elements are denoted and the way in which the operation of combination is denoted. Hence all properties which are expressible strictly in terms of the group axioms (i)–(iv) and which hold for one group, are also satisfied by isomorphic groups. Thus isomorphic groups are not to be viewed as different for group-theoretic investigations.

Now let $\mathfrak{G}$ be a group. In the following its elements are to be denoted by capital Latin letters. The *product* of two elements of $\mathfrak{G}$ is defined by the existence of the composition according to (i). We now define the product of k elements by complete induction.

Definition. Suppose we have already defined the element $A_1 \cdot A_2 \cdots A_n$ of S which is to denote the product of n arbitrary elements $A_1, A_2, \ldots, A_n$. Then we define the product of $n + 1$ arbitrary elements $A_1, \ldots, A_{n+1}$ of $\mathfrak{G}$ by the equation

$$A_1 \cdot A_2 \cdots A_{n+1} = (A_1 \cdot A_2 \cdots A_n) \cdot A_{n+1}.$$

We now prove the

Lemma. *For an arbitrary integer* $k \geq 3$

$$A_1 \cdot A_2 \cdots A_k = A_1 \cdot (A_2 \cdot A_3 \cdots A_k).$$

For $k = 3$ this is obviously true, according to the associative law (ii). If however the theorem is true for $k = n$, then also for $k = n + 1$ as we have

$$A_1 \cdot A_2 \cdots A_{n+1} = (A_1 \cdot A_2 \cdots A_n) \cdot A_{n+1} = A_1 \cdot (A_2 \cdot A_3 \cdots A_n) \cdot A_{n+1}$$
$$= A_1 \cdot (A_2 \cdot A_3 \cdots A_{n+1}).$$

Thus the lemma is proved in general.

Moreover it follows for $1 < l < k$

$$(A_1 \cdot A_2 \cdots A_l)(A_{l+1} \cdots A_k) = [(A_1 \cdot A_2 \cdots A_{l-1}) \cdot A_l](A_{l+1} \cdots A_k)$$
$$= (A_1 \cdot A_2 \cdots A_{l-1})(A_l A_{l+1} \cdots A_k),$$

that is, the two inner parentheses may be shifted one place to the left in the original product without the result being changed. Consequently the inner parentheses can also be shifted as many places as desired to the right or to the left and thus

$$(A_1 A_2 \cdots A_l)(A_{l+1} \cdots A_k) = A_1 \cdot A_2 \cdots A_k$$

entirely independently of where the parentheses stand. Hence in a product of two expressions in parentheses, the parentheses may be omitted without the result being changed and one can easily prove the theorem for several expressions in parentheses by complete induction:

Theorem 16. *A product of* $r + 1$ *expressions in parentheses*

$$(A_1 \cdots A_n)(A_{n_1+1} \cdots A_{n_2}) \cdot (A_{n_2+1} \cdots A_{n_3}) \cdots (A_{n_r+1} \cdots A_k)$$

does not change if the parentheses are removed and is thus independent of the position in which the parentheses stand and therefore is equal to $A_1 \cdot A_2 \cdots A_k$.

Theorem 17. *In every group there is exactly one element* E *such that*

$$AE = EA = A$$

for every element of the group. E is called the unit (identity) element.

By (iv), to each A there is an E such that

$$AE = A, \quad \text{thus also} \quad YAE = YA.$$

If Y runs through all elements of the group, then, by (iv) this also holds for $YA = B$, hence $BE = B$ holds for each B, and E is independent of B.

Moreover there likewise exists an E' such that for each A

$$E'A = A.$$

For $A = E$ it follows that

$$E'E = E,$$

and from $AE = A$ it follows that for $A = E'$

$$E'E = E', \quad \text{hence} \quad E = E',$$

and the theorem is proved. This unit element may be omitted as a component of a product. Thus it plays the role of the number 1 in ordinary multiplication and it will also be denoted by 1.

Finally, again by (iv), for each A there is again an X and a Y such that

$$AX = E, \quad YA = E.$$

From this it follows by composition with Y that

$$YAX = YE, \quad \text{hence} \quad EX = YE, X = Y.$$

We call the element X uniquely defined in this way by A the *inverse element* (or *inverse*) of A and we denote it by A^{-1}. It is defined by

$$A \cdot A^{-1} = A^{-1} \cdot A = E.$$

We can now introduce the powers of an element A:

By A^m we understand a "product" of m elements, for positive m, each of which is $= A$. Then by Theorem 16 for positive integers m, n

$$A^{m+n} = A^m \cdot A^n = A^n \cdot A^m.$$

Furthermore by Theorem 16

$$A^m \cdot (A^{-1})^m = E,$$

that is, $(A^{-1})^m$ is the reciprocal of A^m, thus $= (A^m)^{-1}$. We denote this element by

$$A^{-m} = (A^{-1})^m = (A^m)^{-1}.$$

Finally for each A we set

$$A^0 = E.$$

Exactly as in elementary algebra one proves for these powers with arbitrary integral exponents:

Theorem 18. *For all integers m, n*

$$A^m \cdot A^n = A^n \cdot A^m = A^{m+n},$$

and

$$(A^m)^n = (A^n)^m = A^{nm}.$$

An equation between elements of a group in one unknown can be solved with the help of the inverse. By multiplication by A^{-1} it follows that

$$\text{if } AX = B, \text{ then } X = A^{-1}B$$

and

$$\text{if } YA = B, \text{ then } Y = BA^{-1}.$$

§6 Subgroups and Division of a Group by a Subgroup

Now a subset of the elements of $\mathfrak{G}$ may form a group under the same rule for composition. Such a group is called a *subgroup* of $\mathfrak{G}$. Let a fixed subgroup be denoted by $\mathfrak{U}$; let $U_1, U_2, \ldots$ be the different elements (finitely or infinitely many) belonging to $\mathfrak{U}$. If A is an arbitrary element of $\mathfrak{G}$ then let us denote the totality of elements AU_i $(i = 1, 2, \ldots)$ by

$$A\mathfrak{U} = (AU_1, AU_2, \ldots).$$

The elements of $\mathfrak{G}$ may now be arranged in a sequence of the form $A\mathfrak{U}$; These sequences are called *cosets*. We then have

Lemma. *If two cosets $A\mathfrak{U}$, $B\mathfrak{U}$ have one element in common, then they have all elements in common, thus they agree except for order.*

To prove this let $AU_a = BU_b$ be a common element. Then it follows that $B = AU_aU_b^{-1}$, hence

$$B\mathfrak{U} = (AU_aU_b^{-1}U_1, AU_aU_b^{-1}U_2, \ldots).$$

However $U_aU_b^{-1}U_i$ runs through all elements of $\mathfrak{U}$ for $i = 1, 2, \ldots$ because of the group property (iv) of $\mathfrak{U}$, hence in fact $A\mathfrak{U}$ and $B\mathfrak{U}$ agree.

The number of different elements occuring in a coset $A\mathfrak{U}$ is obviously independent of A; it is equal to the order of $\mathfrak{U}$. Let this order be called N (where N may also be $= \infty$). Each element A of $\mathfrak{G}$ actually appears in one such coset, e.g., A occurs in $A\mathfrak{U}$ because in any case the unit element must belong to $\mathfrak{U}$, since it is a group, and $AE = A$. Thus we obtain each element of $\mathfrak{G}$ exactly once if we run through all elements of the different sequences. In symbols we express this by the equation

$$\mathfrak{G} = A_1\mathfrak{U} + A_2\mathfrak{U} + \cdots$$

where $A_1\mathfrak{U}, A_2\mathfrak{U}, \ldots$ denote the distinct cosets of this kind.

Now in case $\mathfrak{G}$ is a *finite* group of order h, then the order N of $\mathfrak{U}$ is also finite and then the number of different cosets is also finite, say $= j$. Since each element of $\mathfrak{G}$ occurs in exactly one coset and exactly N different elements are contained in each coset, we have

$$h = j \cdot N$$

and thus we have shown

Theorem 19. *In a finite group of order h, the order N of each subgroup is a divisor of h.*

The quotient $h/N = j$ is called the *index* of the subgroup relative to $\mathfrak{G}$.

In case $\mathfrak{G}$ is an *infinite* group, then the order of $\mathfrak{U}$ as well as the number of different cosets can be infinite and at least one of these cases must obviously occur. Furthermore, the number of different cosets is called the *index* of $\mathfrak{U}$ relative to $\mathfrak{G}$ whether this index is finite or not.

Our further investigations deal first with finite groups.

A system $S = (U_1, U_2, \ldots)$ of elements which belong to a finite group forms a subgroup of $\mathfrak{G}$ as soon as it is known that each product of two elements U again belongs to S. For the group axioms (ii) and (iii) are satisfied automatically, (i) holds by assumption, and with finite groups (iv) is a consequence of the remaining axioms.

For example, all the powers of an element A with a positive exponent always form a subgroup of $\mathfrak{G}$. These powers cannot all be different, since $\mathfrak{G}$ contains only finitely many elements. From $A^m = A^n$ it follows that $A^{m-n} = E$. Hence a certain power of A with exponent different from zero is always $= E$.

In order to gain an overview of those exponents q for which $A^q = E$, we note that these exponents obviously form a *module* since from $A^q = E$ and $A^r = E$ it follows that $A^{q \pm r} = E$. Hence by Theorem 1 these q are identical with all multiples of an integer $a\,(> 0)$. This exponent a, uniquely determined by A, is called the *order of* A. This exponent has the property:

$$A^r = E \quad \text{if and only if} \quad r \equiv 0 \,(\text{mod } a).$$

The only element of order 1 is E. More generally

Theorem 20. *If a is the order of A, then*

$$A^m = A^n,$$

if and only if

$$m \equiv n \,(\text{mod } a).$$

Consequently among the powers of A there are only a distinct ones, say $A^0 = E, A^1, \ldots, A^{a-1}$, and by the above these form a subgroup of $\mathfrak{G}$ of order a. Moreover from Theorem 19 we have

Theorem 21. *The order a of each element of $\mathfrak{G}$ is a divisor of the order h of $\mathfrak{G}$ and hence*

$$A^h = E$$

for each element A.

§7 Abelian Groups and the Product of Two Abelian Groups

The groups which occur in number theory are almost exclusively those whose composition laws are *commutative*: $AB = BA$ for all of its elements. Groups of this kind are called *Abelian groups*. In this and the next section we will undertake a more precise investigation of the structure of an arbitrary finite Abelian group. In the following, $\mathfrak{G}$ denotes a finite Abelian group of order h.

Theorem 22. *If a prime number p divides the order h of $\mathfrak{G}$, then there is an element of order p in $\mathfrak{G}$.*

Let $C_1, C_2, \ldots, C_h$ be the h elements of $\mathfrak{G}$ and let $c_1, c_2, \ldots, c_h$ be their respective orders. We form all products

$$C_1^{x_1} C_2^{x_2} \cdots C_h^{x_h}, \tag{8}$$

in which each x_i runs through a complete residue system mod c_i. Then we obtain $c_1 \cdot c_2 \cdots c_h$ formally different products, among which are all elements of $\mathfrak{G}$. Since a representation of the unit element is at once obtained from two different representations of the same element all elements occur equally frequently, say Q times in the form (8). Hence

$$c_1 c_2 \cdots c_h = h \cdot Q.$$

The prime number p, which divides h, must therefore divide at least one c_i, say c_1. Then

$$A = C_1^{c_1/p}$$

is an element of order p by Theorem 20.

Theorem 23. *Let $h = a_1 \cdot a_2 \cdots a_r$ and suppose that the integers $a_1, \ldots, a_r$ are pairwise relatively prime. Then each element C of $\mathfrak{G}$ can be represented in one and only one way in the form*

$$C = A_1 \cdot A_2 \cdots A_r$$

with the conditions

$$A_1^{a_1} = A_2^{a_2} = \cdots = A_r^{a_r} = E.$$

For let r integers $n_1, \ldots, n_r$ be determined so that

$$\frac{h}{a_1} n_1 + \frac{h}{a_2} n_2 + \cdots + \frac{h}{a_r} n_r = 1,$$

which is always possible by Theorem 3 because of the assumption about the a_i. If we then set

$$A_i = C^{(h/a_i)n_i},$$

then by Theorem 21

$$A_i^{a_i} = C^{hn_i} = E$$

and with this

$$C = A_1 \cdot A_2 \cdots A_l.$$

is represented in the required form. To see the uniqueness of the representation let $C = B_1 \cdot B_2 \cdots B_r$ be yet another representation of this type. Then

$$(B_1 \cdot B_2 \cdots B_r)^{h/a_1} = (A_1 \cdot A_2 \cdots A_r)^{h/a_1}. \tag{9}$$

However, since composition is commutative, a fact which is used at this point for the first time, it follows from (9) that

$$B_1^{h/a_1} \cdot B_2^{h/a_1} \cdots B_r^{h/a_1} = A_1^{h/a_1} \cdot A_2^{h/a_1} \cdots A_r^{h/a_1}.$$

Now since h/a_1 is a multiple of each $a_2, a_3, \ldots, a_r$, the factors with the indices $2, 3, \ldots, r$ must be equal to E by the hypotheses about the A_i, B_i, hence

$$B^{h/a_1} = A_1^{h/a_1}.$$

Since $(a_1, h/a_1) = 1$, there are integers x, y with $a_1 x + (h/a_1)y = 1$ and recalling that

$$E = B_1^{a_1} = A_1^{a_1},$$

we have

$$B_1 = B_1^{a_1 x + (h/a_1)y} = A_1^{a_1 x + (h/a_1)y} = A_1.$$

In general, it follows in this fashion that $A_i = B_i$ and with this the uniqueness of the representation of C.

If a_i' is the number of different elements A with the property

$$A^{a_i} = E,$$

then obviously the totality of these forms a subgroup of $\mathfrak{G}$ of order a_i' because the product of two elements of this kind again has the same property. In any case by Theorem 23 we have

$$h = a_1' a_2' \cdots a_r' = a_1 a_2 \cdots a_r. \tag{10}$$

We see that we must have $a_i' = a_i$, for if p is a prime, and $p | a_i'$, then by Theorem 22 there exists among the elements A with $A^{a_i} = 1$ one of order p, hence $p | a_i$. Therefore a_i' has no prime factors other than those of a_i. Since the a_i are pairwise relatively prime, we must have, by Equation (10), $a_i' = a_i$.

With this we have proved:

Theorem 24. *If* $c | h$, $(h/c, c) = 1$ $(c > 0)$, *then the totality of elements of* $\mathfrak{G}$ *with the property*

$$A^c = 1$$

forms a subgroup of $\mathfrak{G}$ *of order* c.

Theorem 23 makes plain the necessity to introduce a special notation for the relation of the group $\mathfrak{G}$ to the r subgroups $A_1, \ldots A_r$ from which

$\mathfrak{G}$ can be built up by this theorem. One can define $\mathfrak{G}$ simply as a "product" of these subgroups. However, if starting out from two groups $\mathfrak{G}_1$ and $\mathfrak{G}_2$ one merely wishes to define a group $\mathfrak{G}$ which has $\mathfrak{G}_1$ and $\mathfrak{G}_2$ as subgroups and which is then to be called the product of these groups, one must consider that at the outset the product of an element of $\mathfrak{G}_1$ with an element of $\mathfrak{G}_2$ has no meaning at all yet.

For this reason we proceed as follows: We denote the elements of the Abelian group $\mathfrak{G}_i$ $(i = 1, 2)$ with the subscript i. We now define a new group whose elements are pairs (A_1, A_2) and we set

(1) $(A_1, A_2) = (B_1, B_2)$ means $A_1 = B_1$ and $A_2 = B_2$.
(2) The rule of composition for these pairs is to be $(A_1, A_2) \cdot (B_1, B_2) = (A_1 B_1, A_2 B_2)$.

In this way the $h_1 \cdot h_2$ new elements (h_i is the order of $\mathfrak{G}_i$) are combined to form an Abelian group $\mathfrak{G}$. The unit element of this group is (E_1, E_2), where E_i is the unit element of $\mathfrak{G}_i$. The h_1 elements (A_1, E_2), where A_1 runs through the group $\mathfrak{G}_1$ obviously form a subgroup of $\mathfrak{G}$ and this group is isomorphic to $\mathfrak{G}_1$; likewise the group of elements (E_1, A_2) is isomorphic to $\mathfrak{G}_2$. The two subgroups have only the one element (E_1, E_2) in common. Each element from $\mathfrak{G}$ can be represented in exactly one way as a product of two elements of the two subgroups:

$$(A_1, A_2) = (A_1, E_2) \cdot (E_1, A_2).$$

Finally we define

(3) $(A_1, E_2) = A_1, (E_1, A_2) = A_2$, thus in particular $E_1 = E_2$.

This use of the symbol "$=$" is permissible, since the relation "$=$" is still not defined between elements of $\mathfrak{G}$, $\mathfrak{G}_1$, and $\mathfrak{G}_2$, and composition of elements defined as equal yields again equal elements. We call the group $\mathfrak{G}$ defined in this way by (1), (2), (3), with the $h_1 h_2$ elements $A_1 A_2$ the *product of the two groups $\mathfrak{G}_1$ and $\mathfrak{G}_2$* and we write

$$\mathfrak{G} = \mathfrak{G}_1 \cdot \mathfrak{G}_2 = \mathfrak{G}_2 \cdot \mathfrak{G}_1.$$

With this terminology it then follows immediately from Theorem 23 that the formation of products is associative:

Theorem 25. *Each finite Abelian group can be represented as a product of Abelian groups whose orders are powers of primes.*

§8 Basis of an Abelian Group

Now we can prove the following theorem which gives us full information about the structure of the most general finite Abelian group.

Theorem 26 (Fundamental Theorem of Abelian Groups). *In each Abelian group $\mathfrak{G}$ of order h (>1) there are certain elements $B_1, \ldots, B_r$, with orders*

$h_1, \ldots, h_r$ respectively $(h_i > 1)$ *such that each element of $\mathfrak{G}$ is obtained in exactly one way in the form*

$$C = B_1^{x_1} B_2^{x_2} \cdots B_r^{x_r},$$

where the integers x_i each run through a complete system of residues mod h_i independently of one another. Moreover the $h_i = p^{k_i}$ are prime powers and $h = h_1 \cdot h_2 \cdots h_r$.

r elements of this kind are called *a basis for $\mathfrak{G}$.*

By our previous results the truth of this theorem is obtained at once for arbitrary h, as soon as it is proved for all Abelian groups of prime-power order.

Hence let $h = p^k$ be the order of $\mathfrak{G}$, where p is a prime and k is an integer ≥ 1. Then the order of each element of $\mathfrak{G}$ has a value p^α, where $0 \leq \alpha \leq k$, α integral.

A system of m elements $A_1, \ldots, A_m$ with orders $a_1, \ldots, a_m$ is called *independent* if from $A_1^{x_1} \cdot A_2^{x_2} \cdots A_m^{x_m} = E$ it follows that

$$x_i \equiv 0 \pmod{a_i} \quad \text{for } i = 1, 2, \ldots, m.$$

For example, each element A is an independent element. The product of powers of m independent elements obviously forms a group which contains exactly $a_1 \cdot a_2 \cdots a_m$ different elements. If $A_1, \ldots, A_m$ are independent then the $m + 1$ elements $A_1, \ldots, A_m, E$ are always independent and conversely. We now always agree on a numbering of the independent elements, such that the orders form a decreasing sequence:

$$a_1 \geq a_2 \geq a_3 \cdots \geq a_m \geq 1.$$

Let this system of numbers $a_1, a_2, \ldots, a_m$ be called the system of rank numbers of $A_1, \ldots, A_m$ or the *rank R of* $A_1, \ldots, A_m$. We now determine a definite ordering of the systems R. Let two independent systems

$$A_i \text{ of order } a_i = p^{\alpha_i} \quad (i = 1, 2, \ldots, m),$$
$$B_q \text{ of order } b_q = p^{\beta_q} \quad (q = 1, 2, \ldots, n)$$

be given. In case $m \neq n$, and say $m > n$, we define $\beta_{n+1} = \beta_{n+2} = \cdots = \beta_m = 0$. Both systems are said to be of *equal rank* if $\alpha_i = \beta_i$ for all $i = 1, \ldots, m$. Otherwise the rank of $(A, \ldots, A_m)$ is called *higher* or *lower* than the rank of $(B_1, \ldots, B_n)$, according as the first nonvanishing difference $\alpha_i - \beta_i$ is > 0 or < 0. Thus the omission or the addition of elements E does not change the rank. If the rank of $(A_1, \ldots)$ is higher than the rank of $(B_1, \ldots)$ and the rank of $(B_1, \ldots)$ is higher than that of $(C_1, \ldots)$, then the rank of $(A_1, \ldots)$ is higher than the rank of $(C_1, \ldots)$. Obviously there are at most h^h possibilities for the ranks of systems of elements independent of one another and distinct from E; consequently there are systems of independent elements of highest rank. We will call such systems *maximal systems* for short. Let $B_1, \ldots, B_r$ be a maximal system in which there is no element $= E$. We show that $B_1, \ldots, B_r$ is a system of basis elements. For this we must only verify that each element of $\mathfrak{G}$ is representable as a product of powers of the B_i—and for this the following lemmas suffice:

Lemma (a). *No element among the elements* $B_1, \ldots, B_r$ *can be a* pth *power of an element of* $\mathfrak{G}$.

If we had $B_m = C^p$ then the system obtained from the $B_1, \ldots, B_r$ by replacing B_m with C and possibly changing the numbering would also be independent, but obviously of higher rank than the maximal system $B_1, \ldots, B_r$, which is impossible.

Lemma (b). *If we replace one of the* B, *say* B_m, *in the system* $B_1, \ldots, B_r$ *by*

$$A = B_m^u B_{m+1}^{x_{m+1}} \cdots B_r^{x_r},$$

where $u \neq 0$ (mod p), *but the* x_i *are arbitrary integers, then the rank does not change and the new system is again a maximal system.*

A has the same order as B_m, since the orders of $B_{m+1}, \ldots, B_r$ are not larger than that of B_m, and thus are divisors of the order of B_m. Moreover, each product of powers from $A, B_{m+1}, \ldots, B_r$ is representable as a product of powers of $B_m, B_{m+1}, \ldots, B_r$, and conversely. Consequently the new system is also independent and thus it is a maximal system.

Lemma (c). *If an element* C^p *is representable as a product of powers of the* B_i, *then the same holds for* C.

If, in fact,

$$C^p = B_1^{x_1} \cdots B_r^{x_r}, \tag{11}$$

then all x_i are $\equiv 0$ (mod p). For if $x_m = u$ were the first exponent which is not divisible by p, then let B_m be replaced by

$$A = B_m^u B_{m+1}^{x_{m+1}} \cdots B_r^{x_r} = C^p B_1^{-x_1} \cdots B_{m-1}^{-x_{m-1}}$$

in the system of the B_i. This new system would be again a maximal system by (b), but it would contain the pth power of one of its elements, namely A, in contradiction to (a). Consequently, in (11), we may set $x = py_i$ with integral y_i and hence

$$(C^{-1} B_1^{y_1} \cdots B_r^{y_r})^p = 1.$$

If C were not representable as a product of powers of the B_i, then this would also hold for all C^n with $n \neq 0$ (mod p) and we would also have in the parenthesis above

$$C' = C^{-1} B_1^{y_1} \cdots B_r^{y_r} \neq 1;$$

hence C' would be an element of order p. Consequently the $r + 1$ elements $B_1, B_2, \ldots, B_r, C'$ would also be independent, correctly arranged according to decreasing order (as the order of B is greater than 1 and hence $\geq p$). However they would have a higher rank than the maximal system $B_1, \ldots, B_r$, which is impossible. Hence the assumption is false and (c) is proved.

By repeated application of (c) however, the representability of each element A of $\mathfrak{G}$ through the B_i is obtained. For if A is of order p^m, then

$$A^{p^m} = 1$$

is certainly representable by the B_i. Hence, by (c), $A^{p^{m-1}}$ is also representable by the B_i, and thus also $A^{p^{m-2}}$ if $m > 1$ and so on until we arrive at $A^{p^0} = A$ itself.

The elements of a basis for $\mathfrak{G}$ are not uniquely determined by $\mathfrak{G}$. Certain properties of the basis are nevertheless characteristic of $\mathfrak{G}$ itself. The number $e = e(p)$ of those basis elements whose order is divisible by the prime p is considered the most important constant determined by $\mathfrak{G}$ alone; we call e *the basis number belonging to* p. Its independence of the choice of basis elements is shown by

Theorem 27. *If p is a prime, then the number of different elements of $\mathfrak{G}$ with the property*

$$A^p = 1$$

is equal to p^e, where e is the basis number belonging to p.

If $B_1, B_2, \ldots, B_e$ are those basis elements whose orders are powers of p, then from

$$A = B_1^{x_1} B_2^{x_2} \cdots B_e^{x_e} B_{e+1}^{x_{e+1}} \cdots B_r^{x_r} \quad \text{and} \quad A^p = 1$$

we have the sequence of congruences

$$px_i \equiv 0 \ (\text{mod } h_i) \qquad i = 1, 2, \ldots, r,$$

hence for $i = e + 1, \ldots, r$ since $(h, p^i) = 1$,

$$x_i \equiv 0 \ (\text{mod } h_i)$$

and for $i = 1, 2, \ldots, e$, since $h_i = p^{k_i}$,

$$x_i \equiv 0 \left(\text{mod } \frac{h_i}{p}\right).$$

Conversely the latter congruence has as a consequence the equation $A^p = 1$. The number of solutions of each of these congruences which are incongruent mod h_i is 1 for $i = e + 1, \ldots, r$ and p for $i = 1, 2, \ldots, e$. Consequently the number of incongruent systems of solutions is p^e.

The statement is also correct if p does not divide the order h of the group, for then $e = 0$.

The simplest Abelian groups are obtained by raising one element to a power: $A^0 = 1, A, A^2, \ldots$ and $A^{-1}, A^{-2}, \ldots$ If all elements of an Abelian group are powers of a single element A, the group is called *cyclic*, and A is called a *generator* of the group. Here we have

Theorem 28. *An Abelian group $\mathfrak{G}$ of order h is cyclic if and only if for each prime p dividing h, the number of elements A with $A^p = 1$ is equal to p.*

By the preceding theorem the condition is equivalent to: the basis number belonging to p should be $= 1$.

The condition is necessary. Namely if

$$C, \; C^2, \; \ldots, \; C^{h-1}, \; C^h = 1$$

are the h elements of $\mathfrak{G}$, then from $A^p = 1$ it follows that for $A = C^x$

$$px \equiv 0 \,(\mathrm{mod}\; h) \quad \text{and} \quad x \equiv 0 \left(\mathrm{mod}\; \frac{h}{p}\right),$$

that is, x has one of the p values $h/p, 2h/p, \ldots, ph/p \bmod h$, and conversely we thus also obtain p different elements A with $A^p = 1$.

The condition, however, is also sufficient; for if $h = p_1^{k_1} \cdots p_r^{k_r}$ is the decomposition of h into different prime factors then, by hypothesis, only one basis element belongs to each p_i; hence all elements of $\mathfrak{G}$ are of the form

$$A = B_1^{x_1} \cdots B_r^{x_r},$$

where

$$B_i^{h_i} = 1 \quad \text{with } h = p_i^{k_i}.$$

One then obtains h different elements, hence all elements of $\mathfrak{G}$, if one forms the successive powers of

$$C = B_1 \cdot B_2 \cdots B_r.$$

If u is the order of C, then by the basis property of the B it follows that

$$u \equiv 0 \,(\mathrm{mod}\; h_i) \quad \text{for } i = 1, 2, \ldots, r,$$

and since the h_i are pairwise relatively prime, u is divisible by $h = h_1 \cdots h_r$, hence $= h$, since u cannot be greater than h.

§9 Composition of Cosets and the Factor Group

If $\mathfrak{U}$ is a subgroup of the Abelian group $\mathfrak{G}$, hence itself Abelian, then $\mathfrak{U}$ gives rise to another group as follows. By §6 the cosets $A\mathfrak{U}$ are uniquely determined along with $\mathfrak{U}$. The number of cosets is h/N where N is the order of $\mathfrak{U}$; we denote them by $R_1, R_2, \ldots$. We now set up a law of composition between the R's with the following observation. If A_1 and A_1' are elements of R_1, A_2 and A_2' are elements of R_2, then $A_1 A_2$ and $A_1' A_2'$ belong to the same coset R_3. Since

$$A_1' = A_1 U_1, \qquad A_2' = A_2 U_2,$$

where U_1, U_2 are elements of $\mathfrak{U}$, then $A_1' A_2' = A_1 A_2 U_1 U_2$ (here we use the fact that the composition of elements of $\mathfrak{G}$ is commutative). Since $A_1 A_2$ and $A_1' A_2'$ differ only by a factor from $\mathfrak{U}$ they therefore belong to the same coset R_3. Hence R_3 is uniquely determined by R_1 and R_2. We write

$$R_1 \cdot R_2 = R_3.$$

The group axioms (i)–(iii) are obviously satisfied with this composition. Furthermore this composition is obviously commutative. Consequently the cosets R form an Abelian group $\mathfrak{R}$ of order h/N.

Definition. The group $\mathfrak{R}$ defined in this way is called the *factor* (*quotient*) *group* of $\mathfrak{U}$. Its order is equal to the index of $\mathfrak{U}$. One writes

$$\mathfrak{R} = \mathfrak{G}/\mathfrak{U}.$$

We can also describe it as follows: the factor group is obtained from $\mathfrak{G}$ if one considers two elements of $\mathfrak{G}$ as not being different whenever they differ only by an element of $\mathfrak{U}$, where moreover we retain the composition rules of $\mathfrak{U}$.

We will apply these concepts to advantage in the case where $\mathfrak{U}$ is the group of those elements of $\mathfrak{G}$ which can be represented as the pth power of elements of $\mathfrak{G}$, where p is a prime dividing h. In particular this subgroup $\mathfrak{U}$ may now be denoted by $\mathfrak{U}_p$. We have

Theorem 29. *The order of $\mathfrak{G}/\mathfrak{U}_p$ is p^e if e is the basis number of $\mathfrak{G}$ belonging to p. The group $\mathfrak{G}/\mathfrak{U}_p$ is isomorphic to the group of elements C of $\mathfrak{G}$ for which $C^p = 1$.*

In fact we see from Theorem 26 that each element X of $\mathfrak{G}$ can be represented in the form

$$X = B_1^{x_1} B_2^{x_2} \cdots B_e^{x_e} A^p$$

where $B_1, \ldots, B_e$ are the basis elements belonging to the prime p and the e numbers $x_1, \ldots, x_e$ are uniquely determined mod p by X, while A^p is a suitably chosen pth power, i.e., an element from $\mathfrak{U}_p$. Such an element X is a pth power if and only if all x_i are $\equiv 0 \pmod{p}$. Consequently the number of cosets determined by $\mathfrak{U}_p$ is equal to the number of different systems x_i mod p, i.e., $= p^e$. The pth power of each coset is identical with the system $\mathfrak{U}_p$, i.e., in the group $\mathfrak{G}/\mathfrak{R}_p$ of order p^e, each element, if it is not the unit element, has order p. Hence $\mathfrak{G}/\mathfrak{U}_p$ must contain exactly e basis elements, each of order p. By Theorem 27 the group of all C with $C^p = 1$ has the same structure. Moreover it is seen that the e cosets

$$B_i \mathfrak{U} \qquad i = 1, 2, \ldots, e$$

form a system of basis elements in the factor group, and the e elements

$$B_i^{h_i/p} \qquad i = 1, 2, \ldots, e$$

are basis elements in the group of those C with $C^p = 1$. Hence the two groups are isomorphic.

§10 Characters of Abelian Groups

Since the law of composition in an Abelian group, like ordinary multiplication, is commutative, those elements which satisfy the symbolic equation $A^h = 1$ behave formally like the hth roots of unity, thus like certain numbers. The question arises whether it is not possible to transform the investigation of Abelian groups entirely into a problem about numbers, perhaps of the following kind:

To each element A of a given Abelian group $\mathfrak{G}$ there is to be assigned a number, denoted by $\chi(A)$, in such a way that for every two elements A, B from $\mathfrak{G}$

$$\chi(A) \cdot \chi(B) = \chi(AB). \tag{12}$$

The composition of the elements thus corresponds to multiplication of the assigned numbers.

The construction of all these "functions" $\chi(A)$ is obtained according to the fundamental theorem in the following way.

Let the trivial solution "$\chi(A) = 0$ for all A" be discarded.

First we must have

$$\chi(E) = 1$$

for the unit element since for each A

$$\chi(A)\chi(E) = \chi(AE) = \chi(A).$$

Next, if $B_1, \ldots, B_r$ is a basis for $\mathfrak{G}$, then by repeated application of (12) it follows that for

$$A = B_1^{x_1} \cdot B_2^{x_2} \cdots B_r^{x_r},$$
$$\chi(A) = \chi(B_1)^{x_1} \cdots \chi(B_r)^{x_r}. \tag{13}$$

Consequently $\chi(A)$ is known for each element A as soon as it is known for the r basis elements B_i. However these values $\chi(B_i)$ are not arbitrary, but rather they must be chosen in such a way that all systems of exponents x_i which lead to the same A also yield the same value $\chi(A)$ in (13). That is, $\chi(B_i)$ must be a number such that

$$\chi(B_i)^{x_i}$$

depends only on the value of x_i mod h_i. Since $1 = \chi(E) = \chi(B_i^{h_i}) = \chi(B_i)^{h_i}$, we have $\chi(B_i) \neq 0$ and thus it is an h_ith root of unity.

However this condition is also sufficient. To prove this let

$$\chi(B_m) = \zeta_m,$$

$m = 1, \ldots, r$, be any h_mth roots of unity

$$\zeta_m = e^{(2\pi i/h_m)a_m} \qquad (a_m \text{ an arbitrary integer}).$$

Then we define

$$\chi(A) = \zeta_1^{x_1} \cdots \zeta_r^{x_r} \quad \text{if } A = B_1^{x_1} \cdots B_r^{x_r}. \tag{14}$$

Since, in fact, the expression $\chi(A)$ only depends on which residue class mod h_m the x_m are contained in, and since this is uniquely determined by the element A, $\chi(A)$ is therefore uniquely defined by these x_m and also satisfies the requirement (12). Now there are exactly h_m different roots of unity of degree h_m corresponding to the values $a_m = 1, 2, \ldots, h_m$. Consequently there exist exactly $h = h_1 \cdot h_2 \cdots h_r$ formally different functions $\chi(A)$, for which no two are identical for all elements, since they differ for at least one basis element. With this we have proved:

Theorem 30. *There are exactly h distinct functions $\chi(A)$ which have the property: $\chi(AB) = \chi(A) \cdot \chi(B)$ and $\chi(A)$ is not $= 0$ for all elements A of $\mathfrak{G}$. Each χ is an hth-root of unity.*

Each such function $\chi(A)$ is called a *group character* or *character of* $\mathfrak{G}$.

Among the characters $\chi(A)$ there is one which is $= 1$ for all A; it is called the *principal character*. Conversely, there exists exactly one element, namely E, such that $\chi(E) = 1$ for every character.

The characters themselves can be combined again to form a group of order h. For if $\chi_1(A)$ and $\chi_2(A)$ are characters, then $f(A) = \chi_1(A) \cdot \chi_2(A)$ also satisfies the defining equation of a χ, hence it is also a character of $\mathfrak{G}$. If $\chi(A)$ runs through all characters and if $\chi_1(A)$ is a fixed character, then $\chi(A)\chi_1(A)$ also runs through all characters of $\mathfrak{G}$. If we understand by $\sum_A$ a sum extended over all h elements A of $\mathfrak{G}$ and by $\sum_\chi$ a sum extended over all h characters χ, then we have

Theorem 31.

$$\sum_A \chi(A) = \begin{cases} h & \text{if } \chi \text{ is the principal character,} \\ 0 & \text{if } \chi \text{ is not the principal character,} \end{cases}$$

$$\sum_\chi \chi(A) = \begin{cases} h & \text{if } A = E, \\ 0 & \text{if } A \neq E. \end{cases}$$

The first half of each statement is trivial, as each summand $= 1$. If B is an arbitrary element, then along with A, AB also runs through h all elements of $\mathfrak{G}$, hence

$$\sum_A \chi(A) = \sum_A \chi(AB) = \chi(B) \sum_A \chi(A), \quad \text{thus} \quad (1 - \chi(B)) \sum_A \chi(A) = 0.$$

Now if χ is not the principal character, then $\chi(B) \neq 1$ for at least one B, hence $\sum_A$ is equal to 0.

Likewise let χ_1 be an arbitrary character. Then we have

$$\sum_\chi \chi(A) = \sum_\chi \chi_1(A)\chi(A) = \chi_1(A) \sum_\chi \chi(A)$$

$$(1 - \chi_1(A)) \sum_\chi \chi(A) = 0.$$

If $A \neq E$, then for at least one character $\chi_1(A) \neq 1$, hence $\sum_\chi$ is equal to 0.

The element A is determined uniquely by the h numbers $\chi_n(A)$, where χ_n are the h characters for $n = 1, 2, \ldots, h$. For if a second element B had the same values $\chi_n(B)$, then we would have $\chi_n(AB^{-1}) = 1$ for all n, and AB^{-1} would be the unit element, thus $A = B$.

The h numbers $\chi_n(A)$ are, however, not arbitrary. On the contrary the following holds:

Theorem 32. *If A is an element of order f, then $\chi_n(A)$ is an fth root of unity. Among the h numbers $\chi_n(A)$, $n = 1, \ldots, h$, all fth roots of unity occur equally often, namely h/f times.*

To begin with, since $A^f = 1$: $\chi_n(A)^f = \chi_n(A^f) = \chi_n(1) = 1$. Thus the first part of the theorem is true. Now if ζ is an arbitrary fth root of unity, let us consider the sum

$$\sum_{n=1}^{h} (\zeta^{-1}\chi_n(A) + \zeta^{-2}\chi_n(A^2) + \cdots + \zeta^{-f}\chi_n(A^f)) = S.$$

Since by hypothesis A^m is not the unit element for $1 < m < f$—if we exclude the trivial case $f = 1$, that is, $A = E$—, it follows by Theorem 31, that if we split the sum into f individual sums, then $S = h$.

On the other hand the term inside each set of parentheses is equal to $\varepsilon + \varepsilon^2 + \cdots + \varepsilon^f$, where

$$\varepsilon = \zeta^{-1}\chi_n(A), \qquad \varepsilon^f = 1.$$

Hence it is equal to 0 or f, depending on whether $\varepsilon \neq$ or $= 1$, that is, according to whether $\chi_n(A) \neq \zeta$ or $= \zeta$. If k denotes the number of characters $\chi_n(A)$ for which $\chi_n(A) = \zeta$ it follows that $S = kf$. Therefore if we combine this with the first result we get

$$kf = h, \qquad k = \frac{h}{f},$$

independently of ζ, which was to be proved.

Moreover the group of characters is isomorphic with the group $\mathfrak{G}$ itself. To see this we assign to the basis element B_q a primitive h_qth root of unity, say

$$\zeta_q = e^{2\pi i/h_q}.$$

Then each character $\chi(A)$ is represented uniquely in the form

$$\chi(A) = \chi_1^{y_1}(A)\chi_2^{y_2}(A) \cdots \chi_r^{y_r}(A),$$

where

$$A = B_1^{x_1} \cdots B_r^{x_r},$$

by the r basis characters

$$\chi_q(A) = \zeta_q^{x_q} \qquad (q = 1, 2, \ldots, r),$$

where the y_q are uniquely determined integers mod h_q. If we now assign the element

$$B_1^{y_1} B_2^{y_2} \cdots B_r^{y_r}$$

to the character

$$\chi = \chi_1^{y_1}\chi_2^{y_2} \cdots \chi_r^{y_r}$$

then an isomorphism between the group of characters and the group $\mathfrak{G}$ is obviously determined.

Every subgroup can be determined with the help of the characters of an Abelian group. If one takes some distinct characters $\chi_1, \chi_2, \ldots, \chi_k$ of $\mathfrak{G}$, then the totality of elements U for which $\chi_1(U) = \chi_2(U) = \cdots = \chi_k(U) = 1$ obviously forms a subgroup $\mathfrak{U}$ of $\mathfrak{G}$, since along with two elements U_1 and U_2 the product $U_1 \cdot U_2$ also has this property.

Moreover it can be seen, as follows, that each subgroup $\mathfrak{U}$ of $\mathfrak{G}$ can be obtained in this way: let $\mathfrak{U}$ be an arbitrary subgroup of $\mathfrak{G}$; the factor group $\mathfrak{G}/\mathfrak{U}$ whose elements are the different cosets $A\mathfrak{U}$ is also an Abelian group, and accordingly it has exactly j characters which are denoted by $\lambda_1(A\mathfrak{U})$, $\lambda_2(A\mathfrak{U}), \ldots, \lambda_j(A\mathfrak{U})$. With the help of these we define a character by fixing

$$\chi_k(A) = \lambda_k(A\mathfrak{U}) \quad \text{for } k = 1, 2, \ldots, j.$$

For each k this determination is unique since each element A belongs to only one coset. Moreover for any two elements A and B of $\mathfrak{G}$ we always have

$$\chi_k(A) \cdot \chi_k(B) = \lambda_k(A\mathfrak{U}) \cdot \lambda_k(B\mathfrak{U}) = \lambda_k(AB\mathfrak{U}) = \chi_k(AB);$$

consequently $\chi_k(A)$ is actually a character of the group $\mathfrak{G}$. The various characters $\lambda_k(A\mathfrak{U})$, $k = 1, 2, \ldots, j$, have the value 1 simultaneously only for the unit element of the group $\mathfrak{G}/\mathfrak{U}$, that is, only for the coset which is identical with $\mathfrak{U}$. Hence the j characters $\chi_k(A)$ are all equal to 1 precisely for those elements A which belong to $\mathfrak{U}$. That is, the subgroup $\mathfrak{U}$ is to be defined as the totality of those elements A for which the j conditions

$$\chi_k(A) = 1 \quad \text{for } k = 1, 2, \ldots, j \tag{15}$$

are satisfied.

However these j conditions, which each single element A from $\mathfrak{U}$ must satisfy, are not independent of one another as, along with χ_1 and χ_2, $\chi_1 \cdot \chi_2 = \chi_3$ also occurs among the χ_k; thus the condition $\chi_3(A) = 1$ already follows from the two conditions $\chi_1(A) = \chi_2(A) = 1$. In order to find the number of mutually independent conditions among the j conditions (15), we consider that the λ_k, which define the χ_k uniquely, form a group isomorphic to $\mathfrak{G}/\mathfrak{U}$ since they are exactly the characters of $\mathfrak{G}/\mathfrak{U}$. Hence they are representable by a basis, say $\lambda_1, \ldots, \lambda_{r_0}$, where r_0 is the number of basis elements of $\mathfrak{G}/\mathfrak{U}$. That is, each character λ_k is a product of powers of these r_0 characters. Hence it follows from the r_0 conditions

$$\chi_1(A) = \chi_2(A) = \cdots = \chi_{r_0}(A) = 1$$

that all j conditions (15) are satisfied for A and with this it follows that A belongs to $\mathfrak{U}$. If h_i is the order of the basis character λ_i and ζ_i $(i = 1, 2, \ldots, r_0)$ are arbitrarily given h_ith roots of unity, then moreover there is always a coset $A\mathfrak{U}$ such that $\lambda_i(A\mathfrak{U}) = \zeta_i$ for $i = 1, 2, \ldots, r_0$. Thus we have proved

Theorem 33. *If $\mathfrak{U}$ is a subgroup of $\mathfrak{G}$ and if the factor group $\mathfrak{G}/\mathfrak{U}$ has r_0 basis elements, then among the h characters of $\mathfrak{G}$ there are r_0 characters χ_i with order h_i, a power of a prime, $(i = 1, 2, \ldots, r_0)$ such that the r_0 conditions*

$$\chi_i(A) = 1 \qquad (i = 1, 2, \ldots, r_0)$$

are satisfied for all elements A of $\mathfrak{U}$ and only for elements A of $\mathfrak{U}$, while on the other hand there always exist elements B in $\mathfrak{G}$ for which those r_0 characters $\chi_i(B)$ are arbitrarily prescribed h_ith roots of unity.

§11 Infinite Abelian Groups

The theory of infinite Abelian groups has still not been developed in any direction as completely as the theory of finite Abelian groups developed above. The few theorems on infinite Abelian groups which exist refer to groups which are specialized still further. The concepts and facts which have an application to arithmetic in the further course of our presentation will be explained in this section. Moreover the theory of infinite Abelian groups will be used only later from Chapter IV on in the theory of fields.

In an infinite group $\mathfrak{G}$ we distinguish elements of finite order and those of *infinite order*, according as some power of the element is equal to E or not—of course the zeroth power is excluded. As will be shown later with examples, it may happen that an infinite Abelian group has only elements of infinite order (except E) or only elements of finite order.

We call a system of finitely many elements of $\mathfrak{G}$, $A_1, A_2, \ldots, A_r, T_1, T_2, \ldots, T_q$, *independent* if a relation

$$A_1^{x_1} A_2^{x_2} \cdots A_r^{x_r} T_1^{y_1} \cdots T_q^{y_q} = 1$$

with integral x, y implies that all $x_i = 0$ and each $y_i \equiv 0 \pmod{h_i}$, where each A has infinite order and each T_i has finite order h_i. In this case the expression on the left obviously represents different elements if each x runs through all integers (positive and negative) and each y_i runs through a complete residue system mod h_i.

A system of finitely or infinitely many elements of $\mathfrak{G}$: A_i $(i = 1, 2, \ldots)$, T_k $(k = 1, 2, \ldots)$ $(A_i$ of infinite order, T_k of finite order), is called a *basis for* $\mathfrak{G}$ if each element of $\mathfrak{G}$ can be represented in the form

$$A_1^{x_1} A_2^{x_2} \cdots T_1^{y_1} T_2^{y_2} \cdots = C,$$

where

(1) the exponents x_i and y_k are integers and only finitely many are $\neq 0$,
(2) the exponents x_i are determined uniquely and the exponents y_k are determined uniquely mod h_k by C.

Obviously any finite set of elements of a basis must be independent.

The requirement that the h_k are powers of a prime will not be imposed here for the sake of simplicity.

A basis is called finite if it consists of finitely many elements.

Theorem 34. *If an infinite Abelian group $\mathfrak{G}$ has a finite basis, then each subgroup of $\mathfrak{G}$ also has a finite basis.*

Let $B_1, B_2, \ldots, B_m$ be a basis of $\mathfrak{G}$ where $B_1, \ldots, B_r$ are the elements of infinite order and $B_{r+1}, \ldots, B_m$ are those of order $h_1, \ldots, h_{m-r}$. We consider the systems of exponents of all products of powers

$$U = B_1^{u_1} \cdots B_m^{u_m}$$

which belong to $\mathfrak{U}$, where, in addition, the last $u_{r+1}, \ldots, u_m$ are to run through all numbers, not just the numbers which are distinct mod h_i, as long as the product belongs to $\mathfrak{U}$. By the group property of $\mathfrak{U}$, however, we obviously have that along with the system of exponents $(u_1, \ldots, u_m)$ and $(u_1', \ldots, u_m')$, the systems $(u_1 + u_1', \ldots, u_m + u_m')$ and $(u_1 - u_1', \ldots, u_m - u_m')$ also correspond to elements U. In particular we keep in mind the elements

$$U = B_k^{z_k} B_{k+1}^{z_{k+1}} \cdots B_m^{z_m} \qquad (1 \le k \le m) \tag{16}$$

belonging to $\mathfrak{U}$ for a definite k, thus for which $u_1 = \cdots u_{k-1} = 0$—there are such elements, since if all $u_i = 0$ the unit element of $\mathfrak{U}$ is obtained—then the totality of possible first exponents z_k in (16) forms a module of integers in the sense of §1, as long as we do not always have $z_k = 0$. However, all numbers of this module are identical with the multiples of a certain integer; consequently, if we do not always have $z_k = 0$, there is an element U_k in $\mathfrak{U}$ with one such $r_k \neq 0$,

$$U_k = B_k^{r_k} B_{k+1}^{r_{k+1}} \cdots,$$

such that z_k in (16) is a multiple of this r_k. From the U_k with this r_k—possibly infinite in number—we pick out a definite one for each $k = 1, \ldots, m$, where we set $U_k = E$ and $r_k = 0$ in case we always have $z_k = 0$ for this k in (16).

We show that each element in $\mathfrak{U}$ is representable as a product of these elements $U_1, \ldots, U_m$. Let

$$U = B_1^{u_1} \cdots B_m^{u_m}$$

be an element of $\mathfrak{U}$. By the preceding discussion u_1 is a multiple of r_1, $u_1 = v_1 r_1$, and hence

$$UU_1^{-v_1} = B_2^{u_2'}B_3^{u_3'} \cdots B_m^{u_m'} \tag{17}$$

is a product only of powers of $B_2, \ldots, B_m$, which also belongs to $\mathfrak{U}$ by the group property. If we should have $r_1 = 0$ and $U_1 = E$, then we should take $v_1 = 0$. Likewise, in (17), u_2' must be a multiple of r_2 in case this element is $\neq 0$, $u_2' = v_2 r_2$. Moreover if $r_2 = 0$ then u_2' must be $= 0$ and we take $v_2 = 0$. In any case then $UU_1^{-v_1}U_2^{-v_2}$ is an element of $\mathfrak{U}$ and representable as product of powers only of $B_3, \ldots, B_m$ etc. until we arrive at the unit element and obtain a representation

$$U = U_1^{v_1}U_2^{v_2} \cdots U_m^{v_m}.$$

The $U_1, \ldots, U_r$ are of infinite order if they are $\neq E$, the other U's are of finite order.

The products of powers of the $U_{r+1}, \ldots, U_m$ form a finite Abelian group and can hence be represented by a basis $C_1, \ldots, C_q$, by Theorem 26. We assert that $U_1, \ldots, U_r, C_1, \ldots, C_q$ form a basis for $\mathfrak{U}$ if we omit the elements $U_i = E$. First, each element U can be represented by the $U_1, \ldots, U_m$, hence also by the $U_1, \ldots, U_r, C_1, \ldots, C_q$. Now if

$$U_1^{v_1}U_2^{v_2} \cdots U_r^{v_r}C_1^{c_1} \cdots C_q^{c_q} = 1 \tag{18}$$

is a representation of the unit element where $v_i = 0$ is assumed for $U_i = E$ (i.e., $r_i = 0$), then by substitution of the B_i in place of the U_i and C_k, it follows that

$$v_1 r_1 = 0;$$

hence either $v_1 = 0$ or $r_1 = 0$. However, in the latter case we also have $v_1 = 0$ as a consequence of our convention. Likewise $v_2 = 0, \ldots, v_r = 0$. Furthermore, since the C_k form a basis of the finite group, then in (18) each c_k must be a multiple of the order of C_k. Now since each element is represented the same number of times by the U_i as by the C_i, hence the same number of times as the unit element, these elements actually form a basis for $\mathfrak{U}$ as was to be proved.

Those infinite Abelian groups in which no element of finite order except E appears are of chief interest. We call such groups *torsion-free groups*, the others *mixed* groups.

Along with a torsion-free group $\mathfrak{G}$ each subgroup of $\mathfrak{G}$ is also torsion-free. In particular, let $\mathfrak{U}$ be a subgroup of $\mathfrak{G}$ of finite index (§6). Then a certain power of each element of $\mathfrak{G}$ with exponent different from zero must always belong to $\mathfrak{U}$. For if A is an element of $\mathfrak{G}$, then the cosets

$$A\mathfrak{U}, \quad A^2\mathfrak{U}, \quad \ldots, \quad A^m\mathfrak{U}$$

are not all distinct, since the index is assumed to be finite. Thus for some n $A^m\mathfrak{U} = A^n\mathfrak{U}$, that is, A^{m-n} must belong to $\mathfrak{U}$, with $m - n \neq 0$. Hence in the above proof applied to $\mathfrak{G}$ and $\mathfrak{U}$ the case $r_k = 0$, $U_k = E$ can obviously never occur, since, in fact, a system of values $z_k \neq 0$, $z_{k+1} = \cdots = z_m = 0$ always exists, so that

$$U_k = B_k^{z_k} \text{ belongs to } \mathfrak{U}.$$

From this we have immediately

Theorem 35. *If $\mathfrak{G}$ is a torsion-free Abelian group with finite basis $B_1, \ldots, B_h$, then every subgroup $\mathfrak{U}$ of $\mathfrak{G}$ with finite index has a basis $U_1, \ldots, U_n$ of the form*

$$
\begin{aligned}
U_1 &= B_1^{r_{11}} B_2^{r_{12}} \cdots B_n^{r_{1n}}, \\
U_2 &= \phantom{B_1^{r_{11}}} B_2^{r_{22}} \cdots B_n^{r_{2n}}, \\
&\vdots \\
U_n &= \phantom{B_1^{r_{11}} B_2^{r_{22}} \cdots} B_n^{r_{nn}},
\end{aligned}
$$

with $r_{ii} \neq 0$ for $i = 1, 2, \ldots, n$.

Theorem 36. *The index of $\mathfrak{U}$ in $\mathfrak{G}$ is $j = |r_{11} \cdot r_{22} \cdots r_{nn}|$.*

For the proof we must determine the maximum number of elements which can exist in $\mathfrak{G}$ such that no two differ by a factor in $\mathfrak{U}$. We first show that an element

$$B_1^{x_1} B_2^{x_2} \cdots B_n^{x_n},$$

where all $|x_i| < r_{ii}$, belongs to $\mathfrak{U}$ only if all $x_i = 0$. By the definition of the U_i in the preceding proof x_1 must be divisible by r_{11} and since $|x_1| < r_{11}$ we must have $x_1 = 0$. However then x_2 must be divisible by r_{22} and must consequently also be $= 0$ etc.

From this it follows that among the $j = |r_{11} \cdot r_{22} \cdots r_{nn}|$ elements

$$B_1^{z_1} \cdot B_2^{z_2} \cdots B_n^{z_n}, \qquad 0 \leq z_i < r_{ii} \tag{19}$$

no two can differ by a factor in $\mathfrak{U}$. Hence there are at least j different cosets—each represented by one of these elements. On the other hand, however, we obtain all elements of $\mathfrak{G}$ from these elements if we multiply them by all elements of $\mathfrak{U}$, and hence j is the exact value of the index. To see this note that for an arbitrary product of the $B_k, B_{k+1}, \ldots, B_n$,

$$P = B_k^{x_k} B_{k+1}^{x_{k+1}} \cdots B_n^{x_n},$$

we can always determine an integer b_k such that

$$P U_k^{-b_k} = B_k^{z_k} B_{k+1}^{z_{k+1}} \cdots,$$

where the first index z_k satisfies the condition $0 \leq z_k < r_{kk}$. Obviously z_k is the smallest positive remainder of $x_k \bmod r_{kk}$. By applying this conclusion repeatedly we see that for each A in $\mathfrak{G}$ a sequence of exponents $b_1, \ldots, b_n$ can be found such that

$$A U_1^{-b_1} U_2^{-b_2} \cdots U_n^{-b_n}$$

is an element of the system (19). Consequently, A differs from this element only by a factor in $\mathfrak{U}$.

We now investigate the connection between different systems of bases of a group $\mathfrak{G}$ in order to find properties of bases which are determined by $\mathfrak{G}$ alone.

Theorem 37. *If a torsion-free Abelian group $\mathfrak{G}$ has a finite basis of n elements $B_1, \ldots, B_n$, then n is the maximal number of independent elements of $\mathfrak{G}$, independent of the choice of basis.*

Since the $B_1, \ldots, B_n$ are independent in any case, there are n independent elements in $\mathfrak{G}$ and thus we need only show that $n + 1$ elements in $\mathfrak{G}$ are not independent. In fact, between $n + 1$ arbitrary elements

$$A_i = B_1^{c_{i1}} \cdot B_2^{c_{i2}} \cdots B_n^{c_{in}} \qquad (i = 1, 2, \ldots, n + 1)$$

there is the relation

$$A_1^{x_1} A_2^{x_2} \cdots A_{n+1}^{x_{n+1}} = 1,$$

if we choose the $n + 1$ integers x_i so that they satisfy the n linear homogeneous equations

$$\sum_{i=1}^{n+1} x_i c_{ik} = 0 \qquad (k = 1, 2, \ldots, n).$$

As is known this is always possible since the coefficients c_{ik} are integers.

Theorem 38. *From a basis $B_1, \ldots, B_n$ of a torsion-free Abelian group $\mathfrak{G}$ one can obtain all systems of bases $B'_1, \ldots, B'_n$ of $\mathfrak{G}$ in the form*

$$B'_i = B_1^{a_{i1}} B_2^{a_{i2}} \cdots B_n^{a_{in}}, \qquad (i = 1, 2, \ldots, n)$$

where the system of exponents are arbitrary integers a_{ik} with determinant ± 1.

To begin with, the B'_i always form a basis. To see this we need only show that the B_i can be represented through the B'_i. The equation

$$B_m = B_1'^{x_1} \cdot B_2'^{x_2} \cdots B_n'^{x_n}$$

is satisfied if the integers x are chosen so that the n equations

$$x_1 a_{1i} + x_2 a_{2i} + \cdots + x_n a_{ni} = \begin{cases} 0, & \text{if } i \neq m, \\ 1, & \text{if } i = m, \end{cases}$$

hold. Since the determinant of the (integral) coefficients is $= \pm 1$ and the right side is also integral, the x_i are uniquely determined integers.

Secondly, if n elements

$$B'_i = B_1^{c_{i1}} \cdots B_n^{c_{in}} \qquad (i = 1, 2, \ldots, n)$$

form a basis, then B_q must be representable through the B'_i,

$$B_q = B_1'^{b_{q1}} B_2'^{b_{q2}} \cdots B_n'^{b_{qn}}, \qquad (q = 1, \ldots, n)$$

and if the B's are substituted for the B''s, then the n^2 equations

$$\sum_{i=1}^{n} b_{qi}c_{ik} \quad \begin{cases} 0 & \text{if } q \neq k, \\ 1 & \text{if } q = k, \end{cases}$$

are obtained, by the basis property of the B's. The determinant of this array is thus $= 1$; on the other hand, however, by the multiplication theorem of determinant theory, the determinant is equal to the product of the two determinants $|b_{ik}|$ and $|c_{ik}|$. Hence each of these integers must divide 1, and therefore each integer is itself $= \pm 1$; thus $|c_{ik}| = \pm 1$.

Finally by combining the last three theorems we obtain

Theorem 39. *If $\mathfrak{G}$ is a torsion-free Abelian group with a finite basis $B_1, \ldots, B_n$, $\mathfrak{U}$ a subgroup of finite index j, then $\mathfrak{U}$ also has a finite basis $U_1, \ldots, U_n$, and the determinant $|a_{ik}|$ in the n equations*

$$U_i = B_1^{a_{i1}} B_2^{a_{i2}} \cdots B_n^{a_{in}} \qquad (i = 1, 2, \ldots, n)$$

is always equal to j in absolute value.

The last assertion holds for the special basis mentioned in Theorem 36. The passage from the special basis U' to an arbitrary basis U is done by Theorem 38 using an array of exponents with determinant ± 1. However, in the passage from B to U we obviously obtain an array of exponents whose determinant is equal to the product of the determinants which appear in the passage from B to U' and from U' to U, and hence which is equal to $\pm j$.

Finally we formulate a simple criterion for $\mathfrak{U}$ to be of finite index.

Theorem 40. *If $\mathfrak{G}$ is a group with a finite basis $B_1, \ldots, B_m$, then a subgroup $\mathfrak{U}$ is of finite index if and only if a power of each element of $\mathfrak{G}$ belongs to $\mathfrak{U}$.*

If the N_hth power ($N_h > 0$) of B_h belongs to $\mathfrak{U}$ and if we set

$$N = N_1 N_2 \cdots N_m,$$

then B_h^N also belongs to $\mathfrak{U}$ and consequently the Nth power of each element likewise belongs to $\mathfrak{U}$. Hence each element of $\mathfrak{G}$ differs from some

$$B_1^{x_1} \cdots B_m^{x_m} \qquad (0 \leq x_i < N)$$

by a factor in $\mathfrak{U}$; therefore there are at most N^m different cosets, represented by the above elements. Thus the index of $\mathfrak{U}$ is finite.

Conversely in the case of a finite index the infinitely many cosets

$$A\mathfrak{U}, \quad A^2\mathfrak{U}, \quad A^3\mathfrak{U}, \ldots$$

cannot all be distinct, thus a power of A must belong to $\mathfrak{U}$.

We also note that the definition of a factor group $\mathfrak{G}/\mathfrak{U}$ carries over without change from finite groups to infinite Abelian groups, where it is of no concern whether the group $\mathfrak{G}$ has a basis.

CHAPTER III

Abelian Groups in Rational Number Theory

§12 Groups of Integers under Addition and Multiplication

In the elementary theory of rational numbers we are constantly dealing with Abelian groups. The set of integers has the properties:

  (i) $a + b$ is an integer if a and b are integers; $a + b = b + a$,

 (ii) $a + (b + c) = (a + b) + c$,

(iii) If $a + b = a' + b$, then $a = a'$,

(iv) For each a and b there is an integer x such that $a + x = b$.

Thus under composition by addition, the set of integers (positive and negative) forms an infinite Abelian group $\mathfrak{G}$. The unit element is the number zero: $a + 0 = a$. This group is obtained by composition of the element 1 with itself. Hence we are dealing with a *torsion-free group* with *one* basis element, thus with a *cyclic group*. The integers of a module also obviously form an Abelian group and indeed a subgroup of $\mathfrak{G}$. What we proved earlier about a module in Theorem 2 is expressed as follows in the terminology of group theory: *Every subgroup of an infinite cyclic group is again a cyclic group.*

    The module of those numbers divisible by a fixed number k forms a subgroup $\mathfrak{U}_k$ of $\mathfrak{G}$. The index of $\mathfrak{U}_k$ is the number of distinct integers which differ by an element not in $\mathfrak{U}_k$, that is, which have a difference that is not a multiple of k. Hence the index of $\mathfrak{U}_k$ is equal to the number of integers which are incongruent mod k, that is, $= k$ (k assumed > 0). What we called a coset in group theory is here the system of numbers which arise by composition of a definite number a with all elements of $\mathfrak{U}_k$, thus which arise by adding on all multiples of k. *The cosets are thus simply the different residue classes mod* k. The composition of cosets which led us to the factor group $\mathfrak{G}/\mathfrak{U}_k$ appears here as a composition of residue classes mod k, which will be designated as addition of residue classes.

Thus the k residue classes mod k, with composition by addition, form an Abelian group which is isomorphic to the factor group $\mathfrak{G}/\mathfrak{U}_k$.

In all these cases we are dealing with cyclic groups, thus with very simple groups. The investigation of another kind of composition, multiplication, is more important and more difficult.

We first show that the positive integers *do not form a group* under composition by multiplication, since the group axioms (i)–(iii) hold but (iv) does not: namely, for integers a, b there does not always exist an integer x with $ax = b$. However if we add the fractions then we see:

Under composition by multiplication the positive rational numbers form an infinite Abelian group, and indeed a torsion-free group $\mathfrak{M}$. The unit element is the number 1. The theorem about unique decomposition of integers into prime factors obviously asserts:

The positive primes form an infinite basis in the group $\mathfrak{M}$.

The simplest subgroups of $\mathfrak{M}$ are obtained, say, in the form of rational numbers for whose representation only certain (finitely or infinitely many) primes are needed.

By adding on the negative rational numbers (0 excluded) we obtain an extended group, in which one element of finite order, namely -1, occurs.

We now wish to compose the residue classes mod n by a kind of multiplication. If A and B are two residue classes mod n and $a_1 \equiv a_2 \pmod{n}$, $b_1 \equiv b_2 \pmod{n}$ are two representatives of A and B, then we have $a_1 b_1 \equiv a_2 b_2 \pmod{n}$; the residue class to which $a_1 \cdot b_1$ belongs is determined by the classes A, B, independent of the choice of representatives. We write $A \cdot B$ or more briefly AB for the class defined by A and B in this way. Obviously $AB = BA$ and $A(BC) = (AB)C$. However the residue classes mod n do not form a group, since $R_0 A = R_0 B$ for each A, B, where R_0 denotes the residue class of zero; hence axiom (iii) is not satisfied.

However, if A and B are residue classes mod n which are relatively prime to n, then this also holds for AB. And it follows from $ab = a'b \pmod{n}$ that $a \equiv a' \pmod{n}$, if b and n are relatively prime. With this we have proved:

Theorem 41. *The system of residue classes mod n does not form a group with composition by multiplication. However, the $\varphi(n)$ residue classes prime to n form an Abelian group under composition by multiplication. Let this group be simply called the "group of residue classes mod n" and let it be denoted by* $\mathfrak{R}(n)$. *The unit element is the class which contains 1.*

From this fact we immediately infer Fermat's theorem as a consequence of Theorem 21 on groups: if $(a, n) = 1$ then $A^{\varphi(n)} = E$ or $a^{\varphi(n)} \equiv 1 \pmod{n}$.

We pose the problem of giving the structure of this finite Abelian group.

§13 Structure of the Group $\mathfrak{R}(n)$ of the Residue Classes mod n Relatively Prime to n

First we reduce the investigation of $\mathfrak{R}(n)$ to the case where n is a power of a prime by means of

Theorem 42. *Suppose* $(n_1, n_2) = 1$, $n = n_1 \cdot n_2$. *Then*

$$\mathfrak{R}(n) = \mathfrak{R}(n_1) \cdot \mathfrak{R}(n_2).$$

To prove this we assign to each element A of $\mathfrak{R}(n)$ a pair of elements C_1 from $\mathfrak{R}(n_1)$ and C_2 from $\mathfrak{R}(n_2)$ as follows: If a is a number in A, then choose any two numbers c_1, c_2 according to the conditions

$$c_1 \equiv a \,(\mathrm{mod}\ n_1), \qquad c_2 \equiv a \,(\mathrm{mod}\ n_2). \tag{20}$$

The residue class C_1 of c_1 mod n_1 is determined uniquely by A, likewise the residue class C_2 of c_2 mod n_2. We set

$$A = (C_1, C_2)$$

where C_1 belongs to $\mathfrak{R}(n_1)$ and C_2 belongs to $\mathfrak{R}(n_2)$. Conversely if c_1 and c_2 are two numbers relatively prime to n_1 and n_2 respectively, then by Theorem 15, since $(n_1, n_2) = 1$, there is an a determined uniquely by the modulus $n = n_1 \cdot n_2$ which satisfies (20). Moreover it obviously follows from

$$A = (C_1, C_2), \qquad A' = (C_1', C_2')$$

that

$$AA' = (C_1 C_1', C_2 C_2').$$

Thus the group $\mathfrak{R}(n)$ is represented as a product of the groups $\mathfrak{R}(n_1)$ and $\mathfrak{R}(n_2)$.

It follows by repeated application of the theorem for a product of different primes $p_1, p_2, \ldots, p_k$ that

$$\mathfrak{R}(p_1^{\alpha_1} p_2^{\alpha_2} \cdots p_k^{\alpha_k}) = \mathfrak{R}(p_1^{\alpha_1})\mathfrak{R}(p_2^{\alpha_2})\mathfrak{R}(p_k^{\alpha_k}).$$

Therefore the investigation of $\mathfrak{R}(n)$ is reduced to the case where n is a power of a prime.

Theorem 43. *If p is a prime, then the group $\mathfrak{R}(p)$ of residue classes mod p is a cyclic group of order $p - 1$.*

By Theorem 27, we need only show that if q is a prime dividing $p - 1$, then the number of classes A with $A^q = 1$ is equal to q (by Theorem 22 it must be at least q). However, the number of these classes A is identical with the number of integers a which are incongruent mod p and which satisfy $a^q \equiv 1 \,(\mathrm{mod}\ p)$, that is, with the number of different roots of $x^q - 1 \equiv 0$

(mod p). By Theorem 12 this number is at most equal to the degree q, because the modulus is a prime. Consequently it is precisely equal to q.

Hence there is a generating class mod p. Each number g from this class is called a *primitive root mod p*. Accordingly g is a primitive root mod p if $g, g^2, g^3, \ldots, g^{p-1}$ are all incongruent mod p. The powers g^u, where $(u, p - 1) = 1$, and only these, are again primitive roots. There are $\varphi(p - 1)$ different primitive roots mod p.

Theorem 44. *If p is an odd prime, then the group of residue classes modulo each power p^α is cyclic.*

The order of this group is $h = \varphi(p^\alpha) = p^{\alpha-1}(p - 1)$. Here we may take $\alpha \geq 2$. The primes dividing h are p and the prime divisors q of $p - 1$. If e is the basis number which belongs to p in $\mathfrak{R}(p^\alpha)$, then p^e is the number a of solutions of

$$a^p \equiv 1 \pmod{p^\alpha} \tag{21}$$

which are incongruent mod p^α. By Fermat's theorem each such a is $\equiv 1$ (mod p). We assume $a \neq 1$ and $a = 1 + up^m$, where p^m is the highest power of p dividing $a - 1$; hence we have

$$m \geq 1, \qquad (u, p) = 1. \tag{22}$$

It follows from (21) that

$$(1 + up^m)^p \equiv 1 \pmod{p^\alpha}. \tag{23}$$

We now expand the pth power by the binomial theorem and note that, for a prime p, all binomial coefficients

$$\binom{p}{k} = \frac{p(p - 1)(p - 2) \cdots (p - k + 1)}{1 \cdot 2 \cdot 3 \cdots k} \quad \text{(for } k = 1, 2, \ldots, p - 1)$$

are divisible by p, since the numerators are divisible by p while the denominators are not divisible by the prime p. We now wish to show for m in (23) that $m \geq \alpha - 1$. If we have $m \leq \alpha - 2$, then it would follow from (23) that

$$(1 + up^m)^p \equiv 1 \pmod{p^{m+2}},$$

$$(1 + up^m)^p = 1 + \binom{p}{1}up^m + \cdots + \binom{p}{p-1}u^{p-1}p^{m(p-1)} + u^p p^{mp}. \tag{24}$$

Since $p > 2$ and $m \geq 1$, all terms from the third on are divisible by p^{m+2}, that is,

$$(1 + up^m)^p \equiv 1 + up^{m+1} \pmod{p^{m+2}}.$$

Hence it follows from (24) that

$$up^{m+1} \equiv 0 \pmod{p^{m+2}},$$

i.e.,

$$u \equiv 0 \pmod{p}$$

in contradiction to (22). Therefore in (23), $a = 1 + up^m$ with $m \geq \alpha - 1$. However, among these numbers there are at most p which are incongruent mod p^α.

The basis number e which belongs to p for the group is thus ≤ 1, hence $= 1$. The easiest way to see that the basis number for the primes q is also equal to 1 is the following. By Theorems 23 and 24, the elements of the group of classes mod p^α can be represented in the form

$$A \cdot B$$

where B runs through the $p - 1$ classes with $B^{p-1} = 1$, and A runs through the $p^{\alpha-1}$ classes with $A^{p^{\alpha-1}} = 1$. We thus need only check that the subgroup of the B's is cyclic. Now if a is a primitive root mod p, then since $a \equiv a^p \equiv a^{p^2} \equiv \cdots \equiv a^{p^{\alpha-1}} = b$, b is also such a number. Hence the numbers $b, b^2, \ldots,$ b^{p-1} are different mod p, thus a fortiori mod p^α, while their $(p-1)$th powers are $\equiv 1 \pmod{p^\alpha}$. Hence the group of classes B is represented by the powers of the class of b. Therefore it is cyclic and Theorem 44 is proved.

The exceptional case of the prime 2 is treated by

Theorem 45. *The groups $\Re(2)$ and $\Re(4)$ are cyclic. If $\alpha \geq 3$ then the group $\Re(2^\alpha)$ of order $h = \varphi(2^\alpha) = 2^{\alpha-1}$ has exactly two basis classes. One is of order 2, the other of order $h/2 = 2^{\alpha-2}$.*

The statements are trivial for the moduli 2 and 4. Thus suppose $\alpha \geq 3$. The group of classes mod 2^α has order $h = \varphi(2^\alpha) = 2^{\alpha-1}$. The number of incongruent solutions of $x^2 \equiv 1 \pmod{2^\alpha}$ is 2^2, that is, $e = 2$, because x must be odd in any case, $x = 1 + 2v$, and consequently

$$0 \equiv x^2 - 1 \equiv (1 + 2v)^2 - 1 \equiv 4v(v + 1) \pmod{2^\alpha}$$
$$v(v + 1) \equiv 1 \pmod{2^{\alpha-2}}.$$

Obviously only one of the factors can be even and it must then be divisible by $2^{\alpha-2}$, that is,

$$v = 2^{\alpha-2}w \quad \text{or} \quad v = -1 + 2^{\alpha-2}w$$
$$x = 1 + 2^{\alpha-1}w \quad \text{or} \quad x = -1 + 2^{\alpha-1}w$$

with integral w. Each such x is, in fact, also a solution of $x^2 \equiv 1 \pmod{2^\alpha}$. Exactly four of these numbers are incongruent modulo 2^α, namely for $w = 0$ and 1.

However since there exist two basis classes in this group of order $h = 2^{\alpha-1}$, each class can be of order at most $h/2$. If a class of order $h/2$ exists, then this class must also be a basis class of degree $h/2$; the other class then has order 2. We show that the class represented by the number 5 has order $h/2 = 2^{\alpha-2}$ modulo 2^α. To see this we show that

$$5^{2^k} \not\equiv 1 \pmod{2^\alpha} \quad \text{for } \alpha \geq 3 \text{ and } k < \alpha - 2,$$

but

$$5^{2^{\alpha-2}} \equiv 1 \pmod{2^\alpha}.$$

Obviously this is equivalent to

$$5^{2^{\alpha-2}} = 1 + 2^\alpha u, \quad \text{where } u \text{ is odd } (\alpha \geq 3).$$

Since $25 = 1 + 8 \cdot 3$, the equation is true for $\alpha = 3$. If it is true for α in general, then it follows by squaring that

$$5^{2^{\alpha-1}} = (1 + 2^\alpha u)^2 = 1 + 2^{\alpha+1} u + 2^{2\alpha} u^2 = 1 + 2^{\alpha+1} u (1 + 2^{\alpha-1} u).$$

Therefore we have the validity of the assertion for $\alpha + 1$.

We observe further that for composite moduli n, the group $\mathfrak{R}(n)$ is not cyclic in general. If p is a divisor of $\varphi(n)$, then, by Theorem 42, the basis number $e(p)$ of $\mathfrak{R}(n)$, which belongs to p, is equal to the sum of the basis numbers $e_i(p)$, which belong to p in $\mathfrak{R}(p_i^{\alpha_i})$, where $p = p_1^{\alpha_1} p_2^{\alpha_2} \cdots$ is the decomposition of n into primes. However for odd p_i, 2 is a divisor of $\varphi(p_i^{\alpha_i})$ and consequently $e_i(2) = 1$. Thus if two odd primes divide n, so does $e(2)$ for $\mathfrak{R}(n) \geq 2$. Hence the group is not cyclic.

§14 Power Residues

With the help of the theorems before us, the foundations of the theory of power residues, that is, the solvability of binomial congruences of the form

$$x^q \equiv a \pmod{n} \tag{25}$$

can easily be developed. If we restrict ourselves to the cases where the following hypotheses are satisfied:

q is a positive prime, n is odd and a power of a prime, say p^α, $(a, n) = 1$,

then the solutions x, if any, are likewise relatively prime to the modulus p^α, and the problem of the solvability of (25) in integers can be formulated in group-theoretic fashion as follows:

Let a class A be given in the group of residue classes mod p^α. How many elements X are there in the group such that

$$X^q = A?$$

We distinguish two cases:

1. The prime q does *not* divide the order of the group $h = \varphi(p^\alpha)$. *Then there is exactly one element X of the desired sort.* To see this let the integers m, n be determined so that $qm + hn = 1$, which is possible since $(q, h) = 1$. Since $X^h = 1$, it follows from $X^q = A$ that

$$X = X^{qm+hn} = (X^q)^m = A^m$$

and this element actually satisfies $X^q = A$.

2. q divides $h = \varphi(p^\alpha)$. By Theorem 44 there is an element C (of order h), whose powers yield all elements of the group. We set

$$A = C^{a'}, \qquad X = C^x$$

with integral a' and x, which are completely determined mod h. By Theorem 20, it follows from

$$X^q = A, \qquad C^{xq} = C^{a'}$$

that

$$xq \equiv a' \pmod{h}$$

and conversely. However, since $q \mid h$, this congruence is solvable in integers x only if

$$q \mid a',$$

and then it has exactly q different solutions mod h. *That is, the equation $X^q = A$ has either no solutions or exactly q distinct solutions X.* Since C is a primitive class, the condition $q \mid a'$ is equivalent to

$$A^{h/q} = C^{a'h/q} = (C^h)^{a'/q} = 1.$$

Returning to numbers from residue classes we see that we have proved

Theorem 46. *The congruence*

$$x^q \equiv a \pmod{p^\alpha}$$

where q, p are primes, $p \neq 2$ $(a, p) = 1$, has exactly one solution x in integers if q does not divide $\varphi(p^\alpha)$. However if q divides $\varphi(p^\alpha)$, then the equation has q solutions, and indeed exactly q, if

$$a^{\varphi(p^\alpha)/q} \equiv 1 \pmod{p^\alpha}. \tag{26}$$

If the exponent is also relatively prime to the modulus, that is $q \neq p$, then Condition (26) allows a still simpler formulation. For, since q is a prime, but $q \neq p$, it follows from $q \mid \varphi(p^\alpha)$ that

$$q \mid p - 1, \qquad q' = \frac{p-1}{q} \quad \text{integral,}$$

and (26) reads

$$a^{p^{\alpha-1}q'} \equiv 1 \pmod{p^\alpha}; \tag{26a}$$

hence in particular, because of Fermat's theorem we also have

$$a^{q'} \equiv 1 \pmod{p}. \tag{27}$$

This congruence, which has the solvability of $x^q \equiv a \pmod{p}$ as a consequence, also conversely has (26) as a consequence. Specifically for each prime p, it follows from

$$m \equiv n \pmod{p^r}, \qquad m = n + xp^r \quad \text{(integral } x\text{)},$$

that

$$m^p = (n + xp^r)^p = n^p + \binom{p}{1} xp^r + \cdots \equiv n^p \pmod{p^{r+1}}$$

$$m^p = n^p \pmod{p^{r+1}},$$

since the binomial coefficients $\binom{p}{k}$ are divisible by p for $k = 1, 2, \ldots, p - 1$ as we have already used above once (see p. 43); thus (26) follows from 27.

If $q \mid p - 1$, then (27), which does not depend any more on the exponent α, is also a condition for the solvability of $x^q \equiv a \pmod{p^\alpha}$. Hence

Theorem 46a. *If q is a prime factor of $p - 1$, p an odd prime, and $(a, p) = 1$, then the congruence $x^q \equiv a \pmod{p^\alpha}$ is solvable if and only if it is solvable mod p. For this it is necessary and sufficient that*

$$a^{(p-1)/q} \equiv 1 \pmod{p}.$$

Then there are q solutions incongruent mod p^α.

As moduli, the powers 2^α require special treatment because of Theorem 45.

Theorem 47. *With odd q and a, the congruence $x^q \equiv a \pmod{2^\alpha}$ always has exactly one solution. For $q = 2$ and odd a, $x^2 \equiv a \pmod{2^\alpha}$ is solvable for $\alpha \geq 3$ if and only if it is solvable mod 8, that is, if $a \equiv 1 \pmod{8}$, and indeed the number of incongruent solutions in this case is equal to 4. For $a \equiv 1 \pmod{4}$, $x^2 \equiv a \pmod{4}$ has two solutions, otherwise it has no solutions for odd a, and $x^2 \equiv a \pmod{2}$ always has a solution.*

The first part (q odd) is proved exactly as above in Case 1. Since, by Theorem 45, the classes mod 2^α ($\alpha \geq 3$) can be represented in the form $B_1^{a_1} B_2^{a_2}$, where $B_1^2 = B_2^{2^{\alpha-2}} = 1$, then we see as above in Case 2 that only those classes $A = B^{a_1} B_2^{a_2}$, where $a_1 = 0$, a_2 even, can be represented in the form X^2. And then there are as many classes X with $X^2 = B_2^{a_2}$ as there are classes with $X^2 = 1$, that is, $2^2 = 4$. A simple form of the solvability condition for $x^2 \equiv a \pmod{2^\alpha}$ with $\alpha \geq 3$ $a \equiv 1 \pmod{8}$ arises as follows:

If $x^2 \equiv a \pmod{2^\alpha}$ ($\alpha \geq 3$) is solvable (let $x = x_0$ be a solution), then the congruence is also solvable mod $2^{\alpha+1}$. For let an integer z be determined so that

$$(x_0 + 2^{\alpha-1}z)^2 - a = x_0^2 - a + 2^\alpha x_0 z + 2^{2\alpha-2}z^2 \equiv 0 \pmod{2^{\alpha+1}}$$

which, since

$$2\alpha - 2 = \alpha + (\alpha - 2) \geq \alpha + 1,$$

leads to the solvable congruence

$$\frac{x_0^2 - a}{2^\alpha} + x_0 z \equiv 0 \pmod{2}.$$

If $x^2 \equiv a$ (mod 8) is solvable, then the congruence is consequently also solvable mod 2^α. However, as we see by trying out the other cases, this congruence is solvable only for $a \equiv 1$ (mod 8).

From this we immediately obtain an overview of the solutions of

$$x^q \equiv a \,(\text{mod } n) \tag{28}$$

for composite n. Suppose $(a, n) = 1$. In order that the congruence be solvable mod n it must be solvable modulo each power of a prime which divides n. If $n = p_1^{\alpha_1} p_2^{\alpha_2} \cdots p_r^{\alpha_r}$, where the p_i are distinct primes, and if N_i is the number of different solutions mod $p_i^{\alpha_i}$ of

$$z^q \equiv a \,(\text{mod } p_i^{\alpha_i}),$$

then the number of different solutions of (28) is

$$N = N_1 \cdot N_2 \cdots N_r.$$

To see this assume that the r numbers $z_1, \ldots, z_r$ are solutions of $z_i^q \equiv a$ (mod $p_i^{\alpha_i}$) and then let us determine x from

$$x \equiv z_i \,(\text{mod } p_i^{\alpha_i}), \qquad (i = 1, 2, \ldots, r).$$

Then

$$x^q \equiv z_i^q \equiv a \,(\text{mod } p_i^{\alpha_i}),$$

hence

$$x^q \equiv a \,(\text{mod } n).$$

x is uniquely determined mod n by the z_i. Two different systems z_i and z_i' lead to the same x mod n if and only if $z_i \equiv z_i'$ (mod $p_i^{\alpha_i}$) for $i = 1, 2, \ldots, r$. On the other hand every solution x of (28) is also a system of solutions of the r single congruences, namely $z_i = x$. Consequently $N_1 N_2 \cdots N_r$ is the exact number of solutions of (28) mod n.

§15 Residue Characters of Numbers mod n

In closing these investigations we finally wish to connect the numbers a considered relative to a modulus n with the concepts, developed in §10, related to characters of Abelian groups.

The elements of the group $\mathfrak{R}(n)$ are the different residue classes mod n which are relatively prime to n, and hence, as an Abelian group, there is assigned to these elements a system of $h = \varphi(n)$ characters. Let a be an integer from one such class A. Then corresponding to each character $\chi(A)$ we define a number-theoretic function

$$\chi(a) = \chi(A),$$

for each integer a, relatively prime to n, which has the following properties:

(1) $\chi(a) = \chi(b)$, if $a \equiv b \pmod{n}$.
(2) $\chi(a)\chi(b) = \chi(ab)$.
(3) $\chi(a) \neq 0$ for all a relatively prime to n.

We complete this definition for the remaining integers by fixing

(4) $\chi(a) = 0$, if $(a, n) > 1$.

Statements (1)–(3) are valid for this extended system of arguments where a is allowed to run through all integers.

Each function $\chi(a)$ with the properties (1)–(4) is called a *residue character of a mod n*. There are exactly $h = \varphi(n)$ different residue characters mod n and by Theorem 31 we have

$$\sum_{k \bmod n} \chi(k) = \begin{cases} 0 & \text{if } \chi \text{ is not the principal character,} \\ \varphi(n) & \text{if } \chi \text{ is the principal character.} \end{cases} \tag{29}$$

Here again we call that character which is equal to 1 for all a relatively prime to n the principal character. The summation $k \bmod n$ under $\sum$ signifies that the index k runs through a complete residue system mod n. In an analogous manner we have

$$\sum_{\chi} \chi(k) = \begin{cases} 0 & \text{if } k \not\equiv 1 \pmod{n}, \\ \varphi(n) & \text{if } k \equiv 1 \pmod{n}. \end{cases} \tag{30}$$

With the help of residue characters mod n we now wish to give a different formulation of the conditions for the solvability of a congruence

$$x^q \equiv a \pmod{n}$$

which were developed in the preceding section. Here we will make the hypothesis:

$$(q, n) = 1, \quad q \text{ prime}, \quad \text{and} \quad (a, n) = 1.$$

Thus the class A of a should be a qth power in the group $\mathfrak{R}(n)$. Now the qth powers of all classes form a subgroup $\mathfrak{U}_q$ of $\mathfrak{R}(n)$. By Theorem 29, the order of the factor group is $\mathfrak{R}/\mathfrak{U}_q = q^e$, where $e = e(q)$ is the basis number belonging to q in $\mathfrak{R}(n)$ and e is at the same time the number of basis elements in $\mathfrak{R}/\mathfrak{U}_q$. Consequently, by Theorem 33, there exist exactly e characters for $\mathfrak{R}(n)$ and thus exactly e residue characters mod n

$$\chi_1(a), \ \chi_2(a), \ \ldots, \ \chi_e(a)$$

such that the e equations $\chi_i(a) = 1$ $(i = 1, 2, \ldots, e)$ are the necessary and sufficient conditions for the class A of a to be a qth power. These e characters are independent of one another in the sense that there are always numbers a for which these e characters are arbitrarily given qth roots of unity.

Until now only the fact that $\mathfrak{R}(n)$ is a finite Abelian group was used; the finer structure plays a role only when we try to represent e as a function of

q and n. Now if n is a power of a prime p^z, then $e(q) = 0$ for odd p if q does not divide $\varphi(p^z)$, and $e(q) = 1$ if $q \mid \varphi(p^z)$, since the group $\Re(p^z)$ is cyclic. However if n is composite and odd, $n = p_1^{z_1} \cdots p_r^{z_r}$, then by Theorem 42, $e(q)$ for $\Re(n)$ is equal to the number of those p_i for which $q \mid \varphi(p_i^{z_i})$.

Each residue character $\chi(a)$, which is equal to 1 for all qth powers a, is called a qth *power character of a mod n*. By Theorem 33, each qth power character is representable as a product of powers by the basis characters $\chi_1, \ldots, \chi_e$.

The simplest case, which will concern us exclusively in what follows, is the case with $q = 2$, where we are concerned with the classes which can be represented as squares. The corresponding power characters are then called quadratic characters.

§16 Quadratic Residue Characters mod n

An integer a relatively prime to n is called a *quadratic residue mod n*, or simply a *residue mod n* if the congruence

$$x^2 \equiv a \pmod{n}$$

is solvable in integers x. In the other case a is called a *nonresidue* mod n. By the preceding section the conditions for the solvability are that the $e(2)$ given residue characters mod n, for a, have the value 1. Each of the characters $\chi(a)$ is a square root of unity, hence it can only have the value ± 1.

To begin with, if $n = p$ is an odd prime, then the corresponding $e(2) = 1$, as 2 divides $p - 1$ and the group $\Re(p)$ is cyclic. Thus among the $p - 1$ characters mod n there is exactly one, say $\chi(a)$, which is a square root of unity but not always $= +1$ and $\chi(a) = +1$ is the condition that a is a quadratic residue mod p. We set

$$\chi(a) = \left(\frac{a}{p}\right).$$

By its definition this character is equal to ± 1 for each a not divisible by p. Thus we have

(1) $\left(\frac{a}{p}\right) = \left(\frac{a'}{p}\right)$ if $a \equiv a' \pmod{p}$,
(2) $\left(\frac{ab}{p}\right) = \left(\frac{a}{p}\right)\left(\frac{b}{p}\right)$,
(3) $\left(\frac{a^2}{p}\right) = 1$,
(4) $\left(\frac{a}{p}\right)$ is not equal to 1 for some a,

where a', a, b are integers not divisible by p. The symbol $\left(\frac{a}{p}\right)$ *is defined by these properties alone*, for each a relatively prime to p, since by (1) and (2), it is a residue character mod p, by (3) this character has only the values ± 1, and by (4), it is not always $= +1$. Hence the residue classes A for which it is 1 form a subgroup of $\Re(p)$ to which all squares belong. Thus its index is

≤ 2 but > 1 and therefore exactly $= 2$. Hence $\left(\frac{a}{p}\right) = +1$ for the quadratic residues a mod p and equal to -1 for the nonresidues mod p.

If we recall that since

$$a^{p-1} - 1 \equiv 0 \ (\text{mod } p),$$

$$(a^{(p-1)/2} + 1)(a^{(p-1)/2} - 1) \equiv 0 \ (\text{mod } p),$$

then in view of Theorem 46a, $\left(\frac{a}{p}\right)$ should be defined as the one of the two numbers ± 1 for which

$$\left(\tfrac{a}{p}\right) \equiv a^{(p-1)/2} \ (\text{mod } p). \tag{31}$$

Legendre introduced the residue symbol $\left(\frac{a}{p}\right)$ into number theory in this manner.

The number of incongruent quadratic residues mod p is $(p-1)/2$, so the number of nonresidues $= p - 1 - (p-1)/2 = (p-1)/2$; hence there are just as many residues as nonresidues mod p.

By Theorem 46a, the condition $\left(\frac{a}{p}\right) = +1$ is at the same time the condition that a is a quadratic residue mod p^{α}. The number of residues mod p^{α} is also equal to the number of nonresidues mod p^{α}, namely $= \varphi(p^{\alpha})/2 = p^{\alpha-1}(p-1)/2 \ (\alpha > 1)$.

For a composite, and for the time being, odd $n = p_1^{\alpha_1} p_2^{\alpha_2} \cdots p_r^{\alpha_r}$, the condition that a is a residue mod n is given by $e(2)$ equations for a certain set of $e(2)$ characters mod n. Here $e(2) = r$. The number of quadratic residues mod n is $\varphi(n)/2^r$, hence for $r > 1$ it is *not equal to the number of nonresidues*. By what was done at the end of §14, the conditions that a is a residue mod n are that a is a residue modulo each prime p_i which divides n, that is, that the r equations

$$\left(\frac{a}{p_i}\right) = 1 \qquad (i = 1, 2, \ldots, r)$$

hold. As we know, for the modulus $2^{\alpha} (\alpha \geq 3)$, the group $\mathfrak{R}(2^{\alpha})$ is no longer cyclic, but it has two basis elements. The decision as to whether or not a is a quadratic residue mod 2^{α} cannot be made by statements about *one* residue character mod 2^{α} but rather for this we need two pieces of information. *For the time being we omit the introduction of a residue symbol mod 2^{α}*, and only later, in §46, will we return to it.

On the other hand we define a symbol $\left(\frac{a}{n}\right)$ for composite odd n. Let

$$n = p_1^{\alpha_1} \cdots p_r^{\alpha_r}, \qquad n \text{ odd}.$$

We set

$$\left(\frac{a}{n}\right) = \left(\frac{a}{p_1}\right)^{\alpha_1} \left(\frac{a}{p_2}\right)^{\alpha_2} \cdots \left(\frac{a}{p_r}\right)^{\alpha_r},$$

provided the elements on the right side have a meaning, that is, if $(a, n) = 1$. Finally let

$$\left(\frac{a}{n}\right) = 0 \quad \text{if } (a, n) > 1.$$

For this extended symbol, we have, by definition,

$$\left(\frac{a}{n}\right) = \left(\frac{a'}{n}\right) \quad \text{if } a \equiv a' \ (\text{mod } n),$$

$$\left(\frac{ab}{n}\right) = \left(\frac{a}{n}\right) \cdot \left(\frac{b}{n}\right)$$

for arbitrary integers a, a', b whether they are relatively prime to n or not. Hence this symbol is also a residue character mod n. However we recall once more that for composite n, *no conclusion can be drawn from the value* $\left(\frac{a}{n}\right)$ *as to whether a is a quadratic residue* mod n *or not.* If a is a residue mod n, then $\left(\frac{a}{n}\right) = +1$ but not conversely.

Legendre and before him, in special cases, Euler, had already made the following remarkable discovery about this residue symbol, which has many consequences for all of number theory and which as the *law of quadratic reciprocity* is formulated today as follows:

For positive odd a, n,

$$\left(\frac{a}{n}\right) = \left(\frac{n}{a}\right)(-1)^{((a-1)/2)((n-1)/2)}.$$

Beyond this, the so-called completion theorems hold:

$$\left(\frac{-1}{n}\right) = (-1)^{(n-1)/2}, \qquad n \text{ odd} > 0$$

$$\left(\frac{2}{n}\right) = (-1)^{(n^2-1)/8}, \qquad n \text{ odd}.$$

After *Legendre* published an attempted proof, which was to be sure incomplete in an essential point, Gauss (1796), who was nineteen years old, succeeded in finding the first proof which he published in 1801 in his classical work *Disquisitiones Arithmeticae*. Since then many different proofs have been given for the reciprocity law; the index in Bachmann's book contains 45 entries; eight proofs are due to Gauss alone.

Modern number theory dates from the discovery of the reciprocity law. By its form it still belongs to the theory of rational numbers, as it can be formulated entirely as a simple relation between rational numbers; however its content points beyond the domain of rational numbers. Gauss himself recognized this. He first attempted to carry over the arithmetic concepts to the complex integers $a + b\sqrt{-1}$ where a, b are integers and he succeeded in finding and proving a similar law for fourth power residues. (It was probably this success of complex number theory which induced him to introduce complex numbers, which were viewed at that time with mistrust and used only occasionally as having equal rights with real numbers in the remaining parts of analysis). He recognized that Legendre's reciprocity law represents a special case of a more general and much more encompassing law. For this reason he and many other mathematicians have looked again and again for new proofs whose essential ideas carry over to other number domains in the hope of coming closer to a general law. The last decisive step was taken by *Kummer* through his introduction of ideal prime factors. Then Dedekind laid the foundations for the general theory of algebraic number fields, and, with this available, the formulation and the proof of the most general reciprocity law for qth power residues, where q is a prime, was finally achieved by *Hilbert* and his student *Furtwängler*.

The development of algebraic number theory has now actually shown that the content of the quadratic reciprocity law only becomes understandable if one passes to general algebraic numbers and that a proof appropriate to the nature of the problem can be best carried out with these higher methods. However, it must be said of the elementary proofs that they possess rather the character of supplementary verification.

For this reason we will dispense entirely with a presentation of an elementary proof. Rather we set ourselves the problem of carrying over the concepts of rational number theory, in particular the concept of integer, to other domains of numbers, where new relations between rational integers will also be obtained, e.g., the reciprocity law itself will be presented as a side result.

CHAPTER IV

Algebra of Number Fields

§17 Number Fields, Polynomials over Number Fields, and Irreducibility

Definition. A system of complex numbers is called a *number field* (or, more briefly, a *field*) if it contains more than one number and if along with the numbers α and β it always contains $\alpha + \beta$, $\alpha - \beta$, $\alpha\beta$, and, if $\beta \neq 0$, α/β.

This means that all rational operations can be performed unrestrictedly within the system. Following Kronecker the term *domain of rationality* is also used in place of the term field. The additional condition, that the system is to contain more than one element, only excludes the system which consists of just a zero element, which satisfies the remaining conditions of the definition.

The concept of a field is related to the concept of a group. By definition, the numbers of a field form an infinite Abelian group under composition by addition. Furthermore the numbers of the field, excluding 0, also form an Abelian group under composition by multiplication.

Examples of number fields are:

The system of all rational numbers.
The system of all real numbers.
The system of all complex numbers.
The system of all numbers of the form $R(\omega)$, where $R(x)$ runs through all rational functions of x with rational numbers as coefficients, where ω is a fixed number.

Since $\alpha/\alpha = 1$, each field thus contains the number 1, and thus also $1 + 1 = 2$, $1 - 1 = 0$ etc. Thus it contains all integers and hence also all quotients of these, that is, all rational numbers. For this reason, the field of rational numbers, which we will denote by $k(1)$ is called the *absolute domain of rationality*. This field is contained in every number field.

In this chapter we will be concerned with the *algebra of number fields* while, after introduction of certain numbers of the field as *integral* numbers, the arithmetic properties of number fields will be dealt with in the remaining chapters.

Now let k be an arbitrary number field. By a *polynomial over k* we mean a polynomial all of whose coefficients are numbers from k. The quotient of two polynomials over k is called a *rational function* over k. If $f(x)$ and $g(x)$ are polynomials over k, then, as is known, if $g(x)$ is of degree at least 1, two polynomials $q(x)$ and $r(x)$ can be determined so that

$$f(x) = q(x)g(x) + r(x), \tag{32}$$

where the degree of $r(x)$ is less than that of $g(x)$. We call $r(x)$ the *remainder* of $f(x)$ mod $g(x)$. The coefficients of $q(x)$ and $r(x)$ can be calculated entirely from those of $f(x)$ and $g(x)$ by means of rational operations, and hence these coefficients likewise belong to k. If $r(x)$ is equal to 0, then $f(x)$ is said to be *divisible* by $g(x)$ or $g(x)$ *divides* $f(x)$; in symbols

$$g(x)\,|\,f(x).$$

If, in (32), the degree m of $f(x)$ is less than the degree n of $g(x)$, then $q = 0$ and $r(x) = f(x)$. On the other hand, if $m \geq n$, then the degree of $q(x)$ is equal to $m - n$, $q(x)$ is not 0, and the degree of $r(x)$ is $< n$. Hence if each of the two polynomials $f(x)$ and $g(x)$ is divisible by the other, then they differ only by a constant factor. The trivial factors of any polynomial $f(x)$ are the constants, that is, the polynomials of 0th degree, and the polynomials $cf(x)$. A polynomial, $a(x - \alpha)$, of the first degree has no factors other than these trivial ones. By the fundamental theorem of algebra each polynomial of degree n can be decomposed in exactly one way into n factors of degree 1 such that

$$f(x) = c(x - \alpha_1)(x - \alpha_2) \cdots (x - \alpha_n),$$

where c is a constant different from zero and $\alpha_1, \ldots, \alpha_n$ are n coincident or distinct complex numbers. Thus if we admit arbitrary coefficients for the polynomials, then the polynomials of 0th degree play the same role as the units ± 1 and the polynomials of degree 1 play the same role as the primes in investigations about divisibility.

If we restrict ourselves to polynomials over a fixed number field k, then these relationships are quite different. *We call a polynomial $f(x)$ irreducible over k, or indecomposable over k, if $f(x)$ cannot be represented as a product of two polynomials over k neither of which is a constant.*

Accordingly, for example, every polynomial over k of degree 1 is irreducible over k. However since the fundamental theorem asserts nothing about whether roots α of $f(x)$ belong to k, polynomials of higher degree may also be irreducible over k. For example, $x^2 + 1$ is obviously irreducible over the field of real numbers. Because of this we must leave the problem of the exact nature of polynomials irreducible over k, without discussing it here, and be satisfied with their existence.

The most important fact concerning polynomials over k is stated in the following theorem:

Theorem 48. *Two arbitrary nonzero polynomials $f_1(x)$ and $f_2(x)$ over k have a uniquely determined greatest common divisor $d(x)$, that is, there is a polynomial $d(x)$ with leading coefficient 1, such that*

$$d(x)|f_1(x), \qquad d(x)|f_2(x)$$

and every polynomial which divides $f_1(x)$ and $f_2(x)$, also divides $d(x)$.

Moreover, $d(x)$ can be represented in the form

$$d(x) = g_1(x)f_1(x) + g_2(x)f_2(x), \tag{33}$$

where $g_1(x)$ and $g_2(x)$ are polynomials over k, and thus $d(x)$ is also a polynomial over k.

The proof is well known from elementary algebra, yet no importance is attached there to the nature of the numerical coefficients which appear. For this reason we reproduce quite briefly a proof based on the proof of the analogous fact for rational numbers (Theorems 1 and 2). Among the polynomials

$$L(x) = u_1(x)f_1(x) + u_2(x)f_2(x),$$

where $u_1(x)$ and $u_2(x)$ run through all polynomials over k, we consider such a polynomial with leading coefficient 1 whose degree is as small as possible. Let $d(x)$ be such a polynomial and suppose that (33) holds. If $d(x)$ is of degree 0, then it is $= 1$ and hence it divides $f_1(x)$ and $f_2(x)$. But even if it is of higher degree, it must divide $f_1(x)$, for let the remainder $r(x)$ of $f_1(x)$ mod $d(x)$ be determined

$$f_1(x) = q(x)d(x) + r(x)$$
$$r(x) = f_1(x) - q(x)d(x)$$
$$r = f_1 - qd = f_1 - q(g_1 f_1 + g_2 f_2) = (1 - qg_1)f_1 - qg_2 f_2.$$

Thus this $r(x)$ also has the form $L(x)$, while its degree (as a remainder mod $d(x)$) is less than the degree of $d(x)$. Consequently it cannot have coefficients different from 0, hence it is 0. Thus $d(x)|f_1(x)$; in exactly the same way we see that $d(x)|f_2(x)$.

However by (33) each common divisor of $f_1(x)$ and $f_2(x)$ divides $d(x)$. If a polynomial $d_0(x)$ has the property stated in the first part of the theorem, then $d(x)|d_0(x)$ holds as well as $d_0(x)|d(x)$, consequently $d_0(x)$ and $d(x)$ differ by only a constant factor; since their leading coefficients are 1, $d_0(x) = d(x)$.

We write $(f_1(x), f_2(x)) = d(x)$ and call $f_1(x)$ and $f_2(x)$ *relatively prime* if $d = 1$. The greatest common divisor of two polynomials is completely defined by this alone, not only relative to a definite field k, while, in general, the property of irreducibility of a polynomial is relative to the field k to which it belongs.

We have immediately from Theorem 48:

Theorem 49. *If a polynomial $f(x)$, irreducible over k, has a common zero $x = \alpha$ with a polynomial $g(x)$ over k, then $f(x)$ is a divisor of $g(x)$ and hence all zeros of $f(x)$ are zeros of $g(x)$.*

For $(f(x), g(x))$ is at least divisible by $x - \alpha$ and thus not $= 1$. On the other hand $f(x)$ has no factors over k other than $cf(x)$. Consequently $(f(x), g(x)) = cf(x)$ and $f(x) | g(x)$.

In particular an irreducible polynomial over k, of degree n, has exactly n distinct roots, since otherwise it would have a common zero with the derivative $f'(x)$, which is also a polynomial over k but of degree $n - 1$, and hence it would have to divide $f'(x)$, which cannot be the case.

§18 Algebraic Numbers over k

Suppose that a number θ is a root of a polynomial $P(x)$ over k. Among all polynomials over k, with leading coefficient 1, which have this root θ, there is one of smallest degree. This polynomial is necessarily irreducible over k—since otherwise θ would already be a root of a divisor of this polynomial— and hence by Theorem 49 it is fully determined by θ and k.

The degree n of this polynomial is called the *degree of θ with respect to k*, or the *relative degree of θ*. The n roots of this polynomial, $\theta_1, \theta_2, \ldots, \theta_n$— surely distinct from one another—are called the *conjugates of θ* with respect to k or the *relative conjugates of θ*. Each of the numbers θ_i is called an *algebraic number over k*. If $k = k(1)$ is the field of rational numbers, then in this notation the reference to k is omitted. Thus in particular a number θ is called an *algebraic number* if it is a root of a polynomial with rational coefficients.

Obviously the numbers in k itself are the numbers of relative degree 1. For further investigation we need the *symmetric function theorem* from algebra, which we formulate as follows:

Let $\alpha_1, \alpha_2, \ldots, \alpha_n$ be n independent variables and let $f_1, f_2, \ldots, f_n$ be their n elementary symmetric functions which are the coefficients of the polynomial in x: $(x - \alpha_1)(x - \alpha_2) \cdots (x - \alpha_n)$. *Then every symmetric polynomial $S(\alpha_1, \ldots, \alpha_n)$ in $\alpha_1, \ldots, \alpha_n$ can be represented as a polynomial of $f_1, f_2, \ldots, f_n$:*

$$S(\alpha_1, \ldots, \alpha_n) = G(f_1, \ldots, f_n).$$

The coefficients of G can be calculated from those of S entirely by the operations of addition, subtraction, and multiplication.

If the theorem is applied twice in succession, then we obtain: If $\beta_1, \ldots, \beta_m$ are m additional independent variables and $\varphi_1, \ldots, \varphi_m$ are their elementary symmetric functions, and if $S(\alpha_1, \ldots, \alpha_n; \beta_1, \ldots, \beta_m)$ is a polynomial of the $n + m$ arguments which remains unchanged under each permutation of the α among themselves and under each permutation of the β among themselves, then S can be represented as a polynomial of the $f_1, \ldots, f_n$ and $\varphi_1, \ldots, \varphi_m$:

$$S(\alpha_1, \ldots, \alpha_n; \beta_1, \ldots, \beta_m) = G(f_1, \ldots, f_n, \varphi_1, \ldots, \varphi_m).$$

The coefficients of G can be calculated from those of S by addition, subtraction, and multiplication.

From this we note first of all:

Theorem 50. *If α, β are algebraic numbers over k, then the same is true for $\alpha + \beta, \alpha - \beta, \alpha\beta$, and, if $\beta \neq 0$, for α/β.*

If $\alpha_1, \ldots, \alpha_n$ are the conjugates of α and $\beta_1, \ldots, \beta_m$ are those of β with respect to k, then the elementary symmetric functions of α, as well as those of β, are numbers in k. The product

$$H(x) = \prod_{k=1}^{m} \prod_{i=1}^{n} (x - (\alpha_i + \beta_k))$$

as a symmetric function in the α, and in the β, is then a polynomial over k by reason of the fundamental theorem just stated, and $\alpha + \beta$ is to be found among its roots, which accordingly is an algebraic number over k. This likewise follows for $\alpha - \beta$ and $\alpha\beta$.

With α/β our method breaks down since the analogous product is not a polynomial in the β and hence the fundamental theorem cannot be applied. However if $\beta \neq 0$, then let us set $x = 1/y$ in the irreducible equation for β over k

$$x^m + c_{m-1}x^{m-1} + c_{m-2}x^{m-2} + \cdots + c_1 x + c_0 = 0$$

and let us multiply by y^m. The polynomial in y obtained in this way then has the root $1/\beta$, and this number is thus likewise an algebraic number over k; consequently by what has gone before, the product $\alpha(1/\beta) = \alpha/\beta$ is also an algebraic number over k.

Theorem 51. *If ω is a root of a polynomial*

$$\varphi(x) = x^m + \alpha x^{m-1} + \beta x^{m-2} + \cdots + \lambda$$

whose coefficients are algebraic numbers over k, then ω is also an algebraic number over k.

Suppose α_i runs through the conjugates of α, β_k runs through the conjugates of β etc. Then by the theorem on symmetric functions, the polynomial

$$F(x) = \prod_{i, k, \ldots, s} (x^m + \alpha_i x^{m-1} + \beta_k x^{m-2} + \cdots + \lambda_s)$$

as a symmetric expression in the conjugates has coefficients in k; since $F(\omega) = 0$, ω is an algebraic number over k.

§19 Algebraic Number Fields over k

Each algebraic number θ over k obviously generates a *field*, the totality of all rational functions of θ with coefficients in k. Let this field be denoted by $K(\theta; k)$ or more simply by $K(\theta)$ and let us say of it that it arises by *adjunction* of θ to k. Likewise, by the adjunction of *several algebraic numbers* $\alpha, \beta, \gamma, \ldots$ over k to k, we obtain a field $K(\alpha, \beta, \gamma, \ldots; k)$ whose numbers are the rational functions of $\alpha, \beta, \gamma, \ldots$ with coefficients in k.

Theorem 52. *Every field obtained by adjunction of several algebraic numbers over k can also be generated by adjunction of a single algebraic number over k.*

Obviously, it is enough to prove the theorem for the adjunction of two numbers. Thus let $\alpha_1, \ldots, \alpha_n$ be the n conjugates of a number α_1 of relative degree n and $\beta_1, \ldots, \beta_m$ the m conjugates of β_1 of relative degree m. We will show that with suitable choice of u and v in k the number $u\alpha_1 + v\beta_1 = \omega_{11}$ is a number generating the field $K(\alpha_1, \beta_1; k)$. We must prove that α_1 and β_1 themselves—consequently also every number from $K(\alpha_1, \beta_1; k)$—are representable as rational functions of ω_{11} with coefficients from k.

For this purpose we choose u and v as rational numbers in such a way that the nm numbers

$$\omega_{ik} = u\alpha_i + v\beta_k \qquad (i = 1, 2, \ldots, n; k = 1, 2, \ldots, m)$$

are all distinct. This is possible since what is required is that for all pairs of indices i, k and i', k',

$$u(\alpha_i - \alpha_{i'}) + v(\beta_k - \beta_{k'}) \neq 0$$

is to hold except if $i = i'$ and $k = k'$ hold simultaneously. The coefficients in these linear functions of u, v never both vanish simultaneously as the α_i are distinct from one another and the β_k are distinct from one another. Hence we must choose u/v ($u \neq 0, v \neq 0$) different from the finitely many numbers

$$-\frac{\beta_k - \beta_{k'}}{\alpha_i - \alpha_{i'}}, \qquad i \neq i', k \neq k';$$

then the ω_{ik} are all distinct and are roots of the polynomial over k

$$H(x) = \prod_{i, k} (x - (u\alpha_i + v\beta_k)) = \sum_{h=0}^{nm} c_h x^h.$$

We now try to construct a rational function of x which takes the values β_k for $x = \omega_{1k}$ $(k = 1, 2, \ldots, m)$. Recalling the Lagrange interpolation formula let us consider the expression

$$\Phi(x) = \sum_{i=1}^{n} \sum_{k=1}^{m} \beta_k \frac{H(x)}{x - \omega_{ik}}.$$

This $\Phi(x)$ is a polynomial in k. For since $H(\omega_{ik}) = 0$

$$\frac{H(x)}{x - \omega_{ik}} = \frac{H(x) - H(\omega_{ik})}{x - \omega_{ik}} = \sum_{h=0}^{nm} c_h \frac{x^h - \omega_{ik}^h}{x - \omega_{ik}} = G(x, \omega_{ik})$$

is obviously a polynomial in x and ω_{ik} with coefficients in k, hence

$$\Phi(x) = \sum_{i=1}^{n} \sum_{k=1}^{m} \beta_k G(x, u\alpha_i + v\beta_k)$$

is a polynomial in x, whose coefficients are polynomial expressions in the α_i, β_k with coefficients in k, which are moreover formally symmetric in the quantities $\alpha_1, \ldots, \alpha_n$ as well as the quantities $\beta_1, \ldots, \beta_m$. Consequently these coefficients belong to k and $\Phi(x)$ is a polynomial in k. If we set $x = \omega_{11}$, then $G(\omega_{11}, \omega_{ik})$ vanishes, except if $i = 1$ and $k = 1$, since ω_{11} is different from the remaining ω_{ik} by construction. However from this it follows that

$$\beta_1 = \frac{\Phi(\omega_{11})}{G(\omega_{11}, \omega_{11})}.$$

We show in an analogous manner that α_1 can also be expressed in terms of ω_{11} and with this we have proved

$$K(\alpha_1, \beta_1; k) = K(\omega_{11}; k).$$

Hence it is enough to restrict ourselves to fields which arise by adjoining a single algebraic number over k.

Now let θ be an algebraic number of degree n over k. Then the following holds for numbers in $K(\theta; k)$:

Theorem 53. *Every number in $K(\theta)$ is obtained exactly once in the form*

$$\alpha = c_0 + c_1\theta + c_2\theta^2 + \cdots + c_{n-1}\theta^{n-1} \tag{34}$$

where the $c_0, \ldots, c_{n-1}$ run through all numbers of the ground field k.

To prove this suppose $\alpha = P(\theta)/Q(\theta)$, $Q(\theta) \neq 0$. Then $Q(x)$ does not have the root θ in common with the function $f(x)$ belonging to θ which is irreducible over k; hence by Theorem 49, $Q(x)$ is relatively prime to $f(x)$. Thus there are two polynomials $R(x)$ and $H(x)$ over k such that

$$1 = Q(x)R(x) + f(x)H(x),$$

and since $f(\theta) = 0$,

$$1 = Q(\theta)R(\theta)$$

$$\alpha = \frac{P(\theta)}{Q(\theta)} = P(\theta)R(\theta) = F(\theta),$$

where $F(x) = P(x)R(x)$ is again a polynomial over k. Finally let $g(x)$ be the remainder of $F(x) \bmod f(x)$, which is also a polynomial over k of degree $\leq n - 1$. Then

$$F(x) = q(x)f(x) + g(x),$$
$$F(\theta) = g(\theta),$$

so that in fact α is put into the form (34). If there are two polynomials $g(x)$ and $g_1(x)$ over k, of degree at most $n - 1$, such that $g(\theta) = g_1(\theta)$, then $g(x) - g_1(x)$ is a polynomial over k with the root θ, whose degree is $< n$. Thus $g(x) - g_1(x)$ is identically 0, that is, the coefficients of $g(x)$ and $g_1(x)$ agree.

Theorem 54. *Every number $g(\theta)$ of the field $K(\theta)$ is likewise an algebraic number over k of degree at most n. The relative conjugates of a number $\alpha = g(\theta)$ are the distinct numbers among the numbers $g(\theta_i)$ ($i = 1, 2, \ldots, n$). Each conjugate to α appears equally often among the $g(\theta_i)$.*

For if $\theta_1, \ldots, \theta_n$ are the conjugates of θ with respect to k, we form the product

$$F(x) = \prod_{i=1}^{n} (x - g(\theta_i)).$$

The coefficients of this polynomial are integral rational combinations of $\theta_1, \ldots, \theta_n$, which are moreover symmetric in $\theta_1, \ldots, \theta_n$ and whose coefficients belong to k. Consequently $F(x)$ is a polynomial over k and thus every number $g(\theta_i)$ is an algebraic number over k. Further if $\varphi(x)$ is a polynomial, among whose roots just one of the numbers $\alpha_i = g(\theta_i)$ occurs, then all the α_i are roots of $\varphi(x)$. Namely the polynomial $\varphi(g(y))$ over k has a root $y = \theta_i$ in common with $f(y)$ and hence, by Theorem 49, it vanishes for all $y = \theta_1, \ldots, \theta_n$; consequently, $\varphi(x)$ vanishes for each $x = \alpha_1, \ldots, \alpha_n$.

Moreover, if $\Psi(x)$ is the irreducible polynomial over k with leading coefficient 1 which has α_1 as a root, then $\Psi(x)$ is a divisor of $F(x)$. Let $\Psi(x)^q$ be the highest power of Ψ dividing $F(x)$. Now if $F(x)/\Psi(x)^q$ were not constant, then it would have a α_i, a root of F, as a root; consequently it would still be divisible by $\Psi(x)$, contrary to the assumption about q. Hence for a certain integer q

$$F(x) = \Psi(x)^q;$$

That is, the n numbers

$$\alpha_i = g(\theta_i) \qquad (i = 1, 2, \ldots, n)$$

represent all conjugates to each α_i; however they represent each conjugate q times. Consequently, n is the largest relative degree which a number α in $K(\theta)$ can have with respect to k, and, with this, n is identified as a number determined by the field $K(\theta)$ alone, which is independent of the choice of the generating number θ. Thus n is called the *relative degree of the field* $K(\theta)$ with respect to k. Hence the degree of each number of $K(\theta)$ is a divisor of the degree of the field.

We now modify the concept of the conjugate, keeping in mind the above theorem, by the following:

Definition. If n is the relative degree of $K(\theta)$ with respect to k and if $\alpha = g(\theta)$ is a number of $K(\theta)$, of degree n/q, then the system of n numbers $\alpha_i = g(\theta_i)$ $(i = 1, 2, \ldots, n)$ will be called the *conjugates of α in the field $K(\theta)$ with respect to k*. These are the conjugates of α with respect to k, each one taken q times.

Accordingly the system of these conjugates as a whole depends only on α, the ground field k, and the field K, but it is independent of the choice of the generating θ. Since in the future we will be dealing exclusively with this concept of conjugate, the qualifier "in the field $K(\theta)$ with respect to k" will be omitted in general for the sake of simplicity.

Once we have brought the conjugates of the generating number θ into a definite order by the numbering $\theta_1 \theta_2, \ldots, \theta_n$, then the n conjugates of an arbitrary number α in $K(\theta)$ acquire a definite numbering by representing α in the uniquely determined form $g(\theta)$ by Theorem 53 and then calculating the number $g(\theta_i)$ as the conjugate α_i. We shall consider such a determination as done, then we prove:

Theorem 55. *Each rational equation* $R(\alpha, \beta, \gamma, \ldots) = 0$ *between numbers* $\alpha, \beta, \gamma, \ldots$ *in $K(\theta)$ with coefficients in k remains true if $\alpha, \beta, \gamma, \ldots$ are replaced by the conjugates with the same index.*

As a rational function of $\alpha, \beta, \gamma, \ldots$, R is identical to the quotient of two integeral rational expressions P and Q

$$R(\alpha, \beta, \gamma, \ldots) = \frac{P(\alpha, \beta, \gamma, \ldots)}{Q(\alpha, \beta, \gamma, \ldots)}$$

in $\alpha, \beta, \gamma, \ldots$.

If we substitute for $\alpha, \beta, \gamma, \ldots$ in R their representations as polynomials in θ,

$$\alpha = g(\theta), \qquad \beta = h(\theta), \qquad \gamma = r(\theta), \ldots,$$

then Q becomes a polynomial in θ, which does not vanish for the numerical value θ, as it is equal to the number $Q(\alpha, \beta, \gamma, \ldots)$. Consequently, it does not vanish for any of the conjugates $\theta_1, \ldots, \theta_n$. However, since $R = 0$, the numerical value in the numerator

$$P(g(\theta), h(\theta), r(\theta), \ldots) = 0.$$

Hence this polynomial in θ must vanish for all conjugates θ_i, i.e.,

$$\left.\begin{array}{l} P(\alpha_i, \beta_i, \gamma_i, \ldots) = 0, \\ Q(\alpha_i, \beta_i, \gamma_i, \ldots) \neq 0, \end{array}\right\} \qquad (i = 1, 2, \ldots, n).$$

Thus the n numerical values

$$R(\alpha_i, \beta_i, \gamma_i, \ldots) = 0, \qquad (i = 1, 2, \ldots, n)$$

which was to be proved.

In particular, it follows for each two numbers α, β, in $K(\theta)$

$$\alpha_i \pm \beta_i = (\alpha \pm \beta)_i, \qquad \alpha_i \beta_i = (\alpha\beta)_i, \qquad \frac{\alpha_i}{\beta_i} = \left(\frac{\alpha}{\beta}\right)_i,$$

since, for example, for $\alpha = g(\theta)$ and $\beta = h(\theta)$,

$$g(\theta)h(\theta) = r(\theta),$$

where g, h, r are polynomials of degree $\leq n - 1$. By the above theorem, from this one equation for the numerical value θ the n equations

$$g(\theta_i)h(\theta_i) = r(\theta_i)$$

follow, that is,

$$\alpha_i \beta_i = (\alpha\beta)_i \qquad (i = 1, 2, \ldots, n).$$

§20 Generating Field Elements, Fundamental Systems, and Subfields of $K(\theta)$

Theorem 56. *A number α in $K(\theta)$ belongs to the ground field k if and only if it is equal to its n conjugates. A number α in $K(\theta)$ has degree n with respect to k if and only if it is distinct from all its conjugates. The latter condition is at the same time necessary and sufficient for the number α to generate the field $K(\theta)$.*

The first two statements follow immediately from Theorem 54 and the definition which follows it. Moreover, if α in $K(\theta)$ is to generate the field $K(\theta)$, thus if $K(\theta) = K(\alpha)$ is to hold, then the degree of α must be equal to the degree of $K(\theta)$, that is, it must be $= n$. Therefore the conjugates of α must all be different. However, if $\alpha_i = g(\theta_i)$ are all different for $i = 1, 2, \ldots, n$, then θ can be expressed rationally in terms of α, and hence all numbers from $K(\theta)$ are also contained in $K(\alpha)$.

In order to express θ in terms of α we conclude that

$$H(x) = \prod_{i=1}^{n} (x - \alpha_i) = \prod_{i=1}^{n} (x - g(\theta_i))$$

is a polynomial in k, as in the proof of Theorem 52. Likewise

$$\frac{H(x)}{x - \alpha_i} = G(x, \alpha_i)$$

is a polynomial in the quantities x, α_i with coefficients in k. Hence

$$\Phi(x) = \sum_{i=1}^{n} \theta_i \frac{H(x)}{x - \alpha_i} = \sum_{i=1}^{n} \theta_i G(x, g(\theta_i))$$

as a symmetric expression in $\theta_1, \ldots, \theta_n$, is also a polynomial over k, from which it follows for $x = \alpha_i$ that

$$\theta_i = \frac{\Phi(\alpha_i)}{G(\alpha_i, \alpha_i)},$$

because the denominator is certainly $\neq 0$ by definition.

Until now we have represented each number from $K(\theta)$ as a linear combination of $1, \theta, \theta^2, \ldots, \theta^{n-1}$ with coefficients in k. However for very many purposes more freedom in the choice of these basis elements is desirable.

We call n numbers $\omega^{(1)}, \omega^{(2)}, \ldots, \omega^{(n)}$ a *fundamental system* of $K(\theta)$ if every number α in $K(\theta)$ can be represented in the form

$$\alpha = \sum_{i=1}^{n} x_i \omega^{(i)}$$

with coefficients x_i in k.

Theorem 57. *In order that the numbers*

$$\omega^{(i)} = \sum_{k=1}^{n} c_{ik}\theta^{k-1} \qquad (c_{ik} \text{ numbers in } k) \tag{35}$$

form a fundamental system of $K(\theta)$, it is necessary and sufficient that the determinant $\|c_{ik}\| \neq 0$.

Obviously we need only investigate when the numbers $1, \theta, \ldots, \theta^{n-1}$ can be represented in terms of the $\omega^{(i)}$ as

$$\theta^{p-1} = \sum_{i=1}^{n} a_{pi}\omega^{(i)} \qquad (p = 1, \ldots, n)\, (a_{pi} \text{ numbers in } k). \tag{36}$$

First, if the determinant in (35) is $\neq 0$, then the n equations for the unknowns $1, \theta, \ldots, \theta^{n-1}$ can be solved and these can be obtained as linear combinations of the $\omega^{(i)}$, with coefficients which are derived by rational operations on the c_{ik}, and thereby belong to k.

Secondly, if a representation of the θ^{k-1} in terms of the $\omega^{(i)}$, as in (36), is possible, then let us substitute the expression (35) for $\omega^{(i)}$ into it to obtain

$$\theta^{p-1} = \sum_{i,k=1}^{n} a_{pi}c_{ik}\theta^{k-1} \qquad (p = 1, \ldots, n).$$

However since no linear homogeneous relation holds between $1, \theta, \ldots, \theta^{n-1}$ with coefficients in k, unless all coefficients are 0, we have

$$\delta_{kp} = \sum_{i=1}^{n} a_{ki}c_{ip} = \begin{cases} 0 & \text{if } p \neq k, \\ 1 & \text{if } p = k. \end{cases}$$

Thus the determinant $\|\delta_{kp}\|$ is $= 1$ and on the other hand it is equal to the product $\|a_{ki}\| \cdot \|c_{ip}\|$; hence the determinant of the $c_{ip} \neq 0$.

Theorem 58. *The n numbers $\omega^{(1)}, \ldots, \omega^{(n)}$ in $K(\theta)$ form a fundamental system if and only if there is no linear relation*

$$\sum_{i=1}^{n} u_i \omega^{(i)} = 0 \tag{37}$$

with coefficients in k except if all $u_i = 0$.

The n numbers $\omega^{(i)}$ of this type are said to be *linearly independent.* For, in the notation used above it would follow from (37) that

$$0 = \sum_{i=1}^{n} u_i \sum_{k=1}^{n} c_{ik}\theta^{k-1}$$

and as before, if the u_i belong to k and are not all $= 0$,

$$\sum_{i=1}^{n} u_i c_{ik} = 0, \qquad (k = 1, \ldots, n);$$

thus

$$\|c_{ik}\| = 0.$$

However, if this determinant is $= 0$, then the system is not a fundamental system, and thus, as is known, the n homogeneous equations for u_i

$$\sum_{i=1}^{n} u_i c_{ik} = 0 \qquad (k = 1, \ldots, n)$$

are solvable and indeed there is a solution among the nonvanishing solutions which is obtained by rational operations from the coefficients c_{ik}, hence which also belongs to k. For this solution we then have

$$\sum_{i=1}^{n} u_i \omega^{(i)} = 0.$$

Hence the number α also determines the coefficients in

$$\alpha = \sum_{i=1}^{n} x_i \omega^{(i)}$$

uniquely, if these coefficients also belong to k.

Let the determinant formed by the n numbers $\omega^{(i)}$ and their n conjugates be denoted by

$$\|\omega_k^{(i)}\| = \Delta(\omega^{(1)}, \ldots, \omega^{(n)}).$$

(The index k designates the row of the determinant; the index i designates the column.) It follows from (35) that

$$\Delta(\omega^{(1)}, \ldots, \omega^{(n)}) = \|c_{ik}\| \cdot \Delta(1, \theta, \ldots, \theta^{n-1}).$$

By Theorem 57 this determinant is $\neq 0$ for each fundamental system and only for fundamental systems since by a known formula

$$\Delta(1, \theta, \ldots, \theta^{n-1}) = \begin{vmatrix} 1 & \theta_1 & \theta_1^2 & \cdots & \theta_1^{n-1} \\ 1 & \theta_2 & \theta_2^2 & \cdots & \theta_2^{n-1} \\ 1 & \theta_3 & \theta_3^2 & \cdots & \theta_3^{n-1} \\ \vdots & \vdots & & & \vdots \\ 1 & \theta_n & \theta_n^2 & \cdots & \theta_n^{n-1} \end{vmatrix} = \prod_{1 \leq i < k \leq n} (\theta_i - \theta_k),$$

and therefore $\neq 0$. The determinant is a polynomial function of $\theta_1, \ldots, \theta_n$ with coefficients in k (even in $k(1)$). If any of the θ_i are permuted, then this determinant changes at most by the factor ± 1, thus its square is also symmetric in $\theta_1, \ldots, \theta_n$ and hence is a number of the ground field k. The same also holds for $\Delta^2(\omega_1, \ldots, \omega_n)$. Obviously this number is also independent of the numbering of the conjugates.

As in the second half of the proof of Theorem 58, it is easily obtained that among the $n + 1$ quantities of the field K, say $\beta^{(1)}, \beta^{(2)}, \ldots, \beta^{(n+1)}$, there is always a linear relation

$$\sum_{i=1}^{n+1} u_i \beta^{(i)} = 0,$$

where the u_i denote numbers in the ground field k, which are not all 0. *Thus the degree n of K can also be defined as the maximum number of linearly independent elements in K.*

Finally let us consider the field $K(\theta)$ not relative to k, but relative to another field $K(\alpha)$, which is an algebraic field in its own right, say of degree m over k, generated by the number α which satisfies an irreducible equation of degree m in k. Suppose moreover that α occurs in $K(\theta)$. Accordingly, $K(\theta)$ is an algebraic field of degree $q \leq n$ over $K(\alpha)$, since the generating number θ already satisfies an equation of degree m with coefficients in k, thus a fortiori in $K(\alpha)$. $K(\alpha)$ is called a *subfield* of $K(\theta)$. If we regard $K(\alpha)$ as the ground field, then every quantity in $K(\theta)$ can be brought into the form

$$\omega = \gamma_0 + \gamma_1 \theta + \cdots + \gamma_{q-1} \theta^{q-1}$$

in a unique way, where the quantities γ are numbers in $K(\alpha)$. Likewise, every number in $K(\alpha)$ thus admits a unique representation

$$c_0 + c_1 \alpha + \cdots + c_{m-1} \alpha^{m-1},$$

where the coefficients c_i belong to k. Consequently each ω admits a unique representation as a linear combination of the mq quantities $\alpha^i \theta^k$ ($i = 0, 1, \ldots, m-1$; $k = 0, 1, \ldots, q-1$) with coefficients in k. Hence these mq

numbers also form a fundamental system of $K(\theta)$ (with respect to the ground field k), hence $mq = n$, $q = n/m$. With this the following theorem is proved:

Theorem 59. *If α is a number of degree m over k and if β is a number of degree q over $K(\alpha; k)$, then the field $K(\alpha, \beta; k)$ has degree mq over k. Furthermore if $\theta_1, \ldots, \theta_n$ $(n = mq)$ are the conjugates of a generating number of $K(\alpha, \beta; k)$ with respect to k, then these conjugates decompose into m sequences with q elements in each sequence; here the q numbers of a sequence are always the conjugates with respect to $K(\alpha_i)$, where $\alpha_1, \ldots, \alpha_m$ are the m conjugates of α with respect to k.*

A field $K(\beta; k)$, which is identical with all conjugate fields $K(\beta_i; k)$ ($i = 1, \ldots, n$) is called a *Galois field* or a *normal field with respect to k*. A number field $K(\alpha; k)$ is always contained as a subfield of a Galois field. This follows from the proof of Theorem 52 for the field which arises by adjunction of all relatively conjugate numbers $\alpha_1, \ldots, \alpha_n$ is obviously a Galois field with respect to k.

From now on we will be concerned exclusively with those numbers which are algebraic with respect to $k(1)$; these will simply be called algebraic. Let us merely mention the following about other kinds of numbers:

Numbers which are not algebraic are called *transcendental*. *Liouville*[1] (1851) first proved that there are transcendental numbers, by giving at the same time a method of constructing arbitrarily many such numbers. Later *George Cantor*[2] (1874) furnished a quite different proof which shows that the set of transcendental numbers has an even higher cardinality than the set of algebraic numbers. Until now it has only rarely been possible to decide whether a definite given number is transcendental or not. General methods for this are not known. *Hermite*[3] (1873) has proved transcendency for e, *Lindemann*[4] (1882) for π; later these proofs were greatly simplified by *Hilbert, Hurwitz*, and *Gordan*[5].

[1] Liouville, Sur des classes très étendues de quantités dont la valeur n'est ni algébrique, ni même réductible à des irrationnelles algébriques. *Journal de Mathématiques pures et appliquées*, Sér I.T.**16** (1851).

[2] Cantor, Über eine Eigenschaft des Inbegriffes aller rellen algebraischen Zahlen. *Crelles Journal f. d. reine u. angew. Mathem.* Vol. 77 (1874).

[3] Hermite, Sur la fonction exponentielle. *Comptes rendus* T. 77 (1873).

[4] Lindemann, Über die Zahl π. *Mathem. Annalen* Vol. **20** (1882).

[5] The three papers can be found in *Mathem. Ann.* Vol. **43** (1892).

CHAPTER V

General Arithmetic of Algebraic Number Fields

§21 Definition of Algebraic Integers, Divisibility, and Units

The concepts which were developed in the preceding chapter with respect to a ground field k are now to be understood with respect to the absolute field $k = k(1)$. To develop the foundations of an arithmetic of algebraic numbers we first need a definition of algebraic integer. The following requirements can be reasonably imposed on a concept of integer.

(1) If α and β are algebraic integers, then so are $\alpha + \beta$, $\alpha - \beta$, and $\alpha\beta$.
(2) If an algebraic integer is rational, then it is an ordinary integer.
(3) If α is an algebraic integer, then the conjugates (with respect to $k(1)$) are also algebraic integers.

By (1) each rational integral expression of algebraic integers with rational integral coefficients would be an algebraic integer. In particular, by (3), all elementary symmetric functions of an algebraic integer and its conjugates would then be algebraic integers. On the other hand they are rational and hence, by (2), they are rational integers. If α is an algebraic integer, then the coefficients in the irreducible equation for α in $k(1)$ with leading coefficient 1 would have to be rational integers. Accordingly we define:

Definition. An algebraic number α of degree n is called an *algebraic integer* if in the irreducible equation for α in $k(1)$ with leading coefficient 1, all coefficients are rational integers.

Henceforth we will always understand by "integer" an algebraic integer. Requirements (2) and (3) are obviously satisfied for these integers.

Theorem 60. *If α satisfies any equation at all with integral coefficients whose leading coefficient is equal to* 1, *then α is an integer.*

Let $\varphi(x) = x^N + a_1 x^{N-1} + \cdots + a_N$ with rational integral a's and $\varphi(\alpha) = 0$. Moreover let

$$f(x) = c_0 x^n + c_1 x^{n-1} + \cdots + c_n$$

be the irreducible polynomial in $k(1)$ which has α as a root and in which the c_i are already assumed to be relatively prime rational integers, with $c_0 > 0$. By Theorem 49 we have $f(x) \mid \varphi(x)$. Thus

$$\frac{\varphi(x)}{f(x)} = \frac{b' g(x)}{b}$$

is a rational polynomial over $k(1)$ where we may assume the polynomial $g(x)$ to be integral with relatively prime coefficients, by suitable choice of the rational integers b and b'. It follows from

$$b\varphi(x) = b'f(x)g(x)$$

that $b = b'$, since, by Theorem 13a, $f(x) \cdot g(x)$, as a product of two primitive polynomials is again primitive, and $\varphi(x)$ is also primitive. Moreover, by comparing leading coefficients we learn from $\varphi(x) = f(x)g(x)$ that c_0 must divide the leading coefficient of φ, which is 1; hence $c_0 = 1$ completing the proof.

In order to verify whether an algebraic number is an integer, we will most often use this theorem, which unlike the definition does not require the verification of the irreducibility of a polynomial.

Theorem 61. *The sum, difference, and product of two integers is again an integer. Hence every rational integral function (polynomial) of integers with rational integral coefficients is again an integer.*

For if $\alpha_1, \ldots, \alpha_n$ are the conjugates of a number α and if $\beta_1, \ldots, \beta_m$ are the conjugates of a number β, then

$$F(x) = \prod_{i=1}^{n} \prod_{k=1}^{m} (x - (\alpha_i + \beta_k))$$

is a polynomial in x whose coefficients are symmetric in $\alpha_1, \ldots, \alpha_n$ and $\beta_1, \ldots, \beta_m$. Since the elementary symmetric functions of the α as well as those of the β are rational integral functions then, by hypothesis, $F(x)$ is an integral polynomial over $k(1)$ by the fundamental theorem on symmetric functions. Consequently its root $\alpha + \beta$ is an integer. The assertions about $\alpha - \beta$ and $\alpha\beta$ are proved similarly.

In a manner quite similar to that above and by Theorem 51 we conclude:

Theorem 62. *If ω is a root of an equation*

$$x^m + \alpha x^{m-1} + \beta x^{m-2} + \cdots + \lambda = 0$$

where $\alpha, \beta, \ldots, \lambda$ are integers, then ω is also an integer.

For example the mth root of an integer is again an integer.

Theorem 63. *Every algebraic number α can be transformed into an integer by multiplication by a suitable nonzero rational number.*

To prove this assume that

$$c_0 x^n + c_1 x^{n-1} + \cdots + c_{n-1} x + c_n = 0$$

is an equation for α with rational integral coefficients and $c_0 \neq 0$. Then by multiplication by c_0^{n-1} we obtain an integer equation for $y = c_0 x$ with leading coefficient 1, which has the root $c_0 \alpha$.

The definition of divisibility arises along with the concept of integer.

An integer α is said to be *divisible* by the integer $\beta (\beta \neq 0)$, if α/β is an integer; in symbols we write $\beta \mid \alpha$.

If $\beta \mid \alpha$ and $\beta \mid \gamma$, then $\beta \mid \lambda\alpha + \mu\gamma$ for arbitrary integers λ, μ, for

$$\frac{\lambda\alpha + \mu\gamma}{\beta} = \lambda \frac{\alpha}{\beta} + \mu \frac{\gamma}{\beta}$$

is an integer by Theorem 61.

An integer ε is called a unit if $1/\varepsilon$ is also an integer.

If ε divides 1, then ε also divides $1 \cdot \alpha = \alpha$, that is, ε divides every integer α. The conjugates of each unit (with respect to $k(1)$) are also units, and each divisor of a unit and each product of units is also a unit.

If two integers α, β differ only by a factor which is a unit, then α and β are called *associates*.

In order that an integer ε be a unit it is necessary and sufficient that the product of all conjugates of ε be equal to ± 1.

For the product $\varepsilon_1 \varepsilon_2 \cdots \varepsilon_n$ as a symmetric function, is a rational integer a, and as a product of units it is also a unit, i.e., $a \mid 1$ and thus $a = \pm 1$. However if $\varepsilon_1 \varepsilon_2 \cdots \varepsilon_n = \pm 1$, then $1/\varepsilon_1 = \pm \varepsilon_2 \cdots \varepsilon_n$ is an integer and therefore ε_1 is a unit.

Obviously all roots of unity are units and indeed they have absolute value 1. However, there are infinitely many other units, for example $2 \pm \sqrt{3}$, since

$$\frac{1}{2 + \sqrt{3}} = 2 - \sqrt{3}, \qquad \frac{1}{2 - \sqrt{3}} = 2 + \sqrt{3}$$

are obviously integers. $\varepsilon = 2 - \sqrt{3}$ is < 1 and > 0, and hence there are arbitrarily small numbers among the powers $\varepsilon, \varepsilon^2, \varepsilon^3, \ldots$. Thus the multiples $N\varepsilon^k$ ($N = \pm 1, \pm 2, \ldots, k = 1, 2, \ldots$) of these numbers are obviously everywhere dense in the real line and moreover they are all integers belonging to the field $K(\sqrt{3})$. This fact—if the real algebraic integers are ordered according to magnitude, there exists *no next integer* to a given integer—has as a consequence that the many methods of proof with which we became acquainted in rational number theory cannot be carried over to algebraic numbers.

Every integer α ($\neq 0$) has infinitely many "trivial" divisors namely ε and $\varepsilon\alpha$, where ε runs through all units. But α is also decomposable into integers in a nontrivial way,

$$\alpha = \sqrt{\alpha}\sqrt{\alpha}$$

neither of which factor is a unit, if α is not a unit. Hence there are no irreducible numbers in the domain of all algebraic integers; thus there is surely no analogue to rational primes.

Rather in order to obtain irreducible numbers, we must first restrict the domain of admissible numbers to the point where we operate only with the numbers of a certain number field of degree n.

§22 The Integers of a Field as an Abelian Group: Basis and Discriminant of the Field

We lay the foundations for further investigations of a definite algebraic number field $K(\theta)$, which is generated by the algebraic number θ of degree n. It is no restriction to assume θ is an integer since we can always transform θ into a integer by multiplication by a rational integer. We fix a definite numbering for the conjugates of θ; in this way according to §19 a definite numbering of the conjugates of each number in K is also defined. From now on the conjugates are to be denoted by superscripts.

Moreover, for each number α of the field K we set:

Norm of $\alpha = N(\alpha) = \alpha^{(1)}\alpha^{(2)} \cdots \alpha^{(n)}$;   hence $N(\alpha\beta) = N(\alpha)N(\beta)$.

Trace of $\alpha = S(\alpha) = \alpha^{(1)} + \alpha^{(2)} + \cdots + \alpha^{(n)}$;   hence $S(\alpha + \beta) = S(\alpha) + S(\beta)$.

These quantities are rational numbers and they are rational integers if α is an integer. We have $N(\alpha) = 0$ only if $\alpha = 0$.

Theorem 64. *The integers of K form a (torsion-free) Abelian group under composition by addition. This group has n basis elements. Thus there are n integers $\omega_1, \ldots, \omega_n$ in K such that if the x_i run through all rational integers in the expression*

$$\alpha = x_1\omega_1 + x_2\omega_2 + \cdots + x_n\omega_n,$$

we obtain each integer in K exactly once. The numbers ω *are called a basis of the field.*

The first part follows directly from Theorem 61. In order to prove the second part, we first investigate those integers ρ of the field which have a representation in the form

$$\rho = c_0 + c_1\theta + \cdots + c_{n-1}\theta^{n-1}$$

with rational c. The c's can be determined from the n conjugate equations

$$\rho^{(i)} = c_0 + c_1\theta^{(i)} + \cdots + c_{n-1}\theta^{(i)n-1} \qquad (i = 1, 2, \ldots, n)$$

since the determinant $\Delta(1, \theta, \theta^2, \ldots, \theta^{n-1}) \neq 0$. The solution yields $\Delta \cdot c_k$ equal to a determinant, among whose elements only the $\rho^{(i)}$ and the powers of the $\theta^{(i)}$ occur. In any case this determinant is an algebraic integer A_k, since ρ and θ are algebraic integers. However,

$$c_k = \frac{A^k}{\Delta} = \frac{A_k \Delta}{\Delta^2}$$

implies that $A_k \Delta = \Delta^2 c_k$ is a rational integer, for this number is an integer since A_k and Δ are integers, and rational since c_k and Δ^2 are rational. Consequently c_k is a rational number,

$$c_k = \frac{x_k}{D},$$

where x_k is a rational integer and the denominator $D = |\Delta^2|$ is independent of ρ. The system of all numbers

$$\alpha = x_0 \frac{1}{D} + x_1 \frac{\theta}{D} + x_2 \frac{\theta^2}{D} + \cdots + x_{n-1} \frac{\theta^{n-1}}{D},$$

where the x_i run through all rational integers, thus contains all integral numbers of the field. Moreover the system perhaps contains nonintegers, and in any case forms a (torsion-free) Abelian group (with composition by addition) with a basis of n elements namely $1/D, \theta/D, \ldots, \theta^{n-1}/D$. Hence by Theorem 34 the subgroup of integers of the field contained in this group likewise has a basis. By Theorem 40 this subgroup is of finite index since $D \cdot \alpha$ (that is, in the sense of group theory: the Dth power of each element) is obviously an integer and belongs to the subgroup. Consequently, by Theorem 35 the basis for the integers in the number field also consists of n elements, say $\omega_1, \ldots, \omega_n$. By Theorem 38 two different systems of basis elements, say α_i and ω_i, are connected by a relation

$$\alpha_i = \sum_{k=1}^{n} c_{ik}\omega_k \qquad (i = 1, 2, \ldots, n),$$

with rational integral c_{ik}, whose determinant is ± 1. Consequently $\Delta^2(\omega_1, \ldots, \omega_n)$ is independent of the choice of basis and is determined completely by the field itself. Since in any case the ω_i represent $1, \theta, \ldots, \theta^{n-1}$ by linear combinations, they form a fundamental system and consequently $\Delta^2 \neq 0$.

Definition. The number $\Delta^2(\omega_1, \ldots, \omega_n)$ which is independent of the choice of basis $\omega_1, \ldots, \omega_n$ is called the *discriminant of the field* and will be denoted by d. It is a nonzero rational integer.

We see without difficulty that $|\Delta^2(\alpha_1, \ldots, \alpha_n)|$ is always $\geq |d|$ for a fundamental system of integers α_i, and is equal to $|d|$ if and only if the fundamental system forms a basis for the field. For this reason the basis for the field is also called a *minimal basis*.

It is appropriate to introduce the concept of a module here. By a *module* of integers in a field K we mean a system of integers in K, which along with α and β always contains $\alpha + \beta$ and $\alpha - \beta$, and also contains a number different from 0.

Thus the numbers of a module form a (torsion-free) Abelian group under composition by addition which is a subgroup of the group of all integral elements of the field, and hence by Theorem 34 also possesses a basis of k elements, where $0 < k \leq n$. We call such a module a k-rank module (module of rank k). We will be dealing only with n-rank modules. Such modules are obviously identified by the fact that they contain n linearly independent numbers.

§23 Factorization of Integers in $K(\sqrt{-5})$: Greatest Common Divisors which Do Not Belong to the Field

We now direct our attention to the multiplicative decomposition of the integers of a field. An integer α is called *irreducible* in K if α cannot be represented as a product of two integers, neither of which is a unit. The property of being irreducible thus does not belong to a number in itself but can only be considered with respect to a definite field. Every rational prime is irreducible in $k(1)$, but, for example, 3 is reducible into $\sqrt{3}\,\sqrt{3}$ in $K(\sqrt{3})$.

Are there irreducible numbers also in algebraic fields of degree higher than 1, and can every integer of the field be represented as a product of these numbers in (essentially) one way?

We will determine numerical examples such that the uniqueness of the decomposition does *not* always hold and we will try to find the reason for this.

We consider the field $K(\sqrt{-5})$. The generating number $\theta = \sqrt{-5}$ is a root of $x^2 + 5 = 0$ and as a nonreal number it surely does not satisfy any equation of lower degree in $k(1)$, hence it is of degree 2 over $k(1)$. Hence all numbers in $K(\sqrt{-5})$ have the form

$$\alpha = r_1 + r_2\sqrt{-5}$$

with r_1 and r_2 rational. The conjugate of α will be denoted by α'. Thus

$$\alpha' = r_1 - r_2\sqrt{-5}, \quad \text{hence } (\alpha')' = \alpha.$$

The integers in $K(\sqrt{-5})$ are the numbers $m + n\sqrt{-5}$ with m, n rational integers. In order that α be an integer it is necessary and sufficient that $\alpha + \alpha'$ and $\alpha\alpha'$ be (rational) integers, that is,

$$2r_1 \quad \text{and} \quad r_1^2 + 5r_2^2$$

must be integers.

Accordingly, r_1 and r_2 can have denominators at most 2. We set $r_1 = g_1/2$, $r_2 = g_2/2$. Thus we should have

$$\frac{g_1^2 + 5g_2^2}{4} \text{ integral, that is, } g_1^2 + 5g_2^2 \equiv 0 \ (\text{mod } 4).$$

All squares are $\equiv 0$ or $1 \ (\text{mod } 4)$. From this it follows that g_1 and g_2 must be even, hence r_1 and r_2 must themselves be integers.

There are no units other than ± 1 in the field $K(\sqrt{-5})$. For a unit $\varepsilon = m + n\sqrt{-5}$ we must have

$$\pm 1 = N(\varepsilon) = \varepsilon \cdot \varepsilon' = m^2 + 5n^2.$$

If $n \neq 0$, then the quantity $m^2 + 5n^2 \geq 5$; hence we must have $n = 0, m = \pm 1$.

The following integers are irreducible in $K(\sqrt{-5})$:

$$\alpha = 1 + 2\sqrt{-5},$$
$$\alpha' = 1 - 2\sqrt{-5},$$
$$\beta = 3,$$
$$\rho = 7.$$

If $\beta = 3$ were decomposable into $\gamma\delta$ and γ, δ were not units, then we would have

$$9 = N(3) = N(\gamma) \cdot N(\delta).$$

However a decomposition of 9 into integral rational factors, none of which $= 1$, is only possible as $3 \cdot 3$. Consequently we necessarily have

$$N(\gamma) = N(\delta) = 3$$

and hence for $\gamma = x + y\sqrt{-5}$ with integral x, y,

$$x^2 + 5y^2 = 3, \qquad x^2 \leq 3, \qquad 5y^2 \leq 3,$$

which is obviously impossible. Hence $\beta = 3$ is irreducible and in exactly the same way $\rho = 7$ is shown to be irreducible. Finally if α were decomposable into $\gamma\delta$, $N(\gamma) \neq 1$ and $N(\delta) \neq 1$, then we would have

$$N(\gamma) \cdot N(\delta) = N(\alpha) = 21.$$

Thus either $N(\gamma) = 3$, $N(\delta) = 7$ or vice versa. But we have just shown that there cannot be any γ with $N(\gamma) = 3$. Thus α and hence also its conjugate α' are irreducible.

The number 21 is thus shown to be decomposable in two essentially distinct ways as a product of irreducible numbers in $K(\sqrt{-5})$:

$$21 = \alpha\alpha' = 3 \cdot 7.$$

To understand this fact, that the irreducible number 3 indeed divides the product $\alpha\alpha'$, but divides neither α nor α', we note that the two numbers α and 3 in $K(\sqrt{-5})$ have indeed no factor in common in $K(\sqrt{-5})$ (except ± 1), but that they have a common factor (not a unit) which belongs to another field. For the squares

$$\alpha^2 = -19 + 4\sqrt{-5}$$
$$\beta^2 = 9$$

are divisible by the integer

$$\lambda = 2 + \sqrt{-5}$$

which is not a unit:

$$\alpha^2 = (2 + \sqrt{-5})(-2 + 3\sqrt{-5})$$
$$\beta^2 = (2 + \sqrt{-5})(2 - \sqrt{-5}).$$

Thus α^2/λ, β^2/λ are integers, and hence by Theorem 62 the square roots

$$\frac{a}{\sqrt{\lambda}}, \frac{b}{\sqrt{\lambda}}$$

are also integers. Likewise

$$\alpha'^2 = (-2 + \sqrt{-5})(2 + 3\sqrt{-5})$$
$$\rho^2 = 7^2 = (2 + 3\sqrt{-5})(2 - 3\sqrt{-5})$$

are divisible by

$$\chi = 2 + 3\sqrt{-5};$$

hence

$$\frac{\alpha'}{\sqrt{\chi}}, \frac{\rho}{\sqrt{\chi}}$$

are integers. Furthermore the number $\sqrt{\lambda}$ (which does not belong to the field $K(\sqrt{-5})$) has precisely the properties of a greatest common divisor of α and β: Each integer ω—in $K(\sqrt{-5})$ or not—which divides α and β, also divides $\sqrt{\lambda}$, and any integer which divides $\sqrt{\lambda}$ is also a divisor of α and β. The last fact is self-evidently a direct consequence of the definition of divisibility. In order to prove the first assertion we make use of the fact that the number $\sqrt{\lambda}$ can be represented in the form

$$A\alpha + B\beta = \sqrt{\lambda} \tag{38}$$

with integers A, B (of course not belonging to $K(\sqrt{-5})$), for example,

$$A = -\frac{2\alpha}{\sqrt{\lambda}}, \qquad B = -\frac{(4 - \sqrt{-5})\beta}{\sqrt{\lambda}}.$$

Hence if $\omega | \alpha$ and $\omega | \beta$, then it follows in fact from (38) that $\omega | \sqrt{\lambda}$.

The double decomposition

$$\alpha\alpha' = \beta\rho,$$

into irreducible factors in $K(\sqrt{-5})$, occurs in such a way that

$$\alpha = \sqrt{\lambda}\sqrt{-\chi}, \qquad \beta = \sqrt{\lambda}\sqrt{\lambda'},$$
$$\alpha' = \sqrt{\lambda'}\sqrt{-\chi}, \qquad \rho = \sqrt{\chi}\sqrt{\chi'},$$

and the four factors not belonging to the field in the product

$$21 = \sqrt{\lambda}\sqrt{\lambda'}\sqrt{-\chi}\sqrt{-\chi'}$$

can be put together in several ways so that they yield numbers in K, although every pair of the numbers has no common factor.

We formulate these two most important results as follows:

(i) It may happen that two numbers, irreducible in $K(\sqrt{-5})$, which do not differ only by a unit factor, have a common factor, which then does not belong to the field.

(ii) The totality of integers in $K(\sqrt{-5})$ which are divisible by an irreducible number α in K need not agree with the totality of integers in $K(\sqrt{-5})$ which are divisible by a nonunit factor of α (not belonging to K).

For example, α is irreducible, $\sqrt{\lambda}$ is a factor of α, the number $\beta = 3$ is divisible by $\sqrt{\lambda}$ but not by α, although β belongs to the field $K(\sqrt{-5})$.

Neither of these properties can occur in the field $k(1)$. This is true since two irreducible numbers which do not differ only by a unit factor are always two different (thus relatively prime) prime numbers, say p, q from which 1 can be formed as a combination:

$$1 = px + qy$$

with integral rational x, y. From this it follows that all common factors of p and q must divide 1, and hence are units.

Furthermore if p is again a prime and φ is an arbitrary integer (not a unit and possibly not rational) which divides p then the set of all rational integers which are divisible by φ is a module and hence by Theorem 2 identical with all multiples of a rational integer n. Then p must divide n because otherwise 1 could be formed as a combination of p and n, and φ would then divide 1. Hence $n = \pm p$, that is, *each rational number divisible by φ is divisible by p, provided φ is not a unit and is a divisor of p, where p is prime.*

We have thus arrived at the insight that in the higher algebraic fields the irreducible numbers are not the ultimate building blocks from which all numbers of the field can be put together, that they do not have the property just stated for primes.

It is now a question of extending the domain of numbers so that we consider also those numbers which appear as the GCD of numbers of the field, as $\sqrt{\lambda}$ and $\sqrt{\chi}$ above, without belonging to the field. Indeed we need not consider the individual $\sqrt{\lambda}, \sqrt{\chi}$ themselves exactly, for in the investigation within K we do not have to keep separate two algebraic numbers which have the property that every number divisible by the one number in K is also divisible by the other.

Consequently we will simply seek to characterize a number A, not belonging to K, by giving all numbers of the field which are divisible by A.

Such a system of integers has the property: if α and β belong to the system and λ and μ are arbitrary integers of the field, then $\lambda\alpha + \mu\beta$ belongs to the system. A result arising much later in the presentation of our theory is that the converse also holds: if a set of integers in K has this property, then there is an algebraic integer A, possibly not belonging to the field K, such that the set consists of all numbers of the field divisible by A. Such a set should thus be regarded conceptually as a number and will be called an *ideal* following Dedekind. *Kummer*, who earlier investigated these relations in the case of cyclotomic fields, the first person to do so, and who should be regarded as the creator of ideal theory, called such numbers A, which appear as the GCD of elements of the field without belonging to the field, *ideal numbers of the field*.

In the theory of ideals which is explained in what follows, we should always keep in mind that an ideal serves only to characterize a certain number not belonging to the field, by operations within the field, as this anticipation of results indicates. In the domain extended by ideals the concept of primes and the fact of unique decomposition into primes will be found again, exactly as in rational number theory.

§24 Definition and Basic Properties of Ideals

Definition. A system S of integers of the field K is called an *ideal* in K (for short: an ideal) if whenever α and β belong to S, every combination $\lambda\alpha + \mu\beta$ with arbitrary integer coefficients λ, μ in K, also belongs to S.[1]

Thus the property of being an ideal does not belong in an absolute sense to a system S, but only in reference to a specific field K. Hereafter ideals will

[1] From §31 on, a somewhat more general definition of ideal is used, in which nonintegral numbers are also admitted.

be denoted by German letters $\mathfrak{a}$, $\mathfrak{b}$, $\mathfrak{c}$, ... The ideal which consists of the single number 0 may be denoted by (0); in several respects it plays a special role. Two ideals $\mathfrak{a}$, $\mathfrak{b}$ are said to be equal ($\mathfrak{a} = \mathfrak{b}$) if they contain exactly the same numbers.

Examples of ideals are:

I. The set of numbers S which is represented by a specific linear form $\xi_1\alpha_1 + \cdots + \xi_r\alpha_r$, with $\alpha_1, \ldots, \alpha_r$ integers in K, where $\xi_1, \ldots, \xi_r$ run through all integers in K. This set of numbers is called the range of values of the form. We denote this ideal by $(\alpha_1, \ldots, \alpha_r)$.

II. The set of integers in K which are divisible by a definite integer A, no matter whether A belongs to the field or not.

A final result of our theory, as has already been mentioned, will be that every ideal is of the form I as well as of the form II (§33). For the time being we show:

Theorem 65. *Every ideal $\mathfrak{a}$ can be written in the form $(\alpha_1, \ldots, \alpha_r)$ with the α suitably chosen integers in K. Moreover, we may even take $r \leq n$.*

The numbers of an ideal $\mathfrak{a}$ which is not (0) (the case $\mathfrak{a} = (0)$ is trivial) obviously form an infinite Abelian group, under composition by addition, which is a subgroup of the group of all integers in K. Consequently by Theorem 34 the ideal $\mathfrak{a}$ has a basis, whose size is $\leq n$. On the other hand, by Theorem 37, the number of elements in this basis is equal to the number of independent elements in $\mathfrak{a}$; hence it is $= n$, since, indeed, if α ($\alpha \neq 0$) belongs to $\mathfrak{a}$, the n independent elements α, $\theta\alpha$, $\theta^2\alpha$, $\ldots$, $\theta^{n-1}\alpha$ must also belong to $\mathfrak{a}$. Thus in each ideal $\mathfrak{a} \neq (0)$ there are exactly n numbers $\alpha_1, \ldots, \alpha_n$ such that

$$\alpha = x_1\alpha_1 + \cdots + x_m\alpha_n$$

represents all numbers of the ideal exactly once, if $x_1, \ldots, x_n$ run through all rational integers. Such a system $\alpha_1, \ldots, \alpha_n$ is called a *basis of the ideal* (or *ideal basis*). According to the definition the numbers in $\mathfrak{a}$ simultaneously form the range of values of the form

$$\xi_1\alpha_1 + \cdots + \xi_n\alpha_n, \quad \text{so} \quad \mathfrak{a} = (\alpha_1, \ldots, \alpha_n).$$

We have $(\alpha_1, \ldots, \alpha_r) = (\beta_1, \ldots, \beta_s)$ if and only if each α can be linearly represented by the β and if each β can be linearly represented by the α with integer coefficients in K. Thus in particular, if ω is an arbitrary number in $\mathfrak{a}$, λ an integer in K,

$$\mathfrak{a} = (\alpha_1, \ldots, \alpha_r) = (\alpha_1, \ldots, \alpha_r, \omega) = (\alpha_1 - \lambda\omega, \alpha_2, \ldots, \alpha_r, \omega). \tag{39}$$

An ideal $\mathfrak{a}$ is called a *principal ideal* if there is an integer α such that $\mathfrak{a} = (\alpha)$. Note that for two principal ideals (α) and (β), $(\alpha) = (\beta)$ if and only if α and β are associates, i.e., they differ only by a unit factor.

Every ideal in the field $k(1)$ is a principal ideal by Theorem 2, since it is a module if it is $\neq (0)$. On the other hand, the ideal $(1 + 2\sqrt{-5}, 3)$ in the field $K(\sqrt{-5})$ is not a principal ideal because of what was said in the preceding section. This ideal consists of all numbers divisible by $\sqrt{\lambda}$.

If

$$(\alpha_1, \ldots, \alpha_r) = (A_1, \ldots, A_s) \quad \text{and} \quad (\beta_1, \ldots, \beta_p) = (B_1, \ldots, B_q),$$

then

$$(\alpha_1\beta_1, \ldots, \alpha_i\beta_k, \ldots, \alpha_r\beta_p) = (A_1B_1, \ldots, A_lB_m, \ldots, A_sB_q)$$

since

$$\alpha_i = \sum_l \lambda_{il}A_l, \qquad \beta_k = \sum_m \mu_{km}B_m$$

implies that

$$\alpha_i\beta_k = \sum_{l,m} \lambda_{il}\mu_{km}A_lB_m$$

with integral λ, μ and conversely each A_lB_m is a combination of the $\alpha_i\beta_k$.

By the *product* $\mathfrak{ab}$ of two ideals $\mathfrak{a} = (\alpha_1, \ldots, \alpha_r)$ and $\mathfrak{b} = (\beta_1, \ldots, \beta_p)$ we mean the ideal

$$\mathfrak{ab} = (\alpha_1\beta_1, \ldots, \alpha_i\beta_k, \ldots, \alpha_r\beta_p)$$

thus defined uniquely by $\mathfrak{a}$ and $\mathfrak{b}$.

It follows directly from this definition that multiplication of ideals is commutative and associative:

$$\mathfrak{ab} = \mathfrak{ba}, \qquad \mathfrak{a(bc)} = \mathfrak{(ab)c}.$$

We set $\mathfrak{a} = \mathfrak{a}^1$ and for each positive rational integer m we set $\mathfrak{a}^{m+1} = \mathfrak{a}^m\mathfrak{a}$ so that $\mathfrak{a}^{p+q} = \mathfrak{a}^p\mathfrak{a}^q$ as with ordinary powers.

We call an ideal $\mathfrak{a}$ *divisible* by an ideal $\mathfrak{c}$ or $\mathfrak{c}$ a factor (divisor) of $\mathfrak{a}$ if $\mathfrak{c} \neq (0)$ and there is an ideal $\mathfrak{b}$ such that $\mathfrak{a} = \mathfrak{bc}$. In symbols we write $\mathfrak{c}|\mathfrak{a}$.

The connection between divisibility of numbers and of ideals is made by the following fact: *The principal ideal (α) is divisible by the principal ideal $(\gamma) \neq (0)$ if and only if the number α is divisible by the number γ.*

This follows since $(\alpha) = (\gamma)(\beta_1, \ldots, \beta_r) = (\gamma\beta_1, \ldots, \gamma\beta_r)$ implies $\alpha = \sum_i \lambda_i\gamma\beta_i = \gamma \sum_i \lambda_i\beta_i$ with integers λ_i; hence $\gamma|\alpha$. Conversely, if $\gamma|\alpha$, then for some integer β, $\alpha = \gamma\beta$, and we also have $(\alpha) = (\gamma) \cdot (\beta)$ and $(\gamma)|(\alpha)$.

The unit ideal (1) consists of all integral elements of the field. If an ideal contains the number 1, then it contains all integers, and is thus $= (1)$. For each ideal $\mathfrak{a} \neq (0)$

$$\mathfrak{a} = \mathfrak{a} \cdot (1), \quad \mathfrak{a}|\mathfrak{a}, \quad (1)|\mathfrak{a}, \quad \text{and} \quad \mathfrak{a}|(0).$$

Each ideal $\mathfrak{a}$ has the "trivial" factors $\mathfrak{a}$ and (1).

Definition. An ideal $\mathfrak{p}$ is called a *prime ideal* if it is different from (1) and has no factors other than $\mathfrak{p}$ and (1).

We do not yet know whether there are prime ideals.

Indeed the fact that divisibility of ideals can be reduced to divisibility of numbers, that not only the converse holds, is of basic significance for the foundations of ideal theory by virtue of the following theorem:

Theorem 66. *For each ideal* $\mathfrak{a}$ *there is an ideal* $\mathfrak{b}$ *different from* (0) *such that* $\mathfrak{a}\mathfrak{b}$ *is a principal ideal.*

The different ways of laying the foundations of ideal theory are distinguished in the proof of this theorem. Here we will use a method of Hurwitz which was greatly simplified by Steinitz. It rests on a generalization of the theorem of Gauss about polynomials which have algebraic integers as coefficients:

Theorem 67. *Let*

$$A(x) = \alpha_p x^p + \alpha_{p-1} x^{p-1} + \cdots + \alpha_0, \qquad B(x) = \beta_r x^r + \beta_{r-1} x^{r-1} + \cdots + \beta_0$$

be polynomials with integer coefficients, $\alpha_p, \beta_r \neq 0$. *Then if an integer* δ *divides all coefficients* γ *of*

$$C(x) = A(x) \cdot B(x) = \gamma_s x^s + \gamma_{s-1} x^s + \gamma_{s-1} x^{s-1} + \cdots + \gamma_0$$

it also divides all products $\alpha_i \beta_k$.

In order to prove this assertion, we need the following two lemmas:

Lemma (a). *If*

$$f(x) = \delta_m x^m + \delta_{m-1} x^{m-1} + \cdots + \delta_1 x + \delta_0 \qquad (\delta_m \neq 0)$$

is a polynomial with integral coefficients and ρ *is a root, then* $f(x)/(x - \rho)$ *also has integer coefficients.*

To begin with, $\delta_m \rho$ is an integer in any case, as we see immediately by Theorem 62 in a manner similar to that of the proof of Theorem 63.

Moreover the lemma is true for $m = 1$, in which case $f(x)/(x - \rho) = \delta_1$, where $\rho = -\delta_0/\delta_1$.

Suppose that this lemma has already been proved for all polynomials of degree $\leq m - 1$. Since

$$\varphi(x) = f(x) - \delta_m x^{m-1}(x - \rho)$$

is obviously an integral polynomial of degree $\leq m - 1$ with ρ as a root,

$$\frac{\varphi(x)}{x - \rho} = \frac{f(x)}{x - \rho} - \delta_m x^{m-1}$$

is thus integral. Therefore the same holds for $f(x)/(x - \rho)$, whence Lemma (a) follows by complete induction.

Lemma (b). *If, in the above notation*

$$f(x) = \delta_m(x - \rho_1)(x - \rho_2) \cdots (x - \rho_m),$$

then $\delta_m \rho_1 \rho_2 \cdots \rho_k$ *is an integer for each k with* $1 \le k \le m$.

This follows by repeated application of Lemma (a) from which we obtain

$$\frac{f(x)}{(x - \rho_{k+1})(x - \rho_{k+2}) \cdots (x - \rho_m)} = \delta_m(x - \rho_1) \cdots (x - \rho_k)$$

as an integral polynomial whose constant term is $\pm \delta_m \rho_1 \cdots \rho_k$.

We now arrive at the proof of Theorem 67 as follows: let the decomposition into linear factors be

$$A(x) = \alpha_p(x - \rho_1)(x - \rho_2) \cdots (x - \rho_p)$$
$$B(x) = \beta_r(x - \sigma_1)(x - \sigma_2) \cdots (x - \sigma_r).$$

By hypothesis

$$\frac{C(x)}{\delta} = \frac{\alpha_p \beta_r}{\delta}(x - \rho_1) \cdots (x - \sigma_r)$$

has integral coefficients, hence by Lemma (b) each product

$$\frac{\alpha_p \beta_r}{\delta} \cdot \rho_{n_1} \rho_{n_2} \cdots \rho_{n_i} \sigma_{m_1} \cdots \sigma_{m_k} \tag{40}$$

is an integer, where $n_1, \ldots, n_i$ and likewise $m_1, \ldots, m_k$ are any distinct indices ($i \le p$, $k \le r$). However, since α_i/α_p and β_k/β_r are elementary symmetric functions of the ρ and of the σ, $\alpha_i \beta_k/\delta$ is a sum of terms of the form (40), and consequently an integer, as was to be proved.

We are now finally able to prove Theorem 66 about ideals. Let $\mathfrak{a} = (\alpha_1, \ldots, \alpha_r)$. We form the integral polynomial

$$g(x) = \alpha_1 x + \alpha_2 x^2 + \cdots + \alpha_r x^r$$

and the conjugate polynomials

$$g^{(i)}(x) = \alpha_1^{(i)} x + \alpha_2^{(i)} x^2 + \cdots + \alpha_r^{(i)} x^r, \qquad (i = 1, 2, \ldots, n)$$

among which the original polynomial $g(x)$ occurs, say for $i = 1$. The product

$$F(x) = \prod_{i=1}^{n} g^{(i)}(x) = \sum_p c_p x^p$$

as a symmetric function of the conjugates is a polynomial with integral rational coefficients c_p. $F(x)$ is divisible by $g(x)$ and the quotient

$$h(x) = \frac{F(x)}{g(x)} = \prod_{i=2}^{n} g^{(i)}(x)$$

is thus a polynomial with coefficients in K which are moreover integers, say

$$h(x) = \beta_1 x + \beta_2 x^2 + \cdots + \beta_m x^m$$

with the β_i integers in K. If we denote the GCD of the rational integers c_p by N, so that $f(x)/N$ is a primitive polynomial, and set

$$b = (\beta_1, \ldots, \beta_m),$$

then we assert that the equation

$$ab = (N)$$

is true. Now $ab = (\ldots, \alpha_i \beta_k, \ldots)$. By Theorem 67, N divides all $\alpha_i \beta_k$, since it divides each coefficient of $g(x)h(x)$. Hence

$$\alpha_i \beta_k = \lambda_{ik} N$$

where λ_{ik} is an integer, and thus all $\alpha_i \beta_k$ and consequently all numbers of ab belong to (N). Secondly, however, N is the GCD of all the coefficients c_p of $h(x)g(x)$, and hence there are rational integers x_p, such that

$$N = c_1 x_1 + c_2 x_2 + \cdots .$$

Each c is a sum of products $\alpha_i \beta_k$; consequently N is representable in the form

$$N = \sum_{i,k} u_{ik} \alpha_i \beta_k$$

with μ_{ik} integers (actually rational integers). Thus N and all numbers of (N) belong to ab, that is, $(N) = ab$.

By reason of the preceding theorem we see the uniqueness of division of ideals:

Theorem 68. *If* $ab = ac$, *then if* $a \neq 0, b = c$.

To see this we determine an ideal m such that $am = (\delta)$ is a principal ideal. Then

$$amb = amc, \qquad (\alpha)b = (\alpha)c.$$

The latter equation asserts that α times every number from b is of the form α times a number from c, that is, every number of b belongs to c, and likewise the converse is true; thus $b = c$.

And now we obtain a new definition of divisibility:

Theorem 69. *An ideal* $c = (\gamma_1, \ldots, \gamma_r)$ *is a divisor of* $a = (\alpha_1, \ldots, \alpha_m)$ *if and only if every number of* a *belongs to* c.

If $c \mid a$, then there is a $b = (\beta_1, \ldots, \beta_p)$ for which $b \neq (0)$ and

$$(\alpha_1, \ldots, \alpha_m) = (\beta, \ldots, \beta_p) \cdot (\gamma_1, \ldots, \gamma_r) = (\ldots, \beta_i \gamma_k, \ldots);$$

hence every number α of a can be represented in the form

$$\alpha = \sum_{i,k} \lambda_{ik} \beta_i \gamma_k = \sum_{k=1}^{r} \gamma_k \left(\sum_{i=1}^{p} \lambda_{ik} \beta_i \right)$$

with integral λ_{ik} and thus belongs to c.

Conversely if every number in $\mathfrak{a}$ is also a number in $\mathfrak{c}$, then for all integers λ_{ik} there exist integers μ_{pk} for which

$$\sum_i \lambda_{ik}\alpha_i = \sum_p \mu_{pk}\gamma_p;$$

then for each $\mathfrak{d} = (\delta_1, \ldots, \delta_s)$

$$\sum_k \sum_i \lambda_{ik}\alpha_i \delta_k = \sum_k \sum_p \mu_{pk}\gamma_p \delta_k,$$

that is, each number in $\mathfrak{a}\mathfrak{d}$ belongs to $\mathfrak{c}\mathfrak{d}$. Now let us choose $\mathfrak{d}$ so that $\mathfrak{c}\mathfrak{d} = (\delta)$ is a principal ideal ($\delta \neq 0$). If $\mathfrak{a}\mathfrak{d} = (\rho_1, \rho_2, \ldots)$, then each ρ_i is a number from (δ); thus it is of the form $\lambda_i \delta$ with integral λ_i and hence

$$(\rho_1, \rho_2, \ldots) = (\delta)(\lambda_1, \lambda_2, \ldots),$$
$$\mathfrak{a}\mathfrak{d} = \mathfrak{c}\mathfrak{d} \cdot (\lambda_1, \lambda_2, \ldots),$$
$$\mathfrak{a} = \mathfrak{c} \cdot (\lambda_1, \lambda_2, \ldots), \quad \text{i.e.,} \quad \mathfrak{c}\,|\,\mathfrak{a}.$$

As an immediate consequence of this theorem we emphasize:

Let $\mathfrak{a}$ be an ideal which is not $= (0)$.

The integer α occurs in $\mathfrak{a}$ if and only if $\mathfrak{a}\,|\,(\alpha)$. If $\mathfrak{a}\,|\,(\alpha)$ and $\mathfrak{a}\,|\,(\beta)$, then also $\mathfrak{a}\,|\,(\lambda\alpha + \mu\beta)$ for all integers λ, μ.

It follows from $\mathfrak{a}\mathfrak{b} = (1)$ that $\mathfrak{a} = (1)$ and $\mathfrak{b} = (1)$.

If each of two ideals is a divisor of the other, then they are equal.

§25 The Fundamental Theorem of Ideal Theory

Theorem 70. *For every two ideals* $\mathfrak{a} = (\alpha_1, \ldots, \alpha_r)$, $\mathfrak{b} = (\beta_1, \ldots, \beta_s)$ *which are not both* $= (0)$, *there is a uniquely determined greatest common divisor* $\mathfrak{d} = (\mathfrak{a}, \mathfrak{b})$ *which has the following property:* $\mathfrak{d}$ *is a divisor of* $\mathfrak{a}$ *and* $\mathfrak{b}$. *Furthermore if* $\mathfrak{d}_1\,|\,\mathfrak{a}$ *and* $\mathfrak{d}_1\,|\,\mathfrak{b}$, *then* $\mathfrak{d}_1$ *is a divisor of* $\mathfrak{d}$. *Indeed* $\mathfrak{d} = (\alpha_1, \ldots, \alpha_r, \beta_1, \ldots, \beta_s)$.

We show that $\mathfrak{d} = (\alpha_1, \ldots, \alpha_r, \beta_1, \ldots, \beta_s)$ has the stated properties of divisibility. Since every sum "number in $\mathfrak{a}$ + number in $\mathfrak{b}$" obviously belongs to $\mathfrak{d}$, then all the numbers of $\mathfrak{a}$ and of $\mathfrak{b}$ belong to $\mathfrak{d}$, and consequently by Theorem 69 $\mathfrak{d}\,|\,\mathfrak{a}$ and $\mathfrak{d}\,|\,\mathfrak{b}$.

Moreover if $\mathfrak{d}_1\,|\,\mathfrak{a}$ and $\mathfrak{d}_1\,|\,\mathfrak{b}$, then all numbers of $\mathfrak{a}$ and of $\mathfrak{b}$ and consequently also each sum "number in $\mathfrak{a}$ + number in $\mathfrak{b}$" belong to $\mathfrak{d}_1$, that is, each number of $\mathfrak{d}$ belongs to $\mathfrak{d}_1$. Again we have $\mathfrak{d}_1\,|\,\mathfrak{d}$.

If an ideal $\mathfrak{d}_2$ likewise has this property, then $\mathfrak{d}_2\,|\,\mathfrak{d}$ and $\mathfrak{d}\,|\,\mathfrak{d}_2$ thus $\mathfrak{d} = \mathfrak{d}_2$. Consequently $\mathfrak{d}$ is uniquely determined by this property.

We see, accordingly, that an *ideal* $\mathfrak{a} = (\alpha_1, \ldots, \alpha_r)$ *can be regarded as the GCD of the principal ideals* $(\alpha_1), (\alpha_2), \ldots, (\alpha_r)$.

We conclude immediately from the expression for $\mathfrak{d}$ that

$$\mathfrak{c} \cdot (\mathfrak{a}, \mathfrak{b}) = (\mathfrak{c}\mathfrak{a}, \mathfrak{c}\mathfrak{b}). \tag{41}$$

Thus from this follows a part of the fundamental theorem:

Theorem 71. *If* $\mathfrak{p}$ *is a prime ideal and* $\mathfrak{p} \mid \mathfrak{ab}$, *then* $\mathfrak{p}$ *divides either* $\mathfrak{a}$ *or* $\mathfrak{b}$ *or both.*

For if $\mathfrak{p}$ does not divide the factor $\mathfrak{b}$, then

$$(\mathfrak{p}, \mathfrak{b}) = (1),$$

since, as a prime ideal, $\mathfrak{p}$ has no factors except (1) and $\mathfrak{p}$. It follows from (41) that

$$\mathfrak{a} = \mathfrak{a}(1) = \mathfrak{a}(\mathfrak{p}, \mathfrak{b}) = (\mathfrak{ap}, \mathfrak{ab})$$

and since $\mathfrak{p} \mid \mathfrak{ab}$, $\mathfrak{p}$ must divide $\mathfrak{a}$.

From this we obtain, as in rational number theory (Theorem 5), that a representation of an ideal as a product of prime ideals is possible, if at all, only in one single manner, except of course for the order of the factors.

However, we are still missing the proof that a decomposition into prime ideal factors is always possible. For this we must show:

(a) Every ideal $\mathfrak{a}$ which is not (0) has only finite many divisors.
(b) Every proper divisor of $\mathfrak{a}$ ($\mathfrak{a} \neq (0)$) has fewer divisors than $\mathfrak{a}$.

For the proof of (a) we recall that every ideal $\mathfrak{a}$ which is not (0) divides a certain principal ideal (α), and that each divisor of $\mathfrak{a}$ is also a divisor of (α). Thus it is sufficient to verify the finiteness of the number of divisors of each principal ideal (α), and here we may take α as a rational integer, since $\alpha \mid N(\alpha)$ implies $(\alpha) \mid N(\alpha)$ and $N(\alpha) = N$ is such a number.

By Theorem 69, an ideal (N) is divisible only by those ideals $\mathfrak{a}$ in which N occurs. Now let $\mathfrak{a} = (\alpha_1, \ldots, \alpha_r)$ be a divisor of (N), hence let N occur in $\mathfrak{a}$. It is sufficient to assume $r \leq n$, since, for example, we can indeed choose for the α_i a basis for $\mathfrak{a}$. Now

$$(\alpha_1, \ldots, \alpha_r) = (\alpha_1, \ldots, \alpha_r, N) = (\alpha_1 - N\lambda_1, \alpha_2 - N\lambda_2, \ldots, \alpha_r - N\lambda_r, N)$$

is true for arbitrary integers λ_i. We show that the λ_i can be chosen so that the $\alpha_i - N\lambda_i$ belong to a definite finite range of values. Let $\omega_1, \ldots, \omega_n$ be a basis for the field. To each integer $\alpha = x_1\omega_1 + \cdots + x_n\omega_n$, an integer $\lambda = u_1\omega_1 + \cdots + u_n\omega_n$ (x_i and u_i rational integers) can obviously be determined so that in

$$\alpha - N\lambda = (x_1 - Nu_1)\omega_1 + \cdots + (x_n - Nu_n)\omega_n$$

the n rational integers $x_i - Nu_i$ belong to the interval $0, \ldots, N - 1$. Among these numbers, which we call "reduced mod N" for the moment, there are only $|N|^n$ distinct ones. We now choose the λ_i so that all numbers $\alpha_i - \lambda_i N$ are reduced mod N; then the, at most, n numbers $\alpha_i - \lambda_i N$ belong to a definite finite set of numbers determined only by N, and hence they can give rise to only finitely many distinct ideals $\mathfrak{a}$; that is, (N) has only finitely many divisors and Lemma (a) is proved. Now in order to prove Lemma (b), let

c be a proper divisor of a. Thus $a = bc$ where $b \neq (1)$, $c \neq a$. Then c surely does not have a as a divisor, and consequently c has at least one less divisor than a.

Now at least one prime ideal must occur among the finitely many, say m, divisors of a which are not $=(1)$, unless a itself is (1). Namely the divisor or divisors which have as few divisors as possible are obviously prime ideals by Lemma (b). Consequently, we can split off a prime ideal p_1 from a, $a = p_1 a_1$, where a_1 has at most $m - 1$ divisors which are $\neq (1)$. In case we do not have $a_1 = (1)$, we can again split off a prime ideal p_2 from a_1, where a_2 has at most $m - 2$ divisors $\neq (1)$, $a = p_1 p_2 a_2$ and so on. Since the $a_1, a_2, \ldots$ always have decreasing numbers of divisors, the process must come to an end after finitely many steps, which can only occur if $a_k = (1)$. Then $a = p_1 p_2 \cdots p_k$ is represented as a product of prime ideals, and we have proved

Theorem 72 (Fundamental Theorem of Ideal Theory). *Every ideal in K different from* (0) *and* (1) *can be written in one and only one way* (*except for order*) *as a product of prime ideals.*

§26 First Applications of the Fundamental Theorem

We see at once that this theorem on ideals can be used in the investigation of divisibility properties of numbers, e.g., this theorem gives an entirely new method for deciding whether or not an integer α is divisible by an integer β. By §24 we must investigate whether (α) is divisible by (β). First we decompose both ideals into their distinct prime factors:

$$(\alpha) = p_1^{a_1} p_2^{a_2} \cdots p_k^{a_k} \qquad (a_i \geq 0),$$
$$(\beta) = p_1^{b_1} p_2^{b_2} \cdots p_k^{b_k} \qquad (b_i \geq 0).$$

By the fundamental theorem, β divides α if and only if $a_i - b_i \geq 0$ for $i = 1, 2, \ldots, k$.

Theorem 73. *There are infinitely many prime ideals in each field.*

Each rational prime p defines an ideal (p), and moreover if p and q are distinct positive primes, then $(p, q) = 1$ in the sense of our ideal theory, since the number 1 occurs in the form $px + qy$ in (p, q). Consequently, the same primes never divide (p) and (q); hence there are at least as many prime ideals as there are positive primes p.

We now simplify the notation in that *when designating principal ideals* (α) *we omit the parentheses* whenever there is no danger of misunderstanding; however we must keep in mind that from the equality of the ideals α and β

it only follows that: $\alpha = \beta \times$ unit. Likewise, in all statements which concern the divisibility of an (α), we replace the ideal by the number α. Thus α is divisible by $\mathfrak{a}$ means that (α) is divisible by $\mathfrak{a}$. The statement $\beta|\alpha$ already has meaning, it actually agrees with $(\beta)|(\alpha)$ by what we have done earlier. The greatest common divisor of $\alpha_1, \ldots, \alpha_r$ is accordingly the ideal $\mathfrak{a} = (\alpha_1, \ldots, \alpha_r)$. If this ideal $= (1)$, then we call the numbers $\alpha_1, \ldots, \alpha_r$ *relatively prime*. In order that the numbers be relatively prime it is necessary and sufficient that $\mathfrak{a}$ contains the number 1, that is, that there are integers λ_i in K such that

$$\lambda_1 \alpha_1 + \lambda_2 \alpha_2 + \cdots + \lambda_r \alpha_r = 1.$$

It follows from $\mathfrak{a}|\alpha$ and $\mathfrak{a}|\beta$ that $\mathfrak{a}|\lambda\alpha + \mu\beta$ for all integers λ and μ in K.

Theorem 74. *If $\mathfrak{a}$ and $\mathfrak{b}$ are ideals distinct from (0), then there is always a number ω for which*

$$(\omega, \mathfrak{ab}) = \mathfrak{a}.$$

This ω then obviously has a decomposition $\omega = \mathfrak{ac}$ where $(\mathfrak{c}, \mathfrak{b}) = 1$. Thus the theorem asserts that each $\mathfrak{a}$ can be made into a principal ideal by multiplication with such a $\mathfrak{c}$ which is relatively prime to the given $\mathfrak{b}$.

For a proof, let $\mathfrak{p}_1, \ldots, \mathfrak{p}_r$ be all the distinct prime ideals which divide $\mathfrak{ab}$, and let $\mathfrak{a} = \mathfrak{p}_1^{a_1} \cdots \mathfrak{p}_r^{a_r}$ $(a_i \geq 0)$. We define the r ideals $\mathfrak{d}_1, \ldots, \mathfrak{d}_r$ by

$$\mathfrak{p}_i^{a_i+1}\mathfrak{d}_i = \mathfrak{a}\mathfrak{p}_1 \cdots \mathfrak{p}_r, \qquad (i = 1, \ldots, r)$$

so that $\mathfrak{d}_i$ is relatively prime to $\mathfrak{p}_i$, but contains all remaining prime ideals $\mathfrak{p}$ to a higher power than in $\mathfrak{a}$. Since these $\mathfrak{d}_i$ in their totality are relatively prime, there are numbers δ_i in $\mathfrak{d}_i$ such that

$$\delta_1 + \delta_2 + \cdots + \delta_r = 1.$$

Here δ_i is divisible by $\mathfrak{d}_i$, hence by all $\mathfrak{p}_k$ $(k \neq i)$. Consequently, since 1 is not divisible by $\mathfrak{p}_i$, δ_i is surely not divisible by $\mathfrak{p}_i$.

We now determine r numbers α_i, such that $\mathfrak{p}_i^{a_i}|\alpha_i$ but $\mathfrak{p}_i^{a_i+1}$ does not divide α_i, which is obviously always possible since for this to happen α_i need only be a number from $\mathfrak{p}_i^{a_i}$ which does not occur in $\mathfrak{p}_i^{a_i+1}$. Then the number

$$\omega = \alpha_1 \delta_1 + \alpha_2 \delta_2 + \cdots + \alpha_r \delta_r$$

has the property asserted in Theorem 74. For each of the prime ideals $\mathfrak{p}_i$ occurs in $r - 1$ summands at least to the power $\mathfrak{p}_i^{a_i+1}$; however, it occurs precisely to the power $\mathfrak{p}_i^{a_i}$ in the ith summand; consequently ω is divisible by precisely the a_ith power of $\mathfrak{p}_i$, but no higher power.

By taking $\mathfrak{ab}$ itself as a principal ideal β, which is divisible by $\mathfrak{a}$, we obtain

Theorem 75. *Every ideal $\mathfrak{a}$ can be represented as the greatest common divisor of two elements of the field: $\mathfrak{a} = (\omega, \beta)$.*

§27 Congruences and Residue Classes Modulo Ideals and the Group of Residue Classes under Addition and under Multiplication

We now carry over the concept of congruence in rational number theory to ideal theory. Only slight modifications are needed in the methods of proof used earlier, which we deal with very briefly.

For two integers α, β and an ideal $\mathfrak{a}$, which is always assumed to be different from 0 in this section,

$$\alpha \equiv \beta \;(\text{mod } \mathfrak{a}) \qquad (\alpha \text{ congruent to } \beta \text{ mod } \mathfrak{a})$$

is to mean

$$\mathfrak{a} \mid \alpha - \beta.$$

If $\mathfrak{a}$ does not divide $\alpha - \beta$, then we write $\alpha \not\equiv \beta \;(\text{mod } \mathfrak{a})$.

These congruences satisfy the same rules of calculation given in §2 for congruences in the rational number field and in the case where α, β and $\mathfrak{a}$ are rational numbers, they mean exactly the same things as earlier.

All numbers which are congruent to each other mod $\mathfrak{a}$ form a residue class mod $\mathfrak{a}$.

Theorem 76. *The number of residue classes* mod $\mathfrak{a}$ *is finite. If the number of residue classes is denoted by $N(\mathfrak{a})$ and if $\alpha_1, \ldots, \alpha_n$ is a basis for $\mathfrak{a}$, then $N(\mathfrak{a}) = |\Delta(\alpha_1, \ldots, \alpha_n)/\sqrt{d}|$. For a principal ideal $\mathfrak{a} = \alpha$, $N(\mathfrak{a}) = |N(\alpha)|$.*

The numbers of $\mathfrak{a}$ form a subgroup of the group $\mathfrak{G}$ of all integers of the field. The different cosets in $\mathfrak{G}$ determined by $\mathfrak{a}$ obviously form the different residue classes mod $\mathfrak{a}$. Hence the number of distinct residue classes mod $\mathfrak{a}$ is the index of $\mathfrak{a}$ in $\mathfrak{G}$. *This index is finite.* For if α is any nonzero number in $\mathfrak{a}$, then the positive rational number $a = |N(\alpha)|$ also belongs to $\mathfrak{a}$, since $\alpha \mid N(\alpha)$, and consequently the product $a \times$ arbitrary integral field element belongs to $\mathfrak{a}$. Thus in group-theoretic terms the ath power, in the sense of composition, of each element of $\mathfrak{G}$ belongs to $\mathfrak{a}$. Consequently by Theorem 40 the index of $\mathfrak{a}$ is finite; it is denoted $N(\mathfrak{a})$ (*norm of* $\mathfrak{a}$). If $\alpha_1, \ldots, \alpha_n$ is a basis for $\mathfrak{a}$, $\omega_1, \ldots, \omega_n$ a basis for $\mathfrak{G}$, then there exists a system of equations

$$\alpha_i = \sum_{k=1}^{n} c_{ik}\omega_k \qquad (i = 1, 2, \ldots, n),$$

with rational integers c_{ik}, and by Theorem 39 the absolute value of the determinant $\|c_{ik}\|$ is equal to the index $N(\mathfrak{a})$. On the other hand, by passage to the conjugates

$$\Delta(\alpha_1, \ldots, \alpha_n) = \|c_{ik}\| \cdot \Delta(\omega_1, \ldots, \omega_n),$$

and since

$$\Delta^2(\omega_1, \ldots, \omega_n) = d \neq 0,$$

we thus have

$$N(\mathfrak{a}) = \left| \frac{\Delta(\alpha_1, \dots, \alpha_n)}{\sqrt{d}} \right|.$$

With a principal ideal (α) we obviously obtain a basis of the form $\alpha\omega_1, \dots, \alpha\omega_n$ thus

$$\Delta(\alpha\omega_1, \dots, \alpha\omega_n) = N(\alpha)\,\Delta(\omega_1, \dots, \omega_n), \qquad N(\mathfrak{a}) = |N(\alpha)|.$$

Theorem 78. *For given α and β the congruence*

$$\alpha\xi \equiv \beta \pmod{\mathfrak{a}}$$

can be solved by an integer ξ in K if and only if $(\alpha, \mathfrak{a}) | \beta$. If $(\alpha, \mathfrak{a}) = 1$, then the solution is completely determined mod $\mathfrak{a}$.

If we assume $(\alpha, \mathfrak{a}) = 1$ to begin with, and let ξ run through a system of $N(\mathfrak{a})$ numbers which are incongruent mod $\mathfrak{a}$, then $\alpha\xi$ runs through all the residue classes mod $\mathfrak{a}$, for it follows from $\alpha\xi_1 \equiv \alpha\xi_2 \pmod{\mathfrak{a}}$ that $\mathfrak{a} | \alpha(\xi_1 - \xi_2)$. However, since $(\alpha, \mathfrak{a}) = 1$ we must have $\mathfrak{a} | \xi_1 - \xi_2$, that is, $\xi_1 \equiv \xi_2 \pmod{\mathfrak{a}}$ by the fundamental theorem. Thus, among the numbers $\alpha\xi$, one from the residue class of β also occurs. For the same reason the solution is obviously determined uniquely mod $\mathfrak{a}$.

Moreover if we now have $(\alpha, \mathfrak{a}) = \mathfrak{d}$ and there is an integer ξ_0 with $\alpha\xi_0 \equiv \beta \pmod{\mathfrak{a}}$, then $\alpha\xi_0 = \beta + \rho$, where $\mathfrak{a} | \rho$. Thus $\mathfrak{d} | \rho$ and $\mathfrak{d} | \alpha\xi_0 - \rho$, that is, $\mathfrak{d} | \beta$.

Conversely, if

$$\mathfrak{d} | \beta, \qquad \beta = \mathfrak{d}\mathfrak{b},$$

then let us set $\alpha = \mathfrak{d}\mathfrak{a}_1$, $\mathfrak{a} = \delta\mathfrak{a}_2$ so that $(\mathfrak{a}_1, \mathfrak{a}_2) = 1$, and let us determine a number $\mu = \mathfrak{m}\mathfrak{a}_1$ such that $(\mu, \mathfrak{a}_1\mathfrak{d}\mathfrak{a}_2) = \mathfrak{a}_1$, thus $(\mathfrak{m}, \mathfrak{d}\mathfrak{a}_2) = 1$. This is possible by Theorem 74. Then $\mathfrak{d}\mathfrak{a}_1 | \mathfrak{m}\mathfrak{a}_1\mathfrak{d}\mathfrak{b}$, hence $\alpha | \mu\beta$ and the congruence

$$\mu\xi \equiv \frac{\mu\beta}{\alpha} \pmod{\mathfrak{a}_2}$$

is solvable for ξ by what has just been proved, since $(\mu, \mathfrak{a}_2) = (\mathfrak{m}\mathfrak{a}_1, \mathfrak{a}_2) = 1$ follows from $(\mathfrak{m}, \mathfrak{a}_2) = 1$ and $(\mathfrak{a}_1, \mathfrak{a}_2) = 1$. From $\mathfrak{a}_2 | \mu\xi - (\mu\beta/\alpha)$ it follows that

$$\alpha\mathfrak{a}_2 | (\alpha\mu\xi - \mu\beta),$$

i.e.,

$$\mathfrak{d}\mathfrak{a}_1\mathfrak{a}_2 | (\mu)(\alpha\xi - \beta), \qquad \mathfrak{d}\mathfrak{a}_1\mathfrak{a}_2 | \mathfrak{m}\mathfrak{a}_1(\alpha\xi - \beta)$$
$$\mathfrak{d}\mathfrak{a}_2 | \mathfrak{m}(\alpha\xi - \beta), \qquad \mathfrak{d}\mathfrak{a}_2 | \alpha\xi - \beta$$

(as $(\mathfrak{m}, \mathfrak{d}\mathfrak{a}_2) = 1$), i.e., $\alpha\xi \equiv \beta \pmod{\mathfrak{a}}$.

Two numbers congruent modulo $\mathfrak{a}$ have the same GCD with $\mathfrak{a}$ so this property is thus a property of the whole residue class. *The number of residue classes relatively prime to $\mathfrak{a}$ is denoted by* $\varphi(\mathfrak{a})$.

Theorem 79. *For two ideals* $\mathfrak{a}$ *and* $\mathfrak{b}$, *we always have*

$$N(\mathfrak{ab}) = N(\mathfrak{a}) \cdot N(\mathfrak{b}).$$

Let α be a number divisible by $\mathfrak{a}$ such that $(\alpha, \mathfrak{ab}) = \mathfrak{a}$. If we let ξ_i $(i = 1, 2, \ldots, N(\mathfrak{b}))$ run through a complete system of residues mod $\mathfrak{b}$ and let η_k $(k = 1, 2, \ldots, N(\mathfrak{a}))$ run through a complete system of residues mod $\mathfrak{a}$, then no two of the numbers $\alpha\xi_i + \eta_k$ are congruent mod $\mathfrak{ab}$. On the other hand, each integer ρ is congruent mod $\mathfrak{ab}$ to one of these numbers $\alpha\xi_i + \eta_k$. For let η_k be determined so that

$$\eta_k \equiv \rho \;(\text{mod } \mathfrak{a})$$

and then let ξ be determined so that

$$\alpha\xi \equiv \rho - \eta_k \;(\text{mod } \mathfrak{ab}).$$

Since $(\alpha, \mathfrak{ab}) = \mathfrak{a}$ and $\mathfrak{a} | \rho - \eta_k$ this congruence can be solved by Theorem 78 and ξ can be determined mod $\mathfrak{b}$ so that ξ can be chosen equal to ξ_i. Consequently the $N(\mathfrak{a}) \cdot N(\mathfrak{b})$ numbers $\alpha\xi_i + \eta_k$ form a complete system of residues mod $\mathfrak{ab}$ and thus there must also be $N(\mathfrak{ab})$ of them.

Theorem 80. *If* $(\mathfrak{a}, \mathfrak{b}) = 1$, *then* $\varphi(\mathfrak{ab}) = \varphi(\mathfrak{a}) \cdot \varphi(\mathfrak{b})$ *and in general*

$$\varphi(\mathfrak{a}) = N(\mathfrak{a}) \prod_{\mathfrak{p} | \mathfrak{a}} \left(1 - \frac{1}{N(\mathfrak{p})} \right)$$

where $\mathfrak{p}$ *runs through the distinct prime divisors of* $\mathfrak{a}$.

To see this let α be chosen so that $(\alpha, \mathfrak{ab}) = \mathfrak{a}$ and β so that $(\beta, \mathfrak{ab}) = \mathfrak{b}$. Then if ξ runs through a complete system of residues mod $\mathfrak{b}$ and η runs through such a system mod $\mathfrak{a}$ in $\alpha\xi + \beta\eta$ we obtain a complete system of residues mod $\mathfrak{ab}$. These numbers are relatively prime to $\mathfrak{ab}$ if and only if $(\xi, \mathfrak{b}) = 1$ and $(\eta, \mathfrak{a}) = 1$.

For a power $\mathfrak{p}^a$ of a prime ideal $\mathfrak{p}$, the numbers which are not relatively prime to $\mathfrak{p}^a$ are those which are divisible by $\mathfrak{p}$. Among these there are $N(\mathfrak{p}^{a-1}) = (N(\mathfrak{p}))^{a-1}$ incongruent modulo $\mathfrak{p}^a$. Therefore

$$\varphi(\mathfrak{p}^a) = N(\mathfrak{p})^a - N(\mathfrak{p})^{a-1} = N(\mathfrak{p}^a)\left(1 - \frac{1}{N(\mathfrak{p})} \right).$$

Theorem 81. *The norm of a prime ideal* $\mathfrak{p}$ *is a power of a certain rational prime* p, $N(\mathfrak{p}) = p^f$. f *is called the degree of* $\mathfrak{p}$. *Every ideal* (p), *where* p *is a rational prime, can be decomposed into at most n factors.*

For each prime ideal $\mathfrak{p}$ divides certain rational numbers and consequently also certain rational primes p. Suppose that $\mathfrak{p} | p$, $p = \mathfrak{pa}$. Then $N(p) = N(\mathfrak{p}) \cdot N(\mathfrak{a})$ and consequently the rational integer $N(\mathfrak{p})$ divides $N(p) = p^n$; hence $N(\mathfrak{p}) = p^f$ and $f \leq n$. If we think of (p) as decomposed into its prime factors $p = \mathfrak{p}_1 \mathfrak{p}_2 \cdots \mathfrak{p}_r$, then the positive rational integers $N(\mathfrak{p}_1) \cdots N(\mathfrak{p}_r)$ have as product $N(p) = p^n$, while none of these integers is $= 1$; thus their number r must be $\leq n$.

In this way we obtain one of the few statements which connect the degree of a field with other properties of the numbers of the field. If it is known that a rational prime p is decomposable into k ideal factors in a number field, then the degree of the field is at least $= k$.

One proves, as we did Theorem 12 about rational primes:

Theorem 82. *A congruence modulo a prime ideal* $\mathfrak{p}$

$$x^m + \alpha_1 x^{m-1} + \cdots + \alpha_{m-1} x^{m-1} + \alpha_m \equiv 0 \ (\text{mod } \mathfrak{p}),$$

with integer coefficients α, *has at most* m *solutions* x *which are incongruent modulo* $\mathfrak{p}$.

The system of $N(\mathfrak{a})$ residue classes mod $\mathfrak{a}$ again forms an Abelian group under composition by *addition* in that two integers α and β determine by their sum $\alpha + \beta$ another residue class mod $\mathfrak{a}$ which depends only on the classes of α and β. *Let the Abelian group of order* $N(\mathfrak{a})$ *which is defined in this way be called* $\mathfrak{G}(\mathfrak{a})$. Theorem 19 of group theory ($A^h = E$) asserts that for all α

$$\alpha \cdot N(\mathfrak{a}) \equiv 0 \ (\text{mod } \mathfrak{a}),$$

since the unit element is represented by the residue class of 0. In particular it follows that for $\alpha = 1$

$$N(\mathfrak{a}) \equiv 0 \ (\text{mod } \mathfrak{a}) \tag{42}$$

In general the group $\mathfrak{G}(\mathfrak{a})$ is *not cyclic* as it is in the field $K(1)$. For example let $\mathfrak{a} = (a)$ where a is a positive rational integer. Since a number $x_1 \omega_1 + \cdots + x_n \omega_n$ (where the x_i are rational integers and the ω_i a basis of the field) is divisible by a if and only if all x_i are divisible by a, we obtain all residue classes mod a exactly once in the form $x_1 \omega_1 + \cdots + x_n \omega_n$ where $0 \leq x_i < a$. Consequently, for each prime p dividing a there exist exactly n *basis classes* whose order is a power of p. Moreover, for a prime ideal $\mathfrak{p}$ we have:

Theorem 83. *The group of residue classes* mod $\mathfrak{p}$ *is an Abelian group* $\mathfrak{G}(\mathfrak{p})$ *of order* $N(\mathfrak{p}) = p^f$ *under composition by addition and the number of basis elements is equal to the degree* f *of the prime ideal* $\mathfrak{p}$.

For since $\mathfrak{p} \mid p$, the number of residue classes whose elements α satisfy the congruence

$$p\alpha \equiv 0 \ (\text{mod } \mathfrak{p})$$

is equal to the number of all residue classes, thus p^f. Consequently, by Theorem 27, f is equal to the number of basis elements. Therefore there are exactly f elements $\omega_1, \ldots, \omega_f$ such that all residue classes mod $\mathfrak{p}$ are obtained exactly once by the representatives $x_1 \omega_1 + \cdots + x_f \omega_f$, where the rational integers x_i satisfy the inequalities $0 \leq x_i < p$.

Thus the group $\mathfrak{G}(\mathfrak{p})$ *is cyclic for the prime ideals of degree* 1 *and only for these.* The prime ideals of degree 1, of which an infinite number always exist, as will be seen in §43, play a decisive role in investigations of number fields.

Moreover the system of residue classes mod $\mathfrak{a}$ relatively prime to $\mathfrak{a}$, forms a finite Abelian group under *composition by multiplication* in that two numbers α, β relatively prime to $\mathfrak{a}$ determine, by their product, a residue class $\alpha \cdot \beta$ mod $\mathfrak{a}$, which is completely determined by the residue classes of α and β and which is, of course, also relatively prime to $\mathfrak{a}$. Thus we have exactly as before

Theorem 84. *The residue classes* mod $\mathfrak{a}$ *relatively prime to* $\mathfrak{a}$, *under composition by multiplication, form an Abelian group of order* $\varphi(\mathfrak{a})$, *which will be denoted by* $\mathfrak{R}(\mathfrak{a})$. *For each prime ideal* $\mathfrak{p}$, $\mathfrak{R}(\mathfrak{p})$ *is a cyclic group.*

A number ρ whose powers yield all classes of $\mathfrak{R}(\mathfrak{p})$ is called a *primitive root* mod $\mathfrak{p}$.

In particular for a prime ideal $\mathfrak{p}$ and every integer α of the field the generalization of Fermat's theorem

$$\alpha^{N(\mathfrak{p})} \equiv \alpha \ (\text{mod } \mathfrak{p})$$

holds.

On the other hand we *cannot* conclude from this that all groups $\mathfrak{R}(\mathfrak{p}^a)$ are cyclic.

Those classes of $\mathfrak{R}(\mathfrak{p})$ which can be represented by a rational number obviously form a subgroup of $\mathfrak{R}(\mathfrak{p})$; these are the classes of $1, 2, \ldots, p - 1$, if $N(\mathfrak{p}) = p^f$. These classes are also distinct mod $\mathfrak{p}$, since a rational integer a, not divisible by p, is relatively prime to p in $k(1)$; thus the number 1 occurs in the form $ax + py$. Consequently (a) and (p) are also relatively prime in K and therefore $(a, \mathfrak{p}) = 1$ and a is not divisible by $\mathfrak{p}$. For each class A of this subgroup, which thus consists of $p - 1$ elements, A^{p-1} is the unit class. Since the entire group $\mathfrak{R}(\mathfrak{p})$ is cyclic, there are no more than $p - 1$ classes C for which $C^{p-1} = 1$. Thus the subgroup of the rational residue classes of $\mathfrak{R}(\mathfrak{p})$ is identical with the group of classes whose $(p - 1)$th power is the unit class. With this we obtain

Theorem 85. *In order that a number* α *be congruent to a rational number mod* $\mathfrak{p}$, *it is necessary and sufficient that* $\alpha^p \equiv \alpha \ (\text{mod } \mathfrak{p})$.

§28 Polynomials with Integral Algebraic Coefficients

To conclude these elementary considerations about congruences we consider functional congruences. They play a decisive role in the foundations which Kronecker gave for ideal theory. Indeed, even today, certain facts of ideal theory can be proved most easily with these methods.

In this section a polynomial is an integral rational function of an arbitrary number of variables $x_1, \ldots, x_m$, in which the coefficients of the various products of powers are all integers in K.

A polynomial $P(x_1, \ldots, x_m)$ is said to be $\equiv 0 \pmod{\mathfrak{a}}$ if all coefficients are divisible by $\mathfrak{a}$. Moreover two polynomials P and Q are congruent to each other mod $\mathfrak{a}$ if the polynomial $P - Q \equiv 0 \pmod{\mathfrak{a}}$. For polynomials which reduce to constants, this agrees with the definition of congruence of numbers.

Theorem 86. If $\mathfrak{p}$ is a prime ideal, and if for two polynomials P and Q the product

$$P(x_1, \ldots, x_m) \cdot Q(x_1, \ldots, x_m) \equiv 0 \pmod{\mathfrak{p}},$$

then at least one of the polynomials is $\equiv 0 \pmod{\mathfrak{p}}$.

The theorem is true for polynomials of 0 variables, that is, for constants. We show that it is correct in general by passing from m to $m + 1$. Assume it is already proven for all polynomials with m or fewer variables. Each polynomial of $m + 1$ variables can be put into the form

$$P(x_0, \ldots, x_m) = \sum_k x_0^k P_k(x_1, \ldots, x_m)$$

where the P_k are polynomials in $x_1, \ldots, x_m$. Obviously $P \equiv 0 \pmod{\mathfrak{p}}$ means that all $P_k \equiv 0 \pmod{\mathfrak{p}}$. Without loss of generality we may replace P and Q by polynomials which are congruent to them mod $\mathfrak{p}$ in which the terms with the highest powers of x_0 are not congruent to zero, provided not all members are congruent to zero. If the leading terms are $x_0^p P_p(x_1, \ldots, x_m)$ and $x_0^q Q_q(x_1, \ldots, x_m)$, then the highest term of PQ in x_0 is equal to the product $x_0^{p+q} P_p Q_q$ and it follows from

$$P(x_0, \ldots, x_m) \cdot Q(x_0, \ldots, x_m) \equiv 0 \pmod{\mathfrak{p}}$$

that

$$P_p(x_1, \ldots, x_m) \cdot Q_q(x_1, \ldots, x_m) \equiv 0 \pmod{\mathfrak{p}}.$$

However since we are dealing here with polynomials in m variables, at least one of the factors must be $\equiv 0 \pmod{\mathfrak{p}}$. That is, either in $P(x_0, \ldots, x_m)$ or in $Q(x_0, \ldots, x_m)$ there is no term which is not $\equiv 0 \pmod{\mathfrak{p}}$. Thus one of the two polynomials P, Q must be $\equiv 0 \pmod{\mathfrak{p}}$.

Furthermore from this it follows that if $\mathfrak{p}^a$ and $\mathfrak{p}^b$ are the highest powers of a prime ideal $\mathfrak{p}$ which divides all coefficients of the polynomials $A(x_1, \ldots, x_m)$ and $B(x_1, \ldots, x_m)$ respectively, then $\mathfrak{p}^{a+b}$ is the highest power of $\mathfrak{p}$ which divides all coefficients of the product $A(x_1, \ldots, x_m) \cdot B(x_1, \ldots, x_m)$.

To prove this we choose integers, say α_1, α_2, in K such that $(\alpha_1/\alpha_2)A(x_1, \ldots, x_m)$ is a polynomial which has coefficients not all divisible by $\mathfrak{p}$. For this purpose we choose

$$\alpha_2 = \mathfrak{a}\mathfrak{p}^a, \qquad \alpha_1 = \mathfrak{a}\mathfrak{m}, \qquad \text{where } (\mathfrak{a}, \mathfrak{p}) = (\mathfrak{m}, \mathfrak{p}) = 1.$$

In an analogous fashion we choose β_1 and β_2 to be integers such that $(\beta_1/\beta_2)B(x_1, \ldots, x_m)$ also has integer coefficients which are not all divisible by $\mathfrak{p}$. Then, by Theorem 85 the product

$$\frac{\alpha_1}{\alpha_2} \cdot \frac{\beta_1}{\beta_2} A(x_1, \ldots, x_m) \cdot B(x_1, \ldots, x_m) = C(x_1, \ldots, x_m)$$

is a polynomial which is not $\equiv 0 \pmod{\mathfrak{p}}$, while $A \cdot B = (\alpha_2\beta_2/\alpha_1\beta_1)C$ also has integer coefficients. Hence $\mathfrak{p}^{a+b}$ is precisely the highest power of $\mathfrak{p}$ dividing $A \cdot B$ because of the numerical factor $\alpha_2\beta_2/\alpha_1\beta_1$.

We now define the *content*, $J(P)$, *of a polynomial* to be the ideal which is equal to the GCD of the coefficients of P. Then it follows from what has been proved:

Theorem 87. *The content of a product of two polynomials is equal to the product of the contents of the two factors.*

With this we have achieved a considerable strengthening of Kronecker's Theorem 67 and also the generalization of Theorem 13 of Gauss to several variables and arbitrary algebraic number fields.

If in a valid congruence for a polynomial mod $\mathfrak{a}$, we replace the variables $x_1, \ldots$ by integers of the field K to which the ideal $\mathfrak{a}$ belongs, then obviously we obtain a valid numerical congruence mod $\mathfrak{a}$ between the integers in K. Finally from

$$\alpha^{N(\mathfrak{p})} \equiv \alpha \pmod{\mathfrak{p}} \tag{43}$$

for each integer α, it follows that for each polynomial $P(x_1, \ldots, x_m)$

$$P(x_1, \ldots, x_m)^{N(\mathfrak{p})} \equiv P(x_1^{N(\mathfrak{p})}, x_2^{N(\mathfrak{p})}, \ldots, x_m^{N(\mathfrak{p})}) \pmod{\mathfrak{p}}. \tag{44}$$

This statement is obviously true for a polynomial which consists of only a single term by (43). Suppose that it has already been proved for polynomials which contain at most k terms. Now if G is such a polynomial and α is any integer of K, then for each positive rational prime p

$$(G(x_1, \ldots, x_m) + \alpha x_1^{a_1} \cdots x_m^{a_m})^p \equiv G^p + \alpha^p x_1^{pa_1} \cdots x_m^{pa_m} \pmod{p}$$

because, by the properties of the binomial coefficients $\binom{p}{i}$, the difference of the two sides of this equation has only coefficients which are divisible by p.

By repeatedly raising this congruence to powers we obtain

$$(G + \alpha x_1^{a_1} \cdots x_m^{a_m})^{p^f} \equiv G^{p^f} + \alpha^{p^f} x_1^{p^f a_1} \cdots x_m^{p^f a_m} \pmod{p}$$

for each positive rational integer f. If the prime ideal $\mathfrak{p}$ divides (p), then this congruence is also true mod $\mathfrak{p}$. If in addition $N(\mathfrak{p}) = p^f$, then by our assumption about G, the truth of the assertion (44) also follows for the polynomial in parentheses, which has at most $k + 1$ terms. Consequently, (44) holds in general.

§29 First Type of Decomposition Laws for Rational Primes: Decomposition in Quadratic Fields

As we have established the connection between the rational primes and the prime ideals of an algebraic number field in §27, the question of the exact nature of these relations naturally arises. We are interested in the following three points:

(1) How many different prime ideals of a given number field divide a given rational prime?
(2) What are the degrees of these prime ideals?
(3) To what power do they divide $\mathfrak{p}$?

We first mention a result concerning (3) of great generality, for which we are indebted to Dedekind:

The prime ideals dividing the discriminant of the field have the characteristic property that they and only they are divisible by a power of a prime ideal higher than the first. (Compare §§36, 38.)

On the other hand our knowledge about the answers to (1) and (2) is extremely slight. At this time we can make a general and exhaustive statement about the number and degree of a prime ideal dividing some prime p only for quite special kinds of algebraic fields. These fields are completely characterized by a property of their "Galois groups" as defined in algebra[2]. Thus *two formally entirely different types of decomposition laws*, with which we now wish to become acquainted, appear. With all remaining fields we have, at this time, no idea at all even of the approximate nature of the decomposition laws valid in these fields.

Before the investigation of the two known kinds of fields we make a general remark about *Galois fields*.

Each ideal $\mathfrak{a} = (\alpha_1, \ldots, \alpha_r)$ of a field determines a sequence of n ideals $\mathfrak{a}^{(i)}$ ($i = 1, \ldots, n$) which arise from $\mathfrak{a}$ when all numbers of $\mathfrak{a}$ are replaced by the conjugates with the same upper index i; obviously $\mathfrak{a}^{(i)} = (\alpha_1^{(i)}, \ldots, \alpha_r^{(i)})$. These n ideals form the *conjugate ideals* to $\mathfrak{a}$. By Theorem 55 each valid congruence remains valid, if we replace all numbers occuring in the congruence by their conjugates.

In a Galois field (end of §20) the conjugate ideals can be multiplied with one another since these ideals belong to the same field. Hence we have

Theorem 88. *For each ideal $\mathfrak{a}$ of a Galois field the principal ideal $(N(\mathfrak{a})) = \mathfrak{a}^{(1)}\mathfrak{a}^{(2)} \cdots \mathfrak{a}^{(n)}$ (compare with Theorem 107).*

---

[2] These are the fields whose generating numbers can be represented by radical signs laid one upon the other. The corresponding equations are the so-called algebraically solvable equations with rational coefficients.

For the proof, we form the polynomial $P(x) = \alpha_1 x + \alpha_2 x^2 + \cdots + \alpha_r x^r$ from a new variable x and $\mathfrak{a} = (\alpha_1, \ldots, \alpha_r)$ where the GCD of the coefficients $= \mathfrak{a}$. The product of the conjugate polynomials

$$f(x) = \prod_{i=1}^{n} (\alpha_1^{(i)} x + \cdots + \alpha_r^{(i)} x^r)$$

is then a polynomial with rational integral coefficients whose GCD we set $= a$, where a is a rational integer. Since 1 is a linear combination of the coefficients of $(1/a)f(x)$, the ideal (a) is also the GCD of the coefficients as an ideal in the field under consideration. Hence, by Theorem 87,

$$\mathfrak{a}^{(1)} \mathfrak{a}^{(2)} \cdots \mathfrak{a}^{(n)} = (a).$$

Now obviously the conjugates have equal norm. Consequently on applying

$$N(\mathfrak{a}^{(1)}) \cdots N(\mathfrak{a}^{(n)}) = N(\mathfrak{a}^{(i)})^n = N((\mathfrak{a})) = |a|^n$$

we have

$$N(\mathfrak{a}^{(i)}) = \pm a, \qquad (N(\mathfrak{a}^{(i)})) = (a) = \mathfrak{a}^{(1)} \cdots \mathfrak{a}^{(n)},$$

for each i, and the theorem is proved. This relation justifies the name norm for the number of incongruent elements mod $\mathfrak{a}$.

In particular, for a prime ideal of degree f

$$p^f = N(\mathfrak{p}) = \mathfrak{p}^{(1)} \cdots \mathfrak{p}^{(n)}.$$

Consequently, no prime ideals other than the conjugate prime ideals divide p. Furthermore if p is not divisible by the square of any prime ideal, then among the $\mathfrak{p}^{(1)}, \ldots, \mathfrak{p}^{(n)}$ each is repeated f times and p is the product of the $k = n/f$ distinct prime ideals among the n conjugate prime ideals $\mathfrak{p}^{(i)}$.

Hence if a rational prime p in a Galois field is a product of k prime ideals which are distinct, then these prime ideals are conjugate and have the same degree $f = n/k$ which is thus a divisor of n.

We now turn to the *quadratic number field* which may be assumed, without loss of generality, to be generated by the root of an equation $x^2 - D = 0$, where D is a (positive or negative) rational integer which is not divisible by any square except 1. This field $K(\sqrt{D})$ is a Galois field; its numbers can be brought into the form

$$\alpha = x + y\sqrt{D}$$

in a unique way, where x, y are rational. Here $\sqrt{D}$ is an arbitrarily fixed value of the two roots. Let the conjugate of α be denoted by α',

$$\alpha' = x - y\sqrt{D}, \qquad (\alpha')' = \alpha.$$

In order that α be an integer it is necessary and sufficient that

$$\alpha + \alpha' \quad \text{and} \quad \alpha\alpha'$$

are integers.

If $2x$ and $x^2 - Dy^2$ are integers, then since D was assumed squarefree, y as well as x can have denominator at most 2. If we set $x = u/2$, $y = v/2$ with rational integers u, v then

$$u^2 - Dv^2 \equiv 0 \ (\mathrm{mod} \ 4).$$

If $D \equiv 2$ or $3 \ (\mathrm{mod} \ 4)$, then since a square can only be congruent to 0 or 1 mod 4 it obviously follows that u, v are both even; consequently x and y are both integers. However, if $D \equiv 1 \ (\mathrm{mod} \ 4)$, then it follows that $u \equiv v$ $(\mathrm{mod} \ 2)$. Hence α is an integer if

(a) $D \equiv 2, 3 \ (\mathrm{mod} \ 4)$: $\alpha = x + y\sqrt{D}$; x, y both integers; a basis for $K(\sqrt{D})$ is 1, $\sqrt{D}$ and the discriminant $d = 4D$,

(b) $D \equiv 1 \ (\mathrm{mod} \ 4)$: $\alpha = g + v(1 + \sqrt{D})/2$; $g = (u - v)/2$, v an integer; a basis for $K(\sqrt{D})$ is 1, $(1 + \sqrt{D})/2$ and the discriminant $d = D$.

Thus in each case if d is the discriminant

$$1, \ \frac{d + \sqrt{d}}{2} \quad \text{is a basis,}$$

for both these numbers are integers and their discriminant is equal to d. We now prove the decomposition theorem:

Theorem 89. *Let p be a rational prime which does not divide d. Then p splits in the field $K(\sqrt{d})$ into two distinct prime ideals $\mathfrak{p}$, $\mathfrak{p}'$ provided the congruence*

$$x^2 \equiv d \ (\mathrm{mod} \ 4p) \tag{45}$$

can be solved in rational integers x. If, however, the congruence cannot be solved, then p is a prime ideal in $K(\sqrt{d})$.

If the prime p which does not divide d splits in $K(\sqrt{d})$, then p can only split into prime factors $\mathfrak{p}$, $\mathfrak{p}'$ which are of degree 1. By Theorem 85, each integer in K is congruent mod $\mathfrak{p}$ to a rational number, and hence there is a rational integer r such that

$$r \equiv \frac{d + \sqrt{d}}{2} \ (\mathrm{mod} \ \mathfrak{p}).$$

From this it follows that

$$2r - d \equiv \sqrt{d} \ (\mathrm{mod} \ 2\mathfrak{p}),$$
$$(2r - d)^2 \equiv d \ (\mathrm{mod} \ 4\mathfrak{p}).$$

Moreover this congruence between rational numbers is also true mod $4p$. Hence $x = 2r - d$ is a solution of (45). The ideal

$$\mathfrak{a} = \left(p, r - \frac{d + \sqrt{d}}{2} \right)$$

is obviously divisible by p and

$$\mathfrak{a}\mathfrak{a}' = \left(p^2, p\left(r - \frac{d + \sqrt{d}}{2}\right), p\left(r - \frac{d - \sqrt{d}}{2}\right), \frac{(2r - d)^2 - d}{4}\right)$$

$$= (p)\left(p, r - \frac{d - \sqrt{d}}{2}, r - \frac{d + \sqrt{d}}{2}, \frac{(2r - d)^2 - d}{4p}\right).$$

However, the last ideal factor is $= (1)$, for this ideal contains p and the difference between the second and the third numbers, which is $\sqrt{d}$; thus this ideal contains the two relatively prime numbers p and d. Finally we obtain from this

$$\mathfrak{p} = \left(p, r - \frac{d + \sqrt{d}}{2}\right), \qquad \mathfrak{p}' = \left(p, r - \frac{d - \sqrt{d}}{2}\right).$$

The two prime ideals are moreover distinct, thus relatively prime, as $(\mathfrak{p}, \mathfrak{p}')$ contains the two coprime numbers p, d.

Conversely, if x is a solution of (45), then the number

$$\omega = \frac{x + \sqrt{d}}{2}$$

is obviously an integer; moreover ω/p is not an integer, as $((\omega - \omega')/p)^2 = d/p^2$ is not an integer. Thus, since p does not divide ω or ω', but does divide the product $\omega\,\omega'$, p cannot be a prime ideal. Thus it splits in $K(\sqrt{d})$ into two prime factors which are distinct from one another by the above.

Moreover if q is an odd prime factor of d, then the ideal

$$\mathfrak{q} = \left(q, \frac{d + \sqrt{d}}{2}\right) = \left(q, \frac{-d + \sqrt{d}}{2} + d\right) = \left(q, \frac{d - \sqrt{d}}{2}\right) = \mathfrak{q}',$$

$$\mathfrak{q}^2 = \mathfrak{q}\mathfrak{q}' = q\left(q, \frac{d + \sqrt{d}}{2}, \frac{d - \sqrt{d}}{2}, \frac{d(d - 1)}{4q}\right).$$

However, by the definition of the discriminant d, $d(d - 1)/4q$ is certainly not divisible by q, that is, $d(d - 1)/4q$ is relatively prime to q. Consequently $\mathfrak{q}^2 = q$ and $\mathfrak{q}$ is the unique prime ideal dividing q.

Finally, in case d is even, 2 is also the square of a prime ideal, namely the square of $\mathfrak{q} = (2, \sqrt{D})$ for $D \equiv 2 \pmod 4$ or of $\mathfrak{q} = (2, 1 + \sqrt{D})$, if $D \equiv 3 \pmod 4$.

If we now keep in mind that by §14, since $d \equiv 0$ or $1 \pmod 4$, the solvability of (45) for an odd prime p is equivalent to the solvability of $y^2 \equiv d \pmod p$, then we can also formulate this theorem as follows:

Theorem 90. *If p is an odd prime, then in a quadratic field with discriminant d*

p splits into two distinct factors of degree 1, if $\left(\frac{d}{p}\right) = +1$.
p splits into two identical factors of degree 1, if $\left(\frac{d}{p}\right) = 0$.
p is itself a prime ideal (of degree 2), if $\left(\frac{d}{p}\right) = -1$.

The prime 2 splits into two distinct factors, if d is odd and a quadratic residue mod 8; 2 is itself a prime ideal, if d is odd and a quadratic nonresidue mod 8. If d is even, 2 is a square.

§30 Second Type of Decomposition Theorem for Rational Primes: Decomposition in the Field $K(e^{2\pi i/m})$

We now investigate the fields generated by mth roots of unity, where m is a rational integer > 2. The mth roots of unity are the m roots of $x^m - 1 = 0$, hence they are algebraic integers. The *primitive mth roots of unity* are the $\varphi(m)$ numbers $e^{2\pi i a/m}$, where $(a, m) = 1$; these numbers are not roots of unity of lower order. If we form

$$g(x) = \prod_{k=1}^{m-1} (x^k - 1),$$

then a root of $g(x)$ is also a root of $f(x) = x^m - 1$ if and only if it is a non-primitive mth root of unity. Consequently

$$F(x) = \frac{x^m - 1}{d(x)}, \quad \text{where } d(x) = (f(x), g(x)),$$

is a polynomial with rational integral coefficients all of whose roots are primitive mth roots of unity. Finally, since among the primitive mth roots of unity each root is a power of every other root, the field $K(e^{2\pi i/m})$ is a *Galois number field* of degree $h \leq \varphi(m)$. (That the degree is exactly $\varphi(m)$, i.e., $F(x)$ is irreducible, will not be needed in this section and will emerge as a side result in §43.)

We set $\zeta = e^{2\pi i/m}$ and keep in mind, that according to the proof of Theorem 64 all integers of $k(\zeta)$ can be uniquely represented in the form

$$\omega = r_0 + r_1\zeta + \cdots + r_{h-1}\zeta^{h-1},$$

where the r_i are rational numbers such that their denominators are all divisors of a fixed integer D, the discriminant of $F(x)$.

Now let p be a rational prime which does not divide D, and let D' be determined so that $D'D \equiv 1 \pmod{p}$. Then we see that in each residue class mod p in $k(\zeta)$ there exist numbers for which $r_0, r_1, \ldots$ are all rational integers, since for each integer ω

$$\omega \equiv DD'\omega \pmod{p},$$

and by the above the $DD'r_i$ are rational integers. Hence we do not need to first construct a basis for the field in the investigation of p.

Lemma. *If the prime p does not divide D · m, then for each integer ω of the field K(ζ),*

$$\omega^{p^f} \equiv \omega \pmod{p}.$$

Here f is the smallest positive exponent such that $p^f \equiv 1 \pmod{m}$.

For the proof we think of ω as chosen in its residue class so that

$$\omega = a_0 + a_1\zeta + \cdots + a_{h-1}\zeta^{h-1}$$

with rational *integers* a_i. Then, by (44), for the integral polynomial

$$Q(x) = a_0 + a_1 x + \cdots + a_{h-1}x^{h-1}$$

over $k(1)$, we derive the functional congruence

$$Q(x)^p \equiv Q(x^p) \pmod{p}, \quad \text{more generally} \quad (Q(x))^{p^f} \equiv Q(x^{p^f}) \pmod{p}.$$

We obtain a valid numerical congruence from the functional congruence if we replace x by the algebraic number ζ, and thus the lemma is proved.

Theorem 91. *If the prime p does not divide D · m, then p is not divisible by the square of a prime ideal in K(ζ).*

For if $\mathfrak{p}^2 \mid p$, then let us choose a number ω which is divisible by $\mathfrak{p}$ but not by $\mathfrak{p}^2$. It follows from the lemma that

$$\omega^{p^f} \equiv \omega \pmod{\mathfrak{p}^2}.$$

Since $p^f \geq 2$, and therefore $\omega^{p^f} \equiv 0 \pmod{\mathfrak{p}^2}$,

$$\omega \equiv 0 \pmod{\mathfrak{p}^2},$$

contrary to the hypothesis.

Theorem 92. *If the prime p does not divide D · m, and if f is the smallest positive exponent such that $p^f \equiv 1 \pmod{m}$, then p splits into exactly e = h/f distinct prime factors in K(ζ). Each factor has degree f.*

Let $\mathfrak{p}$ be a prime factor of p of degree f_1. Then, by (43), for each integer ω in $K(\zeta)$,

$$\omega^{p^{f_1}} \equiv \omega \pmod{\mathfrak{p}} \tag{46}$$

and this congruence holds for each integer ω with no value smaller than f_1. Hence by the lemma we have $f_1 \leq f$. On the other hand, it follows from (46) for $\omega = \zeta$ that

$$\zeta^{p^{f_1}} \equiv \zeta \pmod{\mathfrak{p}}.$$

Here, however, we must have $p^{f_1} \equiv 1 \pmod{m}$, for otherwise $\zeta^{p^{f_1}}$ would be a primitive mth root of unity different from ζ and $\zeta^{p^{f_1}} - \zeta$ would be a factor of the discriminant D of $F(x)$. Thus $\mathfrak{p}$ would be a factor of D, contrary to hypothesis.

However by the definition of f, the equation $f_1 = f$ follows from $p^{f_1} \equiv 1$ (mod m) and $f_1 \leq f$.

Since, by Theorem 91, the conjugate prime ideals divide p only to the first power, then, by the remark in §29, p splits into exactly h/f factors and everything is thereby proved.

Accordingly, the field $K(\zeta)$ is closely related to the group of residue classes mod m in the field $k(1)$. *Primes which belong to the same residue class* mod m *split in* $K(\zeta)$ *in exactly the same way*—except for finitely many exceptions. Later, in §43, we will also show that the field $K(\zeta)$ has degree $\varphi(m)$, thus the same degree as the group $\mathfrak{R}(m)$ in $k(1)$. Finally we state without proof that the so-called Galois group of $K(\zeta)$ is isomorphic to the group $\mathfrak{R}(m)$.

For these reasons $K(\zeta)$ is called a *class field* which belongs to the classification of rational numbers into residue classes mod m.

It is known from the theory of cyclotomic numbers that $K(\zeta)$ contains one or more quadratic fields and each quadratic field is also always contained in a $K(\zeta)$. Then we see that from the decomposition laws in $K(\zeta)$ we can deduce those in every subfield, and in this way we obtain, for quadratic fields, an entirely different decomposition law than the one we found in the preceding section. The comparison of the two then yields the proof of the *quadratic reciprocity laws*[3] mentioned in §16.

§31 Fractional Ideals

We now introduce fractional ideals—systems of numbers which may also contain nonintegral numbers of the field and, when they contain only integers, agree with the ideals discussed until now.

A system S of integral or fractional numbers of the field is to be called an ideal from now on if:

(1) *Along with* α *and* β, $\lambda\alpha + \mu\beta$ *belongs to S, where* λ *and* μ *are arbitrary integers in K*.
(2) *There exists a fixed non-zero integer* ν *such that the product* $(\nu \times$ *each number of S*) *is an integer*.

Ideals which contain only integers will be designated as *integral ideals*, the other ones will be designated as *fractional ideals*. Two ideals are said to be equal if they contain exactly the same numbers.

---

[3] The idea of this proof of the quadratic reciprocity law originates with *Kronecker*. Compare, say, the representation of this proof in *Hilbert's* Bericht über die Theorie der algebraischen Zahlkorper, §122. This proof is not used in this book. The connection is shown in principle in the field $K(\sqrt{-3})$, of third roots of unity, in which both forms of the decomposition law hold.

Theorem 93. *Each ideal* $\mathfrak{g}$ *is the range of values of a linear form*

$$\xi_1\rho_1 + \cdots + \xi_r\rho_r,$$

where $\rho_1, \ldots, \rho_r$ *are certain integers or fractions in* $\mathfrak{g}$, *while the* ξ_i *run through all integers in K. We write* $\mathfrak{g} = (\rho_1, \ldots, \rho_r)$.

Let v be chosen for $\mathfrak{g}$ according to (2). Then all products of v with the numbers in $\mathfrak{g}$ obviously form an integral ideal $\mathfrak{a} = (\alpha_1, \ldots, \alpha_r)$ and then $\mathfrak{g} = (\alpha_1/v, \ldots, \alpha_r/v)$.

If $\alpha_1, \ldots, \alpha_n$ is a basis for the integral ideal $\mathfrak{a}$, then, if we again regard $\mathfrak{g}$ as an infinite Abelian group, $\alpha_1/v, \ldots, \alpha_n/v$ is obviously a basis for $\mathfrak{g}$.

The product of two ideals $\mathfrak{g} = (\gamma_1, \ldots, \gamma_r)$ and $\mathfrak{r} = (\rho_1, \ldots, \rho_s)$ is defined in the same way as for integral ideals:

$$\mathfrak{g}\mathfrak{r} = (\ldots, \gamma_i\rho_k, \ldots),$$

and this multiplication is also commutative and associative. Each ideal $\mathfrak{g} \neq (0)$ can be made into an integral ideal by multiplication by a suitable integral ideal (v). Consequently it can also be made into a principal ideal (ω) by multiplication by an appropriate integral ideal.

If $\mathfrak{g} \neq (0)$, *then it follows from* $\mathfrak{g}\mathfrak{r} = \mathfrak{g}\mathfrak{n}$ *that* $\mathfrak{r} = \mathfrak{n}$.

The proof is word for word the same as for Theorem 68.

If $\mathfrak{g}_1$ *and* $\mathfrak{g}_2$ *are arbitrary ideals,* $\mathfrak{g}_1 \neq (0)$, *then there is exactly one* $\mathfrak{r}$ *such that*

$$\mathfrak{g}_1\mathfrak{r} = \mathfrak{g}_2.$$

One writes $\mathfrak{r} = \mathfrak{g}_2/\mathfrak{g}_1$, *and calls* $\mathfrak{r}$ *the quotient of* $\mathfrak{g}_2$ *and* $\mathfrak{g}_1$. *This notation is meaningful only for* $\mathfrak{g}_1 \neq (0)$.

Let us choose $\mathfrak{a} \neq (0)$ so that $\mathfrak{a}\mathfrak{g}_1 = (\omega)$ is a principal ideal; thus $(\omega) \neq 0$. If $\mathfrak{a}\mathfrak{g}_2 = (\rho_1, \ldots, \rho_r)$, we set

$$\mathfrak{r} = \left(\frac{\rho_1}{\omega}, \ldots, \frac{\rho_r}{\omega}\right).$$

Then in fact $\mathfrak{a}\mathfrak{g}_2 = (\omega)\mathfrak{r} = \mathfrak{a}\mathfrak{g}_1\mathfrak{r}$, $\mathfrak{g}_2 = \mathfrak{g}_1\mathfrak{r}$, and, by what has been said before, $\mathfrak{r}$ is uniquely determined.

The equation $\mathfrak{a}/\mathfrak{b} = \mathfrak{c}/\mathfrak{d}$ is accordingly equivalent to $\mathfrak{a}\mathfrak{d} = \mathfrak{b}\mathfrak{c}$; in particular for each ideal $\mathfrak{m} \neq (0)$,

$$\frac{\mathfrak{a}}{\mathfrak{b}} = \frac{\mathfrak{a}\mathfrak{m}}{\mathfrak{b}\mathfrak{m}}, \qquad \frac{\mathfrak{a}}{(1)} = \mathfrak{a}, \qquad \frac{\mathfrak{m}}{\mathfrak{m}} = (1).$$

Thus each ideal can be represented as the quotient of two relatively prime integral ideals which we designate, as with numbers, as numerator and denominator. In particular each fractional principal ideal ω can also be represented as a quotient of integral ideals which we again express by an equation

$$\omega = \frac{a}{b},$$

omitting parentheses.

With fractional ideals we also wish to speak of *divisibility* in the sense that $a \mid b$ or a *divides* b is to mean that b/a is an integral ideal. If a and b are integral ideals, then this definition agrees with the earlier definition of divisibility.

Accordingly, an integer ω occurs in an ideal g if and only if (ω) is divisible by g, that is, (ω) has a decomposition

$$(\omega) = mg,$$

with m an integral ideal.

Hence the number 1 occurs in all ideals which are the reciprocals of integral ideals a, that is, equal to $1/a$, and only in such ideals.

If an ideal g is represented as the quotient of two relatively prime ideals a and b, then we define the *norm of* g:

$$N(g) = \frac{N(a)}{N(b)}, \quad \text{if } g = \frac{a}{b}.$$

This equation is also correct if a, b are not relatively prime or if they are fractional ideals. Again we have

$$N(g_1 \cdot g_2) = N(g_1) \cdot N(g_2).$$

Between the basis and the norm there is again the relationship:

If $\alpha_1, \ldots, \alpha_n$ is a basis for g, then

$$N(g) = \left| \frac{\Delta(\alpha_1, \ldots, \alpha_n)}{\sqrt{d}} \right|. \tag{47}$$

To prove this choose an integer $v \neq 0$ so that vg is an integral ideal b, with basis $\beta_1, \ldots, \beta_m$. Then $\beta_1/v, \ldots, \beta_n/v$ is a basis for g and

$$N(g) = \frac{N(b)}{N(v)} = \frac{\Delta(\beta_1, \ldots, \beta_n)}{|N(v)|\sqrt{d}} = \frac{\Delta\left(\dfrac{\beta_1}{v}, \ldots, \dfrac{\beta_n}{v} \right)}{\sqrt{d}}.$$

§32 Minkowski's Theorem on Linear Forms

In the subsequent development of algebraic number theory, the concept of magnitude will now play an essential role whereas earlier everything depended on the concept of divisibility and the formal algebraic processes. The most important method here is a theorem about the solvability of linear inequalities by rational integers which goes back to *Dirichlet* and which was subsequently extended and sharpened considerably by *Minkowski*. This theorem and its proof is quite independent of the theories which were treated earlier. It reads as follows:

Theorem 94. *Assume we are given n linear homogeneous expressions*

$$L_p(x) = \sum_{q=1}^{n} a_{pq} x_q \qquad (p = 1, 2, \ldots, n),$$

with real coefficients a_{pq}, *whose determinant* $D = |a_{pq}|$ *is different from zero, as well as n positive quantities* $\varkappa_1, \ldots, \varkappa_n$, *for which*

$$\varkappa_1 \cdot \varkappa_2 \cdots \varkappa_n \geq |D|.$$

Then there are always n rational integers $x_1, \ldots, x_n$, *not all equal to 0, such that*

$$|L_p(x)| \leq \varkappa_p \qquad (p = 1, \ldots, n). \tag{48}$$

The proof is along the lines of Minkowski's contribution to the geometry of numbers. To begin with we ask: "What can we say about the quantities $\varkappa$ if the n inequalities (48) have no solution in rational integers $x_q \neq 0$?" We show that under these conditions $\varkappa_1 \cdot \varkappa_2 \cdots \varkappa_n < |D|$.

To this end, we consider the parallelotope in the space of n dimensions with Cartesian coordinates $x_1, \ldots, x_n$ such that

$$|L_p(x)| \leq \frac{\varkappa_p}{2} \qquad (p = 1, 2, \ldots, n)$$

and think of the same parallelotope displaced parallel to itself so that its center, that is, the point $0, \ldots, 0$, corresponds to all lattice points $g_1, \ldots, g_n$, where the g_i run through all rational integers. In this way we have infinitely many parallelotopes $\Pi_{g_1, \ldots, g_n}$ given by

$$|L_p(x - g)| \leq \frac{\varkappa_p}{2} \qquad (p = 1, \ldots, n).$$

If (48) cannot be solved no two of the parallelotopes have a point in common. For if a point (x) belongs to the two parallelotopes $\Pi_{g_1, \ldots, g_n}$ and $\Pi_{g'_1, \ldots, g'_n}$, then from

$$-\frac{\varkappa_p}{2} \leq L_p(x - g) \leq \frac{\varkappa_p}{2}$$

and

$$-\frac{\varkappa_p}{2} \leq L_p(x - g') \leq \frac{\varkappa_p}{2}$$

it follows by subtraction that

$$|L_p(g - g')| \leq \varkappa_p,$$

that is (48) would have a solution $x_q = g_q - g'_q$.

Consequently the sum of the volumes of all the Π which belong to a definite square $|x_q| \leq L$ $(q = 1, 2, \ldots, n)$ must be less than the volume $(2L)^n$ of this square, from which the assertion follows immediately. To see this we first let c be a number such that the coordinates of all points of the initial figure $\Pi_{0,\ldots,0}$ are all $\leq c$ in absolute value. Then in any case all $\Pi_{g_1,\ldots,g_n}$ such that

$$|g_q| \leq L \qquad (q = 1, \ldots, n)$$

belong to the square $|x_q| \leq L + c$. Since from $|L_p(x - g)| \leq \varkappa_p/2$ and $|g_q| \leq L$ it follows that $|x_q| = |x_q - g_q + g_q| \leq |x_q - g_q| + |g_q| \leq c + L$. Hence if L is a positive rational integer, then there are $(2L + 1)^n$ such $\Pi_{g_1,\ldots,g_n}$ and their total volume is

$$(2L + 1)^n J \leq (2L + 2c)^n,$$

where J is the volume of a single Π. After division by L^n and passage to the limit as $L \to \infty$ it follows that

$$J \leq 1.$$

On the other hand we have

$$J = \int \cdots \int\limits_{|L_p(x)| \leq \varkappa_p/2} dx_1 \cdots dx_n = \frac{1}{|D|} \int \cdots \int\limits_{|y_p| \leq \varkappa_p/2} dy_1 \cdots dy_n$$

$$= \frac{\varkappa_1 \varkappa_2 \cdots \varkappa_n}{|D|}.$$

Thus if these inequalities cannot be solved in integers except $0, \ldots, 0$, then $\varkappa_1, \ldots, \varkappa_n \leq |D|$. However, in this assertion the sign $<$ necessarily holds since the unsolvability for the values $\varkappa_1, \ldots, \varkappa_n$ implies, by continuity, the unsolvability for sufficiently near larger values of the $\varkappa$ whose product must thus likewise be still $\leq |D|$. Therefore the product of the original $\varkappa$ is necessarily $< |D|$.

Moreover, with this, we have proved that if the product of the $\varkappa$ is equal to $|D|$ or greater, then the inequalities (48) must have a solution in integers.

Later, we will take the $L_p(x)$ to be the conjugates of a linear form and complex coefficients must also be allowed. By a simple modification of the above theorem we obtain in this connection:

Theorem 95. *Let n linear forms $L_p(x) = \sum_{q=1}^{n} a_{pq} x_q$ $(p = 1, \ldots, n)$ be given with real or complex coefficients whose determinant $D \neq 0$. Moreover if one of the forms is not real, we assume the complex conjugate of a form also occurs among the $L_p(x)$. Finally let $\varkappa_1, \ldots, \varkappa_n$ be positive quantities such that if the forms $L_\alpha(x)$ and $L_\beta(x)$ are complex conjugate, $\varkappa_\alpha = \varkappa_\beta$. Then there are rational integral x_q, not all vanishing, such that*

$$|L_p(x)| \leq \varkappa_p \qquad (p = 1, \ldots, n),$$

if

$$\varkappa_1 \cdot \varkappa_2 \cdots \varkappa_n \geq |D|.$$

To prove this we replace the system $L_p(x)$ by that system of real forms $L'(x)$ which arises if the real and imaginary components of the $L_p(x)$ are considered by themselves. We take $L'_p(x) = L_p(x)$ if $L_p(x)$ is a real form; on the other hand if $L_\alpha(x)$ and $L_\beta(x)$ are conjugate imaginary and, say, $\alpha < \beta$, then we set

$$L'_\alpha(x) = \frac{L_\alpha(x) + L_\beta(x)}{2}, \qquad L'_\beta(x) = \frac{L_\alpha(x) - L_\beta(x)}{2i}.$$

In the latter case we define

$$\varkappa'_\alpha = \varkappa'_\beta = \frac{\varkappa_\alpha}{\sqrt{2}},$$

and, on the other hand,

$$\varkappa'_p = \varkappa_p$$

in the first case.

The system of real forms L' now obviously has a determinant D' with

$$|D'| = 2^{-r_2}|D|,$$

where r_2 denotes the number of pairs of complex conjugate forms among the $L_p(x)$. Hence since $\varkappa'_1 \cdots \varkappa'_n \geq |D'|$, there are rational integers x_q, which are not all 0, such that

$$|L'_p(x)| \leq \varkappa'_p \qquad (p = 1, \ldots, n).$$

For a nonreal form $L_\alpha(x)$ we now have

$$|L_\alpha(x)|^2 = L'^2_\alpha(x) + L'^2_\beta(x) \leq \varkappa'^2_\alpha + \varkappa'^2_\beta = \varkappa^2_\alpha$$

from which the stated theorem follows.

§33 Ideal Classes, the Class Group, and Ideal Numbers

We can now attack the problem which we posed in §23, at the beginning of the ideal theory, namely, we can investigate whether all ideals of a field can always be represented by numbers, which perhaps belong to other fields. To this end we introduce the concept of equivalence and with it a partition of all ideals of K into classes as follows:

Definition. Two integral or fractional ideals $\mathfrak{a},\mathfrak{b}$ are said to be *equivalent*, in symbols

$$\mathfrak{a} \sim \mathfrak{b},$$

if they differ only by a factor which is a principal ideal, that is, if there is a (integral or fractional) principal ideal $(\omega) \neq (0)$ such that

$$\mathfrak{a} = \omega\mathfrak{b}.$$

This concept of equivalence has the following properties:

(1) $\mathfrak{a} \sim \mathfrak{a}$.
(2) From $\mathfrak{a} \sim \mathfrak{b}$ it follows that $\mathfrak{b} \sim \mathfrak{a}$.
(3) From $\mathfrak{a} \sim \mathfrak{b}$ and $\mathfrak{b} \sim \mathfrak{c}$ it follows that $\mathfrak{a} \sim \mathfrak{c}$.
(4) From $\mathfrak{a} \sim \mathfrak{b}$ it follows that $\mathfrak{a}\mathfrak{c} \sim \mathfrak{b}\mathfrak{c}$ and if $\mathfrak{c} \neq (0)$ the converse also holds.

The collection of all ideals equivalent to a fixed $\mathfrak{a}$ forms an *ideal class*. In particular all principal ideals ($\neq 0$) are equivalent to each other. They form the *principal class*.

By (4), the classes can be immediately made into an Abelian group. If by $\mathfrak{a}$ and $\mathfrak{b}$ we understand any ideals in the classes A and B respectively, then by (4) the product $\mathfrak{a}\mathfrak{b}$ belongs to a class determined by A and B alone and does not depend on the choice of $\mathfrak{a}$ and $\mathfrak{b}$ within their class. We denote the class of $\mathfrak{a}\mathfrak{b}$ by AB and with this we have defined a composition of ideal classes, under which the ideal classes form a (finite or infinite) Abelian group, the *class group of the field K*. The unit element is the principal class.

The passage from ideals to ideal classes corresponds precisely to the passage from numbers to residue classes with respect to a modulus since the collection of integral and fractional ideals of K which are $\neq (0)$ obviously forms an infinite Abelian group under ordinary multiplication. (This group has a basis of infinitely many elements, in the sense of §11, namely the set of all prime ideals.) This group $\mathfrak{M}$ contains the subgroup of all principal ideals ($\neq 0$). The latter subgroup will be denoted by $\mathfrak{H}$. *Moreover, the class group defined above is obviously the factor group* $\mathfrak{M}/\mathfrak{H}$. Indeed its elements are the different cosets which consist of all ideals which differ only by an element of $\mathfrak{H}$, that is, by a factor which is a principal ideal.

It is one of the principal problems of number theory to investigate the finer structure of these class groups. They play an essential role in almost all statements about the numbers in K. Yet our knowledge about the class group in general fields is still extremely slight. We state the most important general fact in the following theorem:

Theorem 96. *In each ideal class of K there is an integral ideal whose norm is $\leq |\sqrt{d}|$. Thus the number of ideal classes in K is finite.*

To prove this let $\mathfrak{a}$ be an integral ideal in the class B^{-1}, where B is an arbitrarily given class. If $\alpha_1, \ldots, \alpha_n$ denotes a basis of $\mathfrak{a}$, then, by Theorem 95, there are rational integers $x_1, \ldots, x_n$, not all vanishing, such that

$$|\omega^{(i)}| = \left| \sum_{k=1}^{n} \alpha_k^{(i)} x_k \right| \leq |\sqrt[n]{\Delta}| \qquad (i = 1, \ldots, n)$$

where $\Delta = \Delta(\alpha_1, \ldots, \alpha_n) = N(\mathfrak{a})\sqrt{d}$ is the determinant of the $\alpha^{(i)}$. Thus we have for the product of these conjugates $\omega^{(i)}$

$$|N(\omega)| \leq |\Delta| = N(\mathfrak{a})|\sqrt{d}|. \tag{49}$$

Now, by definition, ω is a nonzero integer, which is divisible by $\mathfrak{a}$; hence ω has a decomposition

$$\omega = \mathfrak{a}\mathfrak{b}$$

where $\mathfrak{b}$ is a certain nonzero integral ideal. Obviously $\mathfrak{b}$ lies in the class B, which is reciprocal to B^{-1}, as $\mathfrak{a}\mathfrak{b} \sim (1)$. Then it follows from (49) that

$$N(\mathfrak{b}) \leq \sqrt{d} \qquad (50)$$

whereby the first part is proved.

However there are only finitely many integral ideals whose norms have a given value z, for, by (42) in §27, they must be divisors of the ideal (z). Consequently, there are also only finitely many integral ideals, whose norms lie below a given bound, as the norms are rational integers. Hence there are only finitely many integral ideals $\mathfrak{b}$ which satisfy Condition (50); thus the number of distinct ideal classes in K is finite.

Henceforth the *class number* will be denoted by h. As an immediate consequence of the finiteness of h we obtain from Theorem 21:

Theorem 97. *The hth power of each ideal in K is a principal ideal.*

From this we can finally prove the statement formulated in §24.

Theorem 98. *For each ideal $\mathfrak{a}$ in K there is a number A which generally does not belong to the field K, such that the numbers of $\mathfrak{a}$ are identical with those numbers of the field K which are divisible by A.*

By Theorem 97 $\mathfrak{a}^h$ is equal to a principal ideal (ω). The number $A = \sqrt[h]{\omega}$ has the asserted property for if α is a number in $\mathfrak{a}$, then α^h belongs to $\mathfrak{a}^h$ and therefore α^h/ω is an integer and $\alpha/\sqrt[h]{\omega} = \alpha/A$ is thus also an integer.

Conversely if α is a number of the field such that α/A is an integer, then α^h/ω is an integer, that is, $\alpha^h/\mathfrak{a}^h$ is an integral ideal. By the fundamental theorem $\alpha/\mathfrak{a}$ is also an integral ideal, that is, α occurs in $\mathfrak{a}$.

Because of the group property of the ideal classes, the numbers A which are needed to represent all ideals of the field K can now be chosen in such a way that they all belong to a field of relative degree h over K, and indeed in the following way:

If $h > 1$, then as a finite Abelian group the class group has a basis, say the classes $B_1, \ldots, B_m$ with orders $c_1, \ldots, c_m$ respectively. If we now choose an ideal $\mathfrak{b}_q$ ($q = 1, \ldots, m$) from each class, then by the definition of a basis, each ideal $\mathfrak{a}$ is equivalent to exactly one product of powers

$$\mathfrak{b}_1^{x_1}, \ldots, \mathfrak{b}_m^{x_m} \qquad (0 \leq x_q < c_q; q = 1, \ldots, m). \qquad (51)$$

That is, we obtain all ideals $\mathfrak{g}$ (integral and fractional) exactly once if in

$$\mathfrak{g} = \rho\mathfrak{b}_1^{x_1} \cdots \mathfrak{b}_m^{x_m} \qquad (52)$$

we let the number ρ run through all numbers of the field which are not associated and x_q run through all rational integers with the conditions (51). Thus if we determine the number B_q for each b_q according to Theorem 98, where

$$B_q = \sqrt[c_q]{\beta_q}, \qquad b_q^{c_q} = (\beta_q),$$

then obviously to each g of the form (52) there is assigned the number

$$\Gamma = \rho B_1^{x_1} \cdots B_m^{x_m} \tag{53}$$

such that the numbers of g are identical with those numbers of the field which are divisible by Γ. If, in (53), we let ρ run through all numbers of the field, as well as the associated numbers, then we obtain a system of numbers which is called a *system of ideal numbers for K*. This system splits into h classes of ideal numbers, corresponding to the ideal classes. Each class contains the numbers (53) with the same system of exponents x_q, and the set of all numbers of the same class (0 included) is closed under addition and subtraction. The set of all nonzero ideal numbers is also closed under multiplication and division. In this sense each ideal of K is really representable by a number in the sense of Theorem 98.

This representation has gained a particular significance in the more recent investigations in analytic number theory. Yet, above all, it should be explicitly stated that the number field $K(B_1, \ldots, B_m)$ which has relative degree h with respect to K, is in general *not* identical to the so-called Hilbert class field of K.

§34 Units and an Upper Bound for the Number of Fundamental Units

In this and the following section we will gain a complete overview of the units which exist in a field K by proving a fundamental theorem of *Dirichlet* which is formulated later. The existence in K of infinitely many units is in general, along with the necessity of introducing the concept of an ideal, the second essential criterion which distinguishes the higher algebraic number fields from the field of rational numbers.

First of all, the set of all units of the field K obviously forms an abelian group under composition by multiplication. Let this group of all units be denoted by $\mathfrak{E}$. The group $\mathfrak{W}$, of all roots of unity in K, which contains at least two elements, namely ± 1, is contained as a subgroup in $\mathfrak{E}$.

Lemma (a). *There are at most finitely many integers in K, which together with all their conjugates, do not exceed a given constant in absolute value. If all the conjugates of an integer in K have absolute value 1, then this integer is a root of unity.*

Assume for the integer α in K that the inequalities $|\alpha^{(i)}| \leq C$ hold, for $i = 1, 2, \ldots, n$. Then from this an upper bound for the absolute values of the elementary symmetric functions of the $\alpha^{(i)}$, depending only on C and n, follows immediately. However these functions have integral rational values and they are the coefficients of the equation of degree n with roots $\alpha^{(i)}$; hence only finitely many possibilities exist for these coefficients. Therefore there are only finitely many equations of nth degree whose roots are integers and at the same time are all $\leq C$ in absolute value.

Moreover, if α is an integer in K and $|\alpha^{(i)}| = 1$ for $i = 1, \ldots, n$, then the same holds for all infinitely many powers α^q $(q = 1, 2, \ldots)$. By what has just been proved, these cannot all be distinct. Consequently some power α^q is $= 1$ and α is a root of unity.

Theorem 99. *The group $\mathfrak{W}$ of all roots of unity in K is finite, and indeed it is a cyclic group of order $w \geq 2$.*

Since all roots of unity, including all conjugates, have absolute value 1, the first assertion follows from the lemma. Moreover if p is a prime dividing the order of $\mathfrak{W}$, then the number of solutions of $x^p = 1$ is equal to p^1, and thus, by Theorem 28, the basis number of the group $\mathfrak{W}$ belonging to p is equal to 1. Thus the group is cyclic.

For further investigations we introduce a definite numbering of the conjugate fields $K^{(p)}$. Let θ be a number generating the field K and suppose that among the conjugates $\theta^{(1)}, \theta^{(2)}, \ldots, \theta^{(r_1)}$ are real, and the remaining $2r_2$ of the $\theta^{(p)}$ are nonreal. In fact, assume that

$\theta^{(p+r_2)}$ is complex conjugate to $\theta^{(p)}$ for $p = r_1 + 1, \ldots, r_1 + r_2$.

By §19, this numbering carries over to the conjugates of all numbers in K, and thus we also have for each number α in K, $\alpha^{(1)}, \ldots, \alpha^{(r_1)}$ real and

$$|\alpha^{(p+r_2)}| = |\alpha^{(p)}| \quad \text{for } p = r_1 + 1, \ldots, r_1 + r_2. \tag{54}$$

Finally we define

$$e_p = \begin{cases} 1 & \text{for } p = 1, 2, \ldots, r_1, \\ 2 & \text{for } p = r_1 + 1, \ldots, n; \end{cases}$$

thus

$$\sum_{p=1}^{r_1+r_2} e_p = n.$$

Now our goal is the following fundamental theorem of Dirichlet:

Theorem 100. *The group $\mathfrak{E}$ of all units in K has a finite basis. Furthermore this basis consists of precisely $r = r_1 + r_2 - 1$ elements of infinite order, while the remaining basis elements are roots of unity.*

Thus this means:

There are $r + 1$ units $\zeta, \eta_1, \eta_2, \ldots, \eta_r$, where ζ is a wth root of unity, such that each unit of the field is obtained exactly once in the form

$$\varepsilon = \zeta^a \eta_1^{a_1} \cdots \eta_r^{a_r},$$

where $a_1, \ldots, a_r$ are all rational integers and a can only take the values $0, 1,$ $2, \ldots, w - 1$. *The r units $\eta_1, \ldots, \eta_r$ are called fundamental units of the field.*

As preparation for the proof, which we will treat in this and the next section, we recall that k units $\varepsilon_1, \ldots, \varepsilon_k$ of infinite order (that is, those which do not belong to $\mathfrak{W}$) are said to be independent in the sense of group theory, if a relation

$$\varepsilon_1^{a_1} \varepsilon_2^{a_2} \cdots \varepsilon_k^{a_k} = 1 \tag{55}$$

with rational integers a only exists if all $a_1 = \cdots = a_n = 0$. However, along with the one relation (55) the analogous ones always hold for all conjugates and hence

$$\left|\varepsilon_1^{(i)}\right|^{a_1} \left|\varepsilon_2^{(i)}\right|^{a_2} \cdots \left|\varepsilon_k^{(i)}\right|^{a_k} = 1 \qquad (i = 1, 2, \ldots, n)$$

or

$$\sum_{m=1}^{k} a_m \log\left|\varepsilon_m^{(i)}\right| = 0. \tag{56}$$

(Here we mean the real values of the logarithms). Conversely, by Lemma (a), it follows immediately from the fact that relations (56) hold with rational integers a for all $i = 1, 2, \ldots, n$, that $\varepsilon_1, \ldots, \varepsilon_k$ cannot be independent since then the number

$$\varepsilon_1^{a_1} \cdots \varepsilon_k^{a_k}$$

would be an integer of K, which along with all conjugates would have absolute value 1. Hence it would be a root of unity with its wth power $= 1$. Now, however, from the r equations

$$\sum_{m=1}^{k} \gamma_m \log\left|\varepsilon_m^{(i)}\right| = 0 \quad \text{for } i = 1, 2, \ldots, r_1 + r_2 - 1 \tag{57}$$

(for some γ), the truth of these equations follows automatically for the remaining indices $i = r_1 + r_2, \ldots, n$. For since ε_m is a unit

$$\sum_{p=1}^{r_1 + r_2} e_p \log\left|\varepsilon_m^{(p)}\right| = 0 \qquad (m = 1, 2, \ldots, k),$$

and hence

$$e_{r_1 + r_2} \sum_{m=1}^{k} \gamma_m \log\left|\varepsilon_m^{(r_1 + r_2)}\right| = -\sum_{p=1}^{r_1 + r_2 - 1} e_p \sum_{m=1}^{k} \gamma_m \log\left|\varepsilon_m^{(p)}\right| = 0;$$

hence (57) is also true for $i = r_1 + r_2$ and with this it is true, by (54), for $i = 1, 2, \ldots, n$. Consequently the k units $\varepsilon_1, \ldots, \varepsilon_k$ are independent if and only if the r linear homogeneous equations for the k unknowns $\gamma_1, \ldots, \gamma_k$

$$\sum_{m=1}^{k} \gamma_m \log|\varepsilon_m^{(i)}| = 0 \qquad (i = 1, 2, \ldots, r) \tag{58}$$

have no solutions in rational integers γ except $\gamma_m = 0$.

Next we obtain an upper bound for the number k of independent units by the following

Lemma (b). *If the r relations (58) hold for the k units $\varepsilon_1, \varepsilon_2, \ldots, \varepsilon_k$ with some real γ_m, which are not all zero, then r such relations also hold with rational integers γ_m, which are not all zero.*

Obviously it is enough to prove this for those units which are not roots of unity. Suppose we choose a number q such that the r equations among the units $\varepsilon_1, \varepsilon_2, \ldots, \varepsilon_{q-1}$

$$\sum_{m=1}^{q-1} \alpha_m \log|\varepsilon_m^{(i)}| = 0 \qquad (i = 1, \ldots, r)$$

hold only for $\alpha_1 = \cdots = \alpha_{q-1} = 0$ and on the other such that between the q units such a system

$$\sum_{m=1}^{q} \beta_m \log|\varepsilon_m^{(i)}| = 0 \qquad (i = 1, 2, \ldots, r) \tag{59}$$

holds with real $\beta_1, \ldots, \beta_q$ not all vanishing. Thus $2 \le q \le k$, and by the assumption about q we necessarily have $\beta_q \ne 0$ and the $q-1$ quotients $\beta_1/\beta_q, \ldots, \beta_{q-1}/\beta_q$ in (59) are uniquely determined. Lemma (b) will be proved once we show that these $q-1$ quotients β_m/β_q $(m = 1, 2, \ldots, q-1)$ are rational numbers.

If we set

$$\frac{\beta_m}{\beta_q} = -\alpha_m, \qquad (m = 1, 2, \ldots, q-1),$$

then it is a matter of checking the n equations

$$\log|\varepsilon_q^{(i)}| = \sum_{m=1}^{q-1} \alpha_m \log|\varepsilon_m^{(i)}| \qquad (i = 1, 2, \ldots, n). \tag{60}$$

More generally, we now consider all units η whose logarithms can be represented in the form

$$\log|\eta^{(i)}| = \sum_{m=1}^{q-1} \rho_m \log|\varepsilon_m^{(i)}| \qquad (i = 1, 2, \ldots, n) \tag{61}$$

with some real ρ_m. If this representation is at all possible, the ρ_m are uniquely determined by η (because of the hypothesis about q). Among the systems $(\rho_1, \ldots, \rho_{q-1})$ which appear here there are only finitely many whose elements all have absolute value < 1, since for the corresponding η we have

$$\left|\log|\eta^{(i)}|\right| \leq \sum_{m=1}^{q-1} \left|\log|\varepsilon_m^{(i)}|\right| \qquad (i = 1, 2, \ldots, n)$$

and by Lemma (a) there can only be finitely many integers of the field with this property. Let H be the number of distinct systems ρ with $|\rho_i| \leq 1$. On the other hand, the set of all systems $(\rho_1, \ldots, \rho_{q-1})$ appearing in (61) has the property that along with $(\rho_1, \ldots, \rho_{q-1})$ the system

$$(N\rho_1 - n_1, N\rho_2 - n_2, \ldots, N\rho_{q-1} - n_{q-1}),$$

where $N, n_1, n_2, \ldots, n_{q-1}$ are arbitrary rational integers, also occurs in this set. Now for each N the $n_1, \ldots, n_{q-1}$ can be chosen so that for all numbers $|N\rho_i - n_i| \leq \frac{1}{2}$ and for different values of N, if ρ_1 is irrational, the numbers $N\rho_1 - n_1$ always have different values. Thus, infinitely many systems $(\rho_1, \ldots, \rho_{q-1})$ are obtained, where all $|\rho_i| < 1$, contrary to what was proved above. Hence neither ρ_1 nor $\rho_2, \ldots, \rho_{q-1}$ can be irrational, so all α_m in (60) are rational, and the lemma is proved.

Moreover, we obtain at the same time, with respect to the denominators which may appear in the ρ_m, that there exists a fixed rational integer $M \neq 0$, which depends only on $\varepsilon_1, \ldots, \varepsilon_{q-1}$ but not on η in (61), such that $M\rho_m$ is a rational integer. In abbreviated notation, if ρ_1 has the form a/b with rational integers a, b ($b > 0$), then among the numbers $|N\rho_1 - n_1|$, there are exactly b distinct numbers, namely $0, 1/b, \ldots, b-1/b$, which are < 1. Consequently b is not greater than the number H, defined above, of all systems $(\rho_1, \ldots, \rho_{q-1})$ where all $|\rho_i| < 1$; thus $H!\rho_1$ is an integer, and hence we may choose $M = H!$. With this we have proved:

Lemma (c). *Assume that $\varepsilon_1, \ldots, \varepsilon_k$ are units such that the r equations*

$$\sum_{m=1}^{k} \gamma_m \log|\varepsilon_m^{(i)}| = 0 \qquad (i = 1, 2, \ldots, r),$$

with γ_m real, hold only for $\gamma_m = 0$. Then there is a fixed rational integer $M \neq 0$ such that the n expressions

$$\sum_{m=1}^{k} \rho_m \log|\varepsilon_m^{(i)}|$$

can be $= \log|\eta^{(i)}|$ (for $i = 1, 2, \ldots, n$), where η is a unit in K, only if $M\rho_m$ is a rational integer.

Furthermore, from Lemmas (b) and (c), it follows immediately that the number k of independent units of infinite order is at most r since for $k > r$ the r linear homogeneous equations (58) for the k unknowns $\gamma_1, \ldots, \gamma_k$ can surely be solved by real nonvanishing values, as the coefficients are real.

Moreover, from (c) we have:

Lemma (d). *The group $\mathfrak{G}$ of all units has a finite basis, and the number k of basis elements of infinite order is $\leq r$.*

For the proof let $\varepsilon_1, \ldots, \varepsilon_k$ be k units of infinite order and suppose there do not exist $k + 1$ independent elements of infinite order. Then, by (b) and (c), for each unit η in K a system of equations

$$\log|\eta^{(i)}| = \sum_{m=1}^{k} \frac{g_m}{M} \log|\varepsilon_m^{(i)}| \qquad (i = 1, 2, \ldots, n)$$

holds, for a certain positive rational integer M, where the g_m are rational integers. From this it follows by Lemma (a) that

$$\eta^M = \varepsilon_1^{g_1} \varepsilon_2^{g_2} \cdots \varepsilon_k^{g_k} \zeta,$$

where ζ is a root of unity in K, that is, a wth root of unity. Hence we have

$$\eta = \varepsilon_1^{g_1/M} \varepsilon_2^{g_2/M} \cdots \varepsilon_k^{g_k/M} \zeta_0^x, \quad \text{where } \zeta_0 = e^{2\pi i/Mw},$$

with rational integral x. We now consider[4] the totality of products of powers of the $k + 1$ numbers

$$H_1 = \varepsilon_1^{1/M}, \quad \ldots, H_k = \varepsilon_k^{1/M}, \qquad H_{k+1} = \zeta_0$$

with arbitrary but fixed values of the roots. The set of these numbers form a (mixed) Abelian group for which $H_1, \ldots, H_{k+1}$ is a basis. By what has already been proved, the group $\mathfrak{E}$ of all units of K is contained as a subgroup, and indeed as a subgroup of finite index, since the Mth power of each element belongs to $\mathfrak{E}$. Thus, by Theorem 34, $\mathfrak{E}$ also has a finite basis, and the number of basis elements of infinite order in $\mathfrak{E}$ is $\leq k$. However, in any case, the wth powers of all units, hence also the k independent units $\varepsilon_1^w, \varepsilon_2^w, \ldots, \varepsilon_k^w$, must occur among the products of powers of these base elements of infinite order. Consequently the number of these basis elements is exactly $= k$, and Lemma (d) is proved.

§35 Dirichlet's Theorem about the Exact Number of Fundamental Units

For a complete proof of Dirichlet's Theorem 100, we must still verify that the number k, which so far we have seen to be $\leq r$, is exactly equal to $r = r_1 + r_2 - 1$.

Since $n = r_1 + 2r_2$, $r = \frac{1}{2}(n + r_1) - 1$ and thus $r = 0$ only if $n + r_1 = 2$, that is, $n = 2$, $r_1 = 0$ or $n = 1$, $r_1 = 1$. These are the cases of the imaginary quadratic field and the trivial case of the rational number field.

---

[4] Here we recall the analogous method for the verification of the existence of a basis in §22.

Lemma (a). *If* $r = 0$, *the group* $\mathfrak{E}$ *is identical to the group* $\mathfrak{W}$ *of the roots of unity in* K.

For in the imaginary quadratic field it follows immediately from $N(\varepsilon) = \pm 1$ that $\varepsilon^{(1)} \cdot \varepsilon^{(2)} = +1$, and since $|\varepsilon^{(1)}| = |\varepsilon^{(2)}|$, this unit, along with its conjugate has absolute value 1; hence this unit is a root of unity.

Lemma (b). *If* $r > 0$, *then to each system of real numbers* $c_1, \ldots, c_r$, *not all vanishing, there is associated a unit* ε *such that*

$$L(\varepsilon) = c_1 \log|\varepsilon^{(1)}| + c_2 \log|\varepsilon^{(2)}| + \cdots + c_r \log|\varepsilon^{(r)}| \neq 0.$$

This second important conclusion in Dirichlet's train of thought rests on Minkowski's Theorem 95.

If $\varkappa_1, \ldots, \varkappa_n$ are n positive quantities such that

$$\varkappa_1 \cdot \varkappa_2 \cdots \varkappa_n = |\sqrt{d}|,$$
$$\varkappa_{p+r_2} = \varkappa_p \quad \text{for } p = r_1 + 1, \ldots, r_1 + r_2,$$

then by Theorem 95 there is a nonzero integer α in K (whose norm thus has at least absolute value 1) such that

$$|\alpha^{(i)}| \le \varkappa_i \quad \text{for } i = 1, 2, \ldots, n, \ 1 \le |N(\alpha)| \le \sqrt{d}.$$

From this it follows that

$$|\alpha^{(i)}| \ge \frac{1}{|\alpha^{(1)}| \cdot |\alpha^{(2)}| \cdots |\alpha^{(i-1)}| \cdot |\alpha^{(i+1)}| \cdots |\alpha^{(n)}|}$$

$$\ge \frac{\varkappa_i}{\varkappa_1 \cdots \varkappa_n} = \frac{\varkappa_i}{|\sqrt{d}|}.$$

(Moreover we may conclude from this that $|d| > 1$, for if $|d| = 1$ the equality sign would have to hold in each of these inequalities.) For this number α, the expression

$$L(\alpha) = \sum_{m=1}^{r} c_m \log|\alpha^{(m)}|$$

satisfies

$$\left| L(\alpha) - \sum_{m=1}^{r} c_m \log \varkappa_m \right| \le \sum_{m=1}^{r} |c_m| \log|\sqrt{d}| < A$$

where A is chosen independent of α and the $\varkappa$. The r quantities $\varkappa_1, \ldots, \varkappa_r$ are positive numbers which can be chosen arbitrarily, so we can find a sequence of systems $\varkappa_1^{(h)}, \ldots, \varkappa_r^{(h)}$ $(h = 1, 2, \ldots)$ such that

$$\sum_{m=1}^{r} c_m \log \varkappa_m^{(h)} = 2Ah \quad (h = 1, 2, \ldots),$$

and for the corresponding α_h we have

$$|L(\alpha_h) - 2Ah| < A$$
$$A(2h - 1) < L(\alpha_h) < A(2h + 1).$$

Consequently

$$L(\alpha_1) < L(\alpha_2) < L(\alpha_3) \ldots \tag{62}$$

while at the same time

$$|N(\alpha_h)| \le |\sqrt{d}|.$$

These infinitely many principal ideals (α_h) whose norms are not greater than $\sqrt{d}$, cannot all be distinct; hence there must exist at least two distinct indices h and m such that

$$(\alpha_h) = (\alpha_m), \quad \text{hence} \quad \alpha_m = \varepsilon\alpha_h,$$

where ε is a unit in k. For this ε, by (62),

$$L(\alpha_h) \ne L(\alpha_m) = L(\varepsilon\alpha_h)$$
$$L(\varepsilon) = L(\varepsilon\alpha_h) - L(\alpha_h) \ne 0,$$

and Lemma (b) is proved. From this we obtain

Lemma (c). *If $r \ge 0$, then the number k of independent units of the field is exactly $= r$.*

By (b) there is a unit ε_1 such that

$$\log|\varepsilon_1^{(1)}| \ne 0.$$

Then if $r > 1$, there is likewise a unit ε_2 such that

$$\begin{vmatrix} \log|\varepsilon_1^{(1)}| & \log|\varepsilon_2^{(1)}| \\ \log|\varepsilon_1^{(2)}| & \log|\varepsilon_2^{(2)}| \end{vmatrix} \ne 0$$

and so on. Thus we conclude from (b) the existence of r units $\varepsilon_1, \ldots, \varepsilon_r$ for which the determinant

$$\begin{vmatrix} \log|\varepsilon_1^{(1)}| & \cdots & \log|\varepsilon_r^{(1)}| \\ \log|\varepsilon_1^{(2)}| & \cdots & \log|\varepsilon_r^{(2)}| \\ \vdots & & \vdots \\ \log|\varepsilon_1^{(r)}| & \cdots & \log|\varepsilon_r^{(r)}| \end{vmatrix} \ne 0.$$

It follows immediately from the nonvanishing of this determinant that none of these units is a root of unity and, at the same time, that the r linear homogeneous equations for $\gamma_1, \ldots, \gamma_r$

$$\sum_{m=1}^{r} \gamma_m \log|\varepsilon_m^{(i)}| = 0 \quad (i = 1, 2, \ldots, r)$$

have the single solution $\gamma_1 = \gamma_2 = \cdots = \gamma_r = 0$. Consequently, by the theorems of the preceding section the number k of independent units of infinite order is exactly $= r$, and together with Lemma (d), the validity of Dirichlet's unit theorem, Theorem 100, is established.

By Theorem 38, §11, we see at once that between two systems of units in $K: \eta_1, \ldots, \eta_r$ and $\varepsilon_1, \ldots, \varepsilon_r$, equations of the form

$$\eta_m = \zeta_m \varepsilon_1^{a_{m1}} \cdot \varepsilon_2^{a_{m2}} \cdots \varepsilon_r^{a_{mr}} \qquad (m = 1, 2, \ldots, r)$$

hold, where the ζ_m are roots of unity, while the a_{mk} are rational integers with determinant ± 1. Thus for each system of fundamental units of K the absolute value of the determinant

$$\begin{vmatrix} \log|\eta_1^{(1)}| & \cdots & \log|\eta_r^{(1)}| \\ \vdots & & \vdots \\ \log|\eta_1^{(r)}| & \cdots & \log|\eta_r^{(r)}| \end{vmatrix}$$

has the same nonzero value; thus this value is a constant of the field.

The absolute value R of the determinant

$$\begin{vmatrix} e_1 \log|\eta_1^{(1)}| & \cdots & e_1 \log|\eta_r^{(1)}| \\ \vdots & & \vdots \\ e_r \log|\eta_1^{(r)}| & \cdots & e_r \log|\eta_r^{(r)}| \end{vmatrix} = \pm R$$

is called the *regulator R* of the field K.

§36 Different and Discriminant

In this section we concern ourselves with deeper properties of the discriminant d of the field K. Hitherto d was defined rather formally as the determinant of a basis of the field; we now try to find a definition of d based on intrinsic properties, which then has the advantage that it can be carried over to relative fields (§38).

We first define the *different of the number* $\alpha^{(p)}$ *in* $K^{(p)}$ as the number

$$\delta(\alpha^{(p)}) = \prod_{h \neq p} (\alpha^{(p)} - \alpha^{(h)}).$$

If $F(x)$ is the nth-degree polynomial with rational coefficients and leading coefficient 1 which has the n quantities $\alpha^{(1)}, \ldots, \alpha^{(n)}$ as roots, then obviously

$$\delta(\alpha) = F'(\alpha). \tag{63}$$

Accordingly, $\delta(\alpha^{(p)})$ is a number in $K^{(p)}$ and, by Theorem 54, it vanishes if and only if α is a number of lower degree than n. We then find the value

$$d(\alpha) = \prod_{n \geq i > k \geq 1} (\alpha^{(i)} - \alpha^{(k)})^2$$

$$= (-1)^{n(n-1)/2} N(\delta(\alpha))$$

for the discriminant of the number α.

Now let $\mathfrak{a}$ ($\neq 0$) be an arbitrary ideal in K with the basis $\alpha_1, \ldots, \alpha_n$.

Theorem 101. *The set of numbers λ in K, for which the trace*

$$S(\lambda\alpha) = \sum_{p=1}^{n} \lambda^{(p)}\alpha^{(p)} = \text{integer} \tag{64}$$

for each number α in $\mathfrak{a}$, forms an ideal $\mathfrak{m}$. Here $\mathfrak{ma}$ is an ideal independent of $\mathfrak{a}$, determined only by the field K, and it is the reciprocal of an integral ideal $\mathfrak{d}$. A basis for $\mathfrak{m}$ can be formed from the n numbers $\beta_1, \ldots, \beta_n$ which are determined along with their conjugates by the equations

$$S(\beta_i\alpha_k) = e_{ik} \qquad (i, k = 1, 2, \ldots, n) \tag{65}$$

where $e_{ik} = 1$ if $i = k$, otherwise $e_{ik} = 0$.

PROOF. The numbers λ with the property (64) cannot have arbitrarily large ideal denominators. This is true since the hypothesis is equivalent to n equations

$$S(\lambda\alpha_k) = g_k \qquad (k = 1, 2, \ldots, n),$$

where the g_k are rational integers, and from the n linear equations for $\lambda^{(1)}, \ldots, \lambda^{(n)}$ these $\lambda^{(i)}$ are obtained as quotients of two determinants. The denominator is the fixed determinant of the $\alpha_k^{(i)}$ which is equal to $N(\mathfrak{a})\sqrt{d}$. The numerator is an integral polynomial in the $\alpha_k^{(i)}$. Consequently there is an integer ω depending only on the α, such that $\omega\lambda$ is an integer. Moreover, if λ_1 and λ_2 belong to this set of λ, then for all integers ξ_1, ξ_2

$$S((\lambda_1\xi_1 + \lambda_2\xi_2)\alpha) = S(\lambda_1\xi_1\alpha) + S(\lambda_2\xi_2\alpha)$$

is also an integer, since $\xi_1\alpha$, $\xi_2\alpha$ belong to the ideal $\mathfrak{a}$; thus $\lambda_1\xi_1 + \lambda_2\xi_2$ also belong to the set of λ. By §31 this set is thus an ideal which depends on $\mathfrak{a}$ and is denoted by $\mathfrak{m} = \mathfrak{m}(\mathfrak{a})$. Furthermore we have $\mathfrak{am}(\mathfrak{a}) = \mathfrak{m}(1)$ which is thus independent of $\mathfrak{a}$ since if λ belongs to $\mathfrak{m}(\mathfrak{a})$, then for each ξ, $S(\lambda\alpha_k\xi)$ is also an integer, that is, $\lambda\alpha_k$ belongs to $\mathfrak{m}(1)$. Conversely, if μ belongs to $\mathfrak{m}(1)$ and $\rho_1, \ldots, \rho_n$ denotes a basis for $1/\mathfrak{a}$, then $\alpha\rho_k$ is an integer and hence $S(\mu\rho_k\alpha)$ is an integer, that is, the products of μ with every number in $1/\mathfrak{a}$ belong to $\mathfrak{m}(\mathfrak{a})$, thus μ belongs to $\mathfrak{am}(\mathfrak{a})$.

Moreover $\mathfrak{m}(1)$ is the reciprocal of an integral ideal $\mathfrak{d}$, as the number 1 obviously belongs to $\mathfrak{m}(1)$. Consequently

$$\mathfrak{m} = \mathfrak{m}(\mathfrak{a}) = \frac{1}{\mathfrak{a}\mathfrak{d}},$$

where $\mathfrak{d}$ is an integral ideal independent of $\mathfrak{a}$.

Finally if we define the n^2 numbers $\beta_i^{(p)}$ by the uniquely solvable equations

$$\sum_{p=1}^{n} \beta_i^{(p)}\alpha_k^{(p)} = e_{ik} \qquad (i, k = 1, 2, \ldots, n) \tag{65}$$

and if we set

$$S(\lambda \alpha_k) = g_k \qquad (k = 1, 2, \ldots, n),$$

where λ satisfies (64), then we also have

$$S(\lambda_0 \alpha_k) = g_k \quad \text{for } \lambda_0 = g_1 \beta_1 + \cdots + g_n \beta_n.$$

Consequently

$$\lambda = \lambda_0 = g_1 \beta_1 + \cdots + g_n \beta_n$$

and the $\beta_1, \ldots, \beta_n$ form a basis for $\mathfrak{m}(\mathfrak{a})$, provided they are numbers in K. The latter fact is obtained directly from the representation of the solutions of (65) as a determinant. Alternatively by multiplication by $\alpha_i^{(q)}$ and summation over i we can deduce from (65) the equivalent system of equations

$$\sum_p \alpha_k^{(p)} \sum_i \beta_i^{(p)} \alpha_i^{(q)} = \sum_i e_{ik} \alpha_i^{(q)} = \alpha_k^{(q)} = \sum_p e_{pq} \alpha_k^{(p)}$$

and from this we can deduce

$$\sum_i \beta_i^{(p)} \alpha_i^{(q)} = e_{pq}, \qquad \sum_i \beta_i^{(p)} \sum_q \alpha_i^{(q)} \alpha_k^{(q)} = \sum_q e_{pq} \alpha_k^{(q)}$$

or

$$\sum_{i=1}^{n} \beta_i^{(p)} S(\alpha_i \alpha_k) = \alpha_k^{(p)}.$$

Since the coefficients on the left-hand side are now rational, the $\beta_i^{(p)}$ are numbers in $K^{(p)}$. Hence Theorem 101 is proved.

For later applications (Chapter 8), we formulate this result in yet another way:

Theorem 102. *If* $\alpha_1, \ldots, \alpha_n$ *are basis elements of the ideal* $\mathfrak{a}$, *then the n sequences of numbers* $\beta_1^{(p)}, \ldots, \beta_n^{(p)}$ ($p = 1, \ldots, n$), *which are defined by* (65), *are conjugate sequences of numbers in* K, *and* $\beta_1, \ldots, \beta_n$ *forms a basis for* $1/\mathfrak{a}\mathfrak{d}$.

Since, moreover,

$$\Delta^2(\beta_1, \ldots, \beta_n) = \frac{1}{\Delta^2(\alpha_1, \ldots, \alpha_n)} = \frac{1}{dN^2(\mathfrak{a})}$$

and by (47)

$$\Delta^2(\beta_1, \ldots, \beta_n) = N^2(\mathfrak{m})d = \frac{d}{N^2(\mathfrak{a}\mathfrak{d})},$$

we have

Theorem 103. $N(\mathfrak{d}) = |d|$.

This ideal $\mathfrak{d}$ defined by Theorem 101 is called the *different* or the *basic ideal of the field*.

Now in order to discover the fundamental connection between this different of the field and the differents of the numbers in K, we must investigate the set of numbers in K which can be represented in the form

$$G(\theta) = a_0 + a_1\theta + a_2\theta^2 + \cdots + a_{n-1}\theta^{n-1}$$

with rational integers a_i. Let θ be an integer which generates the field K. Let the set of numbers $g(\theta)$ with rational integers a_i be called a *number ring or an integral domain* and let it be denoted by $R(\theta)$. In the first place, the numbers of the ring certainly form a module with basis elements $1, \theta, \theta^2, \ldots, \theta^{n-1}$, and secondly they are closed under multiplication.

Lemma (a). *Each number α of the field, for which $\mathfrak{d}\alpha$ is integral can be represented in the form*

$$\alpha = \frac{\rho}{F'(\theta)}$$

where ρ is an integer of the ring $R(\theta)$ and $F'(\theta)$ is the different of θ as in (63).

For the proof we consider the polynomial in x

$$G(x) = \sum_{i=1}^{n} \alpha^{(i)} \frac{F(x)}{x - \theta^{(i)}}, \tag{66}$$

where

$$F(x) = \prod_{i=1}^{n} (x - \theta^{(i)}) = c_0 + c_1 x + \cdots + c_{n-1}x^{n-1} + c_n x^n.$$

$G(x)$ is a polynomial with rational integral coefficients since

$$\frac{F(x)}{x - \theta} = \frac{F(x) - F(\theta)}{x - \theta} = \sum_{h=1}^{n} c_h \sum_{0 \le r \le h-1} x^r \theta^{h-r-1}$$

and hence

$$G(x) = \sum_{h=1}^{n} c_h \sum_{0 \le r \le h-1} x^r S(\alpha \theta^{h-r-1}).$$

However since $\alpha\mathfrak{d}$ is integral by hypothesis, the traces appearing here are rational integers by Theorem 101. If we set $x = \theta$ in (66) we obtain

$$\alpha = \frac{G(\theta)}{F'(\theta)},$$

where, in fact, $G(\theta)$ is a number of the ring.

From this it follows that $F'(\theta) \cdot \alpha$ is an integer if $\mathfrak{d}\alpha$ is integral, thus $F'(\theta)$ has the decomposition

$$F'(\theta) = \mathfrak{d}\mathfrak{f} \tag{67}$$

where $\mathfrak{f}$ is an integral ideal.

Lemma (b). *For each number ρ which belongs to the ring $R(\theta)$, we have*

$$S\left(\frac{\rho}{F'(\theta)}\right) = integer.$$

Obviously this assertion needs only to be proved for $\rho = 1, \theta, \ldots, \theta^{n-1}$, where it follows directly from the so-called Euler formulas

$$\sum_{i=1}^{n} \frac{\theta^{(i)k}}{F'(\theta^{(i)})} = \begin{cases} 0 & \text{for } k = 0, 1, 2, \ldots, n-2, \\ 1 & \text{for } k = n-1. \end{cases}$$

These formulas follow, as we mention for the sake of completeness, from the Lagrange interpolation formula

$$\sum_{i=1}^{n} \frac{\theta^{(i)k+1}}{F'(\theta^{(i)})} \frac{F(x)}{x - \theta^{(i)}} = \begin{cases} x^{k+1} & \text{for } k = 0, 1, \ldots, n-2, \\ x^n - F(x) & \text{for } k = n-1, \end{cases}$$

if we set $x = 0$ (or also if we expand in powers of $1/x$ after division by $F(x)$).

Theorem 104. *All numbers of the ideal $\mathfrak{f} = F'(\theta)/\mathfrak{d}$ belong to the ring $R(\theta)$, and if all numbers of an ideal $\mathfrak{a}$ belong to the ring $R(\theta)$, then $\mathfrak{a}$ is divisible by $\mathfrak{f}$.*

If $\omega \equiv 0 \pmod{\mathfrak{f}}$, then $\alpha = \omega/F'(\theta)$ is a number with denominator $\mathfrak{d}$, and by Lemma (a), $\alpha F'(\theta)$ must be a number of this ring. Hence the first part of our theorem is proved.

Conversely, if all numbers of $\mathfrak{a}$ are numbers of the ring, then by Lemma (b) $S(\alpha/F'(\theta))$ is an integer for all numbers α in $\mathfrak{a}$. Consequently, by Theorem 101, $1/F'(\theta)$ is a number of the ideal $\mathfrak{m}(\mathfrak{a}) = 1/\mathfrak{a}\mathfrak{d}$; thus $F'(\theta) = \mathfrak{d}\mathfrak{f}$ divides $\mathfrak{a}\mathfrak{d}$ so $\mathfrak{f} \mid \mathfrak{a}$, which was to be proved.

This theorem thus yields a new definition of $\mathfrak{f}$; $\mathfrak{f}$ *is the* GCD *of all ideals in K which contain only numbers of the ring.* The ideal $\mathfrak{f}$ is called the *conductor of the ring.*

Lemma (c). *There are always rings $R(\theta)$ in K whose conductor $\mathfrak{f}$ is not divisible by an arbitrary prime ideal $\mathfrak{p}$.*

If ω is an integer divisible by $\mathfrak{p}$ but not $\mathfrak{p}^2$, then the expression

$$\gamma_0 + \gamma_1\omega + \gamma_2\omega^2 + \cdots + \gamma_h\omega^h \tag{68}$$

obviously represents all residue classes mod $\mathfrak{p}^{h+1}$, if $\gamma_0, \ldots, \gamma_h$ run independently through a complete system of residues mod $\mathfrak{p}$. Now let θ be a primitive root mod $\mathfrak{p}$ such that for the number,

$$\omega = \theta^{N(\mathfrak{p})} - \theta$$

which is divisible by $\mathfrak{p}$ is not divisible by $\mathfrak{p}^2$. (If θ does not have the latter property, then nonetheless $\theta + \pi$ surely does as long as π is a number which is divisible by $\mathfrak{p}$ but not by $\mathfrak{p}^2$.) Moreover, by a modification mod $\mathfrak{p}^2$ we can arrange that θ is different from all of its conjugates, and in addition

$$\theta \equiv 0 \ (\text{mod } \mathfrak{a}), \quad \text{where } p = \mathfrak{p}^e \mathfrak{a}, \ (\mathfrak{a}, \mathfrak{p}) = 1, \tag{69}$$

and p is the rational prime which is divisible by $\mathfrak{p}$.

If we now let γ_i in (68) run through the $N(\mathfrak{p})$ numbers

$$0, \ \theta, \ \theta^2, \ \ldots, \ \theta^{N(\mathfrak{p})-1}$$

which are incongruent mod $\mathfrak{p}$, we see that each residue class mod $\mathfrak{p}^h$ can be represented by a number of the ring $R(\theta)$. But then, if (69) holds, the conductor $\mathfrak{f}$ of the ring cannot be divisible by $\mathfrak{p}$. For if

$$N(\mathfrak{d}\mathfrak{f}) = p^k a, \quad \text{where } (a, p) = 1,$$

then to begin with, by the above, to each integer ω there is a number ρ of the ring such that

$$\pi = \omega - \rho \equiv 0 \ (\text{mod } \mathfrak{p}^{ek}).$$

Here $\pi a \theta^k$ is divisible by $F'(\theta) = \mathfrak{d}\mathfrak{f}$, as the decomposition

$$\frac{\pi a \theta^k}{F'(\theta)} = \frac{\pi \theta^k N(\mathfrak{d}\mathfrak{f})}{\mathfrak{d}\mathfrak{f} p^k} = \frac{N(\mathfrak{d}\mathfrak{f})}{\mathfrak{d}\mathfrak{f}} \frac{\pi \theta^k}{\mathfrak{p}^{ek} \mathfrak{a}^k}$$

shows by (69). Hence by Lemma (a), this number can be represented in the form

$$\frac{\pi a \theta^k}{F'(\theta)} = \frac{\rho_1}{F'(\theta)}, \quad \text{thus } \pi = \frac{\rho_1}{a \theta^k},$$

where ρ_1 is a number in $R(\theta)$. However, then

$$a \theta^k \omega = a \theta^k (\rho + \pi) = a \theta^k \rho + \rho_1$$

is likewise a number of the ring, and the ideal $a \theta^k$ (which is not divisible by $\mathfrak{p}$) contains only numbers of the ring. Thus by Theorem 104 it is divisible by $\mathfrak{f}$; consequently $\mathfrak{f}$ is also not divisible by $\mathfrak{p}$. From this we immediately obtain the main theorem of this theory:

Theorem 105. *The greatest common divisor of the differents $\delta(\theta)$ of all integers θ in K is equal to the different $\mathfrak{d}$ of the field.*

It is a noteworthy fact that in contrast to the different, the discriminant d of the field is indeed a common divisor of the discriminant $d(\theta)$ of all integers θ in K, but need not be the greatest common divisor of the same[5].

---

[5] R. *Dedekind*, Über den Zusammenhang zwischen der Theorie der Ideale und der Theorie der höheren Kongruenzen, and: Über die Diskriminanten endlicher Körper, *Abh. d. K. Ges. d. Wiss zu Göttingen* 1878 and 1882 and as well the later papers of Hensel in *Crelles Journal*, Vol. 105 (1889) and Vol. 113 (1894).

§37 Relative Fields and Relations between Ideals in Different Fields

We now turn to the problem of how to modify the concepts which were developed in the preceding sections to the case in which the field K is no longer considered relative to $k(1)$, but rather to any algebraic number field k which is a subfield of K. Of course the ideal theory developed until now holds in k just as in K. Can a relationship be found between the ideals in K and the ideals in k?

We agree to denote elements of K (numbers or ideals) by capital letters, while small letters always denote elements in k. Let K have relative degree m with respect to k (compare §20, Theorem 59), while the degrees of K and k with respect to the rational number field are N and n respectively. Then

$$N = n \cdot m.$$

Furthermore, q arbitrary numbers $\alpha_1, \ldots, \alpha_q$ of k define an ideal in K and an ideal in k, both ideals to be denoted by $(\alpha_1, \ldots, \alpha_q)$ which we distinguish by

$$\mathfrak{a} = (\alpha_1, \ldots, \alpha_q)_k \quad \text{and} \quad \mathfrak{A} = (\alpha_1, \ldots, \alpha_q)_K. \tag{70}$$

Moreover, if a number β belongs to $\mathfrak{a}$, then it of course also belongs to $\mathfrak{A}$. The converse is also true:

Lemma (a). *If β belongs to $\mathfrak{A}$, then, provided (70) holds, β also belongs to $\mathfrak{a}$.*

If an equation

$$\beta = \sum_h \Gamma_h \alpha_h$$

with integers Γ_h in K holds, then the equations

$$\beta = \sum_h \Gamma_h^{(i)} \alpha_h \qquad (i = 1, 2, \ldots, m)$$

with the corresponding relative conjugates also hold and by multiplication it follows that

$$\beta^m = \prod_{i=1}^m \left(\sum_h \Gamma_h^{(i)} \alpha_h \right).$$

Obviously, for indeterminates $x_1, \ldots, x_q$, the expression

$$\prod_{i=1}^m \left(\sum_{h=1}^q \Gamma_h^{(i)} x_h \right) = \sum_{n_1, \ldots, n_q} \gamma_{n_1 n_2, \ldots, n_q} x_1^{n_1} x_2^{n_2} \cdots x_q^{n_q} \tag{71}$$

is a homogeneous polynomial in $x_1, \ldots, x_q$ of degree m where

$$n_1 + n_2 + \cdots + n_q = m.$$

These coefficients γ, as symmetric integral expressions in the $\Gamma_h^{(i)}$, are integers in k. For $x_h = \alpha_h$ $(h = 1, 2, \ldots, q)$ Equation (71) is thus a number from the ideal $\mathfrak{a}^m$. Consequently $\beta^m/\mathfrak{a}^m$ is an integer; hence $\beta/\mathfrak{a}$ is also an integer, that is, β occurs in $\mathfrak{a}$.

Thus, by (70), if we have a further pair of corresponding ideals,

$$\mathfrak{b} = (\beta_1, \ldots, \beta_s)_k \quad \text{and} \quad \mathfrak{B} = (\beta_1, \ldots, \beta_s)_K,$$

then we have

Lemma (b). *If* $\mathfrak{a} = \mathfrak{b}$, *then* $\mathfrak{A} = \mathfrak{B}$ *and conversely.*

The first part is obvious. For the converse, if $\mathfrak{A} = \mathfrak{B}$, then each β belongs to $\mathfrak{A}$, and by Lemma (a) it then also belongs to $\mathfrak{a}$. Likewise each α belongs to $\mathfrak{B}$, thus also to $\mathfrak{b}$; consequently $\mathfrak{a} = \mathfrak{b}$.

Let us now assign to each ideal $\mathfrak{a}$ in k an ideal $\mathfrak{A}$ in K by the following prescription: we set $\mathfrak{a} = (\alpha_1, \ldots, \alpha_q)_k$ and define $\mathfrak{A} = (\alpha_1, \ldots, \alpha_q)_K$. By Lemma (b) this prescription yields an ideal $\mathfrak{A}$ fully determined by $\mathfrak{a}$ (independent of the representation of $\mathfrak{a}$) and indeed in this way we obviously arrive at every ideal in K which can be represented as the GCD of numbers of the ground field. This correspondence, expressed by symbols

$$\mathfrak{a} \rightleftarrows \mathfrak{A}, \tag{72}$$

according to Lemma (b) is moreover a unique one-to-one mapping. Consequently, we have

Theorem 106. *By* (72) *there exists a well-defined invertible correspondence between all ideals in k on the one hand and all ideals in K which can be represented as the GCD of numbers in k on the other hand, such that for an arbitrary number α in k the two statements "α belongs to $\mathfrak{a}$" and "α belongs to $\mathfrak{A}$" are true simultaneously, if* (72) *holds. Moreover we also have*

$$\mathfrak{a}\mathfrak{b} \rightleftarrows \mathfrak{A}\mathfrak{B}, \; \textit{if } \mathfrak{a} \rightleftarrows \mathfrak{A} \textit{ and } \mathfrak{b} \rightleftarrows \mathfrak{B}.$$

Definition. We thus call two ideals connected by (72) *equal* to one another and we say that the ideal $\mathfrak{A}$ in K *lies in the field* k.

Since the relation "$=$" between ideals of different fields has not yet been defined, this definition contains no contradiction to earlier stipulations. By Theorem 106 the following rules hold:

(1) From $\mathfrak{a} = \mathfrak{A}$ and $\mathfrak{a} = \mathfrak{b}$ it follows that $\mathfrak{b} = \mathfrak{A}$.
(2) From $\mathfrak{a} = \mathfrak{A}$ and $\mathfrak{b} = \mathfrak{A}$ it follows that $\mathfrak{a} = \mathfrak{b}$.
(3) From $\mathfrak{a} = \mathfrak{A}$ and $\mathfrak{A} = \mathfrak{B}$ it follows that $\mathfrak{a} = \mathfrak{B}$.
(4) From $\mathfrak{a} = \mathfrak{A}$ and $\mathfrak{b} = \mathfrak{B}$ it follows that $\mathfrak{a}\mathfrak{b} = \mathfrak{A}\mathfrak{B}$.
(5) From $\mathfrak{a}^p = \mathfrak{A}^p$ it follows that $\mathfrak{a} = \mathfrak{A}$ (p a rational integer).

The meaning of these assertions is that the relation "=" between ideals in different fields is a generalization of the already defined relation, likewise indicated by the symbol "=," between ideals of the same field.

Thus, by the above definition, we can decide whether two symbols

$$(\alpha_1, \ldots, \alpha_q) \text{ in } k, \qquad (A_1, \ldots, A_s) \text{ in } K$$

are to be regarded as equal or not. Here K is to be regarded as an extension field of k. However in some cases the two symbols can already have a meaning in a subfield K' of K, and now we wish to see whether from equality in one field one can also deduce equality in other fields.

In fact this is the case since if K' is an extension field of k but a subfield of K, such that A belongs to K', then it obviously follows immediately from

$$(\alpha_1, \ldots, \alpha_q)_{K'} = (A_1, \ldots, A_s)_{K'} \tag{73}$$

that

$$(\alpha_1, \ldots, \alpha_q)_K = (A_1, \ldots, A_s)_K.$$

Conversely, however, if the latter equation is valid, then by the second part of Lemma (b) applied to the extension field K of K', Equation (73) is also valid in K'.

Accordingly the symbol $(\alpha_1, \ldots, \alpha_q)$ defines the same ideal in every field in which it has any meaning at all. And now we can decide whether or not two ideals $\mathfrak{a}_1$ and $\mathfrak{a}_2$, which are defined as the GCD of numbers of two arbitrary fields k_1, k_2 respectively, are equal. To do this consider any field K which contains k_1 as well as k_2 and determine whether or not these two GCDs are equal in K in the sense of our very first definition of equality of ideals (§24). The result is the same in all fields. Thus in the notation $\mathfrak{a} = (\alpha_1, \ldots, \alpha_q)$ we need not make reference to a definite field. By rule 4 the product of two ideals $\mathfrak{a}$, $\mathfrak{b}$ is an ideal which is completely determined by $\mathfrak{a}$ and $\mathfrak{b}$; the same holds for the quotient and the GCD.

In particular the statement "The algebraic integers $\alpha_1, \ldots, \alpha_q$ are relatively prime (have the GCD (1))" is independent of reference to a special number field and is equivalent to the statement "There are algebraic integers $\lambda_1, \ldots, \lambda_q$ for which

$$\lambda_1 \alpha_1 + \cdots + \lambda_q \alpha_q = 1.\text{"}$$

It is then a remarkable fact, which follows immediately from our stipulations, that if any integers λ with this property exist, they may always be chosen from the number field, which is generated by $\alpha_1, \ldots, \alpha_q$.

It should be emphasized, however, that although an ideal $\mathfrak{a}$ distinguishes a definite number field in this way, in general $\mathfrak{a}$ does not lie in each number field in the sense of the above definition. For example

$$\mathfrak{a} = (5, \sqrt{10}) = (\sqrt{5})$$

holds because the square is equal to (5). Thus, for example, $\mathfrak{a}$ belongs to the two quadratic number fields $k(\sqrt{10})$ and $k(\sqrt{5})$ but not to the field $k(1)$.

The property of being a prime ideal belongs to an ideal only with respect to a definite field in which it lies.

If we now connect these concepts with the theorems in §33 about ideal numbers, then we obtain the following: If $\mathfrak{a}$ is an ideal in k and $\mathfrak{a}^h$ is equal to the principal ideal (ω) in k, then the equation

$$\mathfrak{a} = (\sqrt[h]{\omega})$$

has meaning according to our present stipulation and indeed it is a valid equation. Furthermore, if the number A in the system of ideal numbers of k is assigned to the ideal $\mathfrak{a}$, then likewise $\mathfrak{a} = (A)$ holds. The set of all ideal numbers belongs to a field of relative degree h over k, and this fact can then also be expressed as follows: if h is the class number of k, then there is a relative field of relative degree h over k in which all ideals of k become principal ideals. The relative field is not uniquely determined by this requirement, by any means. Also its class number need not be 1.

§38 Relative Norms of Numbers and Ideals, Relative Differents, and Relative Discriminants

If A is some number in K and if $A^{(i)}$ $(i = 1, \ldots, m)$ are its relative conjugates with respect to k, then

$$S_k(A) = A^{(1)} + A^{(2)} + \cdots + A^{(m)}$$
$$N_k(A) = A^{(1)} \cdot A^{(2)} \cdots A^{(m)}$$

are called the *relative trace* and the *relative norm* of A respectively (with respect to k). They are numbers in k. If S and s denote the traces in K and in k, with respect to $k(1)$, likewise N and n denote the norms in K and in k, then, by Theorem 59,

$$S(A) = s(S_k(A)); \qquad N(A) = n(N_k(A)). \tag{74}$$

The number

$$\delta_k(A^{(q)}) = \prod_{i=1, i \neq q}^{m} (A^{(q)} - A^{(i)})$$

is called the *relative different* of $A^{(q)}$ in the field $K^{(q)}$ with respect to k; it is a number in the field $K^{(q)}$. If

$$\Phi(x) = \prod_{i=1}^{m} (x - A^{(i)}) = x^m + \alpha_1 x^{m-1} + \cdots + \alpha_{m-1} x + \alpha_m$$

(where the α_r are obviously numbers in k), then

$$\delta_k(A) = \Phi'(A).$$

The product

$$d_k(A) = \prod_{i \le i < q \le m} (A^{(i)} - A^{(q)})^2 = (-1)^{m(m-1)/2} \prod_{i=1}^{m} \Phi'(A^{(i)})$$

$$= (-1)^{m(m-1)/2} N_k(\delta_k(A))$$

is called the *relative discriminant* of A; it is a number in k.

If $\mathfrak{A}$ is an ideal in K, then the relative conjugate ideal $\mathfrak{A}^{(i)}$ arises if one replaces each number A in $\mathfrak{A}$ by $A^{(i)}$. Obviously for two ideals $\mathfrak{A}$ and $\mathfrak{B}$ we have

$$(\mathfrak{A} \cdot \mathfrak{B})^{(i)} = \mathfrak{A}^{(i)} \cdot \mathfrak{B}^{(i)}.$$

Definition. The ideal

$$N_k(\mathfrak{A}) = \mathfrak{A}^{(1)} \cdot \mathfrak{A}^{(2)} \cdots \mathfrak{A}^{(m)}$$

is called the *relative norm of* $\mathfrak{A}$ with respect to k. We have $N_k(\mathfrak{A}\mathfrak{B}) = N_k(\mathfrak{A}) \cdot N_k(\mathfrak{B})$.

Theorem 107. *The ideal $N_k(\mathfrak{A})$ is an ideal in k. If k is the field of rational numbers, then $N_k(\mathfrak{A}) = (N(\mathfrak{A}))$.*

To begin let $\mathfrak{A} = (A_1, \ldots, A_s)$ be an integral ideal, where the A_i are numbers in K. Then, by §28, for any variables $u_1, \ldots, u_s$, the content of the conjugate polynomials

$$F^{(i)}(u) = A_1^{(i)} u_1 + \cdots + A_s^{(i)} u_s$$

is equal to $\mathfrak{A}^{(i)}$. Hence, by Theorem 87,

$$\mathfrak{A}^{(1)} \cdots \mathfrak{A}^{(m)} = J(F^{(1)}) \cdots J(F^{(m)}) = J(F^{(1)} \cdots F^{(m)}).$$

However the polynomial

$$Q(u) = F^{(1)} \cdot F^{(2)} \cdots F^{(m)}$$

is obviously a polynomial over k; thus $J(Q)$ is an ideal in k. Thus the first part of Theorem 107 is proved, if we recall that each ideal can be written as the quotient of two integral ideals, and by definition

$$N_k\left(\frac{\mathfrak{A}}{\mathfrak{B}}\right) = \frac{N_k(\mathfrak{A})}{N_k(\mathfrak{B})}.$$

For the proof of the second part of Theorem 107, let h be the class number of K. Then $\mathfrak{A}^h = (A)$, where A is a certain number in K, and

$$N_k(\mathfrak{A})^h = N_k(\mathfrak{A}^h) = N((A)) = (N(A)).$$

Since

$$\pm N(A) = N(\mathfrak{A}^h) = N(\mathfrak{A})^h = a^h, \quad \text{where } a = N(\mathfrak{A}),$$

we have

$$N_k(\mathfrak{A})^h = (a)^h, \qquad N_k(\mathfrak{A}) = (a).$$

Thus Theorem 88 of §29, which was only proved in the case of Galois fields, is now seen as valid for every number field, and at the same time the terminology "norm of $\mathfrak{A}$" for the number of residue classes mod $\mathfrak{A}$ has received its justification.

Theorem 108. *For each prime ideal $\mathfrak{P}$ of K there is exactly one prime ideal $\mathfrak{p}$ in k which is divisible by $\mathfrak{P}$. Then*

$$N_k(\mathfrak{P}) = \mathfrak{p}^{f_1},$$

where f_1 is a natural number $\leq m$. f_1 is called the relative degree of $\mathfrak{P}$ with respect to k. $\mathfrak{p}$ splits into at most m factors in K.

By Theorem 107 $N_k(\mathfrak{P})$ is an ideal in k which, by definition, is divisible by $\mathfrak{P}$. If $N_k(\mathfrak{P})$ is decomposed into its prime factors, then by the fundamental theorem $\mathfrak{P}$ must divide at least one of these prime ideals in k. If $\mathfrak{P}$ were to divide two distinct prime ideals $\mathfrak{p}_1$, $\mathfrak{p}_2$ in k then it would also have to be a divisor of $(\mathfrak{p}_1, \mathfrak{p}_2) = 1$, which, however, cannot be the case. Thus there exists exactly one prime ideal $\mathfrak{p}$ in k which is divisible by $\mathfrak{P}$. If the decomposition of $\mathfrak{p}$ into prime ideals in K is, say,

$$\mathfrak{p} = \mathfrak{P}_1 \cdot \mathfrak{P}_2 \cdots \mathfrak{P}_v,$$

then it follows that for the relative norm

$$N_k(\mathfrak{P}_1) \cdot N_k(\mathfrak{P}_2) \cdots N_k(\mathfrak{P}_v) = N_k(\mathfrak{p}) = \mathfrak{p}^m.$$

By the preceding theorem, each factor on the left is an ideal in k and by this equation each factor must be a power of $\mathfrak{p}$. Therefore

$$N(\mathfrak{P}_i) = \mathfrak{p}^{f_i} \quad \text{and} \quad f_1 + f_2 + \cdots + f_v = m;$$

hence

$$f_i \leq m \quad \text{and} \quad v \leq m.$$

Theorem 109. *If N denotes the norm in K and n denotes the norm in k, then for each ideal $\mathfrak{A}$ in K*

$$N(\mathfrak{A}) = n(N_k(\mathfrak{A})).$$

To begin with, this assertion follows immediately for each number A in K, by (74). By Theorem 107, or also by consideration of the principal ideal $\mathfrak{A}^h$, the result is also obtained for each ideal in K.

Theorem 110. *If the relative degree of the prime ideal $\mathfrak{P}$ is equal to 1, then each number in K is congruent modulo $\mathfrak{P}$ to a number in k.*

By Theorem 108, $N(\mathfrak{P}) = n(\mathfrak{p})^{f_1}$; hence for $f_1 = 1$ the number of residue classes mod $\mathfrak{P}$ in K is equal to the number of residue classes mod $\mathfrak{p}$ in k. However if a number α in k is divisible by $\mathfrak{P}$, then $(\alpha, \mathfrak{p})$ is divisible at least by $\mathfrak{P}$, so that $(\alpha, \mathfrak{p}) \neq 1$. Hence as an ideal in k, $(\alpha, \mathfrak{p})$ is necessarily $= \mathfrak{p}$. Consequently a system of incongruent numbers mod $\mathfrak{p}$ in k is also incongruent mod $\mathfrak{P}$ and thus there are $n(\mathfrak{p}) = N(\mathfrak{P})$ incongruent numbers modulo $\mathfrak{P}$ in k.

We take special note of the following fact: If a number A in K is *equal to its relative conjugates* then, by Theorem 56, it is a number in k. However the corresponding statement is not valid for ideals. For example the ideal $(\sqrt{5})$ is equal to its conjugates with respect to $k(1)$ in $K(\sqrt{5})$, but $(\sqrt{5})$ is not an ideal in $k(1)$.

Finally the concepts in §36 can be extended to relative fields and lead to a definition of relative discriminant.

Definition. The set of numbers Δ in K, such that for each integer A in K the relative trace $S_k(\Delta A)$ is an integer forms an ideal $\mathfrak{M}$ in K. Furthermore,

$$\frac{1}{\mathfrak{M}} = \mathfrak{D}_k$$

is an integral ideal and is called the *relative different of K with respect to k.*

The proof that $\mathfrak{M}$ and $\mathfrak{D}_k$ are ideals runs parallel to that of Theorem 101.

Theorem 111. *If $\mathfrak{D}$ and $\mathfrak{d}$ are the differents of K and k respectively, then for the relative differents $\mathfrak{D}_k$ the relation*

$$\mathfrak{D} = \mathfrak{D}_k \mathfrak{d} \tag{75}$$

holds.

PROOF. If Δ is a number in K such that $\Delta \mathfrak{D}_k \mathfrak{d}$ is integral, then by the definition of $\mathfrak{D}_k$,

$$\mathfrak{d} S_k(\Delta A) \text{ is integral} \tag{76}$$

for each integer A, since for each number ξ in k which is divisible by $\mathfrak{d}$

$$S_k(\Delta A \xi) = \xi S_k(\Delta A)$$

is an integer. From the definition of $\mathfrak{d}$, $s(S_k(\Delta A))$ is an integer by (76). Hence $S(\Delta A)$ is an integer and thus

$$\mathfrak{D}\Delta \text{ is integral if } \mathfrak{D}_k \cdot \mathfrak{d}\Delta \text{ is integral.}$$

Conversely, if $\mathfrak{D}\Delta$ is integral, then for each integer A in K and each integer ξ in k, $S(\Delta A \xi)$ is integral; hence

$$s(S_k(\Delta A \xi)) = s(\xi S_k(\Delta A))$$

is an integer. Thus

$$\mathfrak{d} S_k(\Delta A) \text{ is integral, that is, } S_k (\rho \Delta A) \text{ is integral}$$

if ρ is any number of $\mathfrak{d}$ in k and hence $\rho \Delta \mathfrak{D}_k$ is integral. Thus we have shown that if $\mathfrak{D}\Delta$ is integral, $\mathfrak{D}_k \mathfrak{d}\Delta$ is also integral, from which Theorem 111 follows.

The meaning of the relative different which already emerges from this simple equation (75) will become yet more evident when we prove the following fact, which can also serve as the definition of $\mathfrak{D}_k$.

Theorem 112. *The relative different of K is the GCD of all relative differents of integers of K with respect to k.*

For the proof of this theorem we must proceed almost exactly as in the proof of Theorem 105.

If θ is an integer generating the field K, then the *relative ring* $R_k(\theta)$ is the set of all numbers

$$\alpha_0 + \alpha_1 \theta + \cdots + \alpha_{m-1} \theta^{m-1}$$

where $\alpha_0, \ldots, \alpha_{m-1}$ run through all integers in k. If $\Phi(x)$ is the irreducible polynomial over k with leading coefficient 1 which has the root θ, then we have the following lemmas which are proved as in §36:

Lemma (a). *If A is an integer in K such that $A\mathfrak{D}_k$ is integral, then A can be represented in the form*

$$A = \frac{B}{\Phi'(\theta)},$$

where B is a number in $R_k(\theta)$. Thus $\Phi'(\theta)$ is divisible by $\mathfrak{D}_k$.

Lemma (b). *For each number B in $R_k(\theta)$*

$$S_k\left(\frac{B}{\Phi'(\theta)}\right) \text{ is integral.}$$

Theorem 113. *The GCD of all ideals in K which contain only numbers in $R_k(\theta)$ is $\mathfrak{F}$, where $\mathfrak{F}\mathfrak{D}_k = \Phi'(\theta)$.*

Lemma (c). *Corresponding to each prime ideal $\mathfrak{P}$ in K there is a relative ring $R_k(\theta)$, where $\mathfrak{P}$ does not divide $\mathfrak{F} = \Phi'(\theta)\mathfrak{D}_k^{-1}$.*

To see this let p be the prime ideal in k which is divisible by $\mathfrak{P}$,

$$p = \mathfrak{P}^e\mathfrak{A}_1, \text{ where } (\mathfrak{A}, \mathfrak{P}) = 1.$$

Let A be a primitive root mod $\mathfrak{P}$ such that each integer in K is congruent to a number in $R_k(A)$ modulo each power of $\mathfrak{P}$, and such that

$$A \equiv 0 \pmod{\mathfrak{A}}.$$

Finally let β be a number in k which is divisible by $\Phi'(A) = \mathfrak{D}_k\mathfrak{F}$ and assume that p^b is the highest power of p dividing β. Then an appropriate power of β, say $\alpha = \beta^h$, furnishes a decomposition into two numerical factors in k,

$$\alpha = \pi \cdot \mu, \text{ where } \pi = p^{hb}, (\mu, p) = 1.$$
$$\alpha \equiv 0 \pmod{\mathfrak{F}\mathfrak{D}_k}.$$

Then let us determine, for an arbitrarily given integer Δ in K, a number Γ in $R_k(A)$, such that

$$\Delta \equiv \Gamma \ (\mathrm{mod} \ \mathfrak{P}^{ehb}).$$

The number $B\mu A^{hb} = (\Delta - \Gamma)\mu A^{hb}$ is then divisible by $\mathfrak{D}_k\mathfrak{F} = \Phi'(A)$, since

$$\frac{B\mu A^{hb}}{\Phi'(A)} = \frac{\pi\mu}{\mathfrak{D}_k\mathfrak{F}} \ \frac{BA^{hb}}{\pi} = \frac{\alpha}{\mathfrak{D}_k\mathfrak{F}} \cdot \frac{BA^{hb}}{\mathfrak{P}^{ehb}\mathfrak{A}^{hb}} \ \text{is integral.}$$

If we apply Lemma (a), we thus obtain a representation

$$\Delta\mu A^{hb} = \text{number in } R_k(A)$$

for each Δ, from which by Theorem 113, μA^{hb} generates an ideal which is divisible by $\mathfrak{F}$. Thus, in any case, it follows that $\mathfrak{F}$ is prime to $\mathfrak{P}$.

With this Theorem 112 is also proved.

We then define the *relative discriminant of K with respect to k* to be the relative norm of the relative different of K. By Theorem 103 the discriminant ideal with respect to $k(1)$, defined in this way, is then the same as the ideal (d), where d is the discriminant of K. However we must distinquish the discriminant of a field, which is a well-defined number d, from the relative discriminant of the same field with respect to $k(1)$, which is an ideal, namely (d).

To finish the investigations about differents we finally prove the following theorem, which is tied to the general problem which we posed at the beginning of §29 for the ground field $k = k(1)$:

Theorem 114. *If a prime ideal $\mathfrak{P}$ in K divides a prime ideal $\mathfrak{p}$ in k to a higher power than the first, then $\mathfrak{P}$ is a factor of the relative different of K with respect to k. Thus there can only exist finitely many prime ideals $\mathfrak{P}$ of this kind.*

For a proof, let the decomposition of $\mathfrak{p}$ in K be

$$\mathfrak{p} = \mathfrak{P}^e\mathfrak{A}, \quad \text{where } (\mathfrak{A}, \mathfrak{P}) = 1, e \geq 2.$$

For each integer A in K we now have, by the often used properties of the binomial coefficients $\binom{p}{n}$,

$$S_k(A)^p \equiv S_k(A^p) \ (\mathrm{mod} \ p), \text{ hence also mod } \mathfrak{p}, \tag{77}$$

if p is the rational prime which is divisible by $\mathfrak{p}$. If we now choose

$$A \equiv 0 \ (\mathrm{mod} \ \mathfrak{P}^{e-1}\mathfrak{A}),$$

then since $e \geq 2$

$$A^p \equiv 0 \ (\mathrm{mod} \ \mathfrak{p}) \quad \text{and} \quad S_k(A^p) \equiv 0 \ (\mathrm{mod} \ \mathfrak{p}). \tag{78}$$

Hence, by (77), it follows from (78) that

$$S_k(A) \equiv 0 \ (\mathrm{mod} \ \mathfrak{p}) \text{ if } A \equiv 0 \ (\mathrm{mod} \ \mathfrak{P}^{e-1}\mathfrak{A}). \tag{79}$$

Now let α be a nonintegral number in k with ideal denominator $\mathfrak{p}$,

$$\alpha = \frac{\mathfrak{a}}{\mathfrak{p}}, \qquad (\mathfrak{a}, \mathfrak{p}) = 1.$$

By (79), if A runs through all numbers of $\mathfrak{P}^{e-1}\mathfrak{A}$, that is, all numbers of $\mathfrak{a}/\mathfrak{P}$, then

$$\alpha S_k(A) = S_k(\alpha A) \text{ is integral.}$$

Thus by definition $\mathfrak{P}$ divides the relative different $\mathfrak{D}_k$.

The converse of Theorem 114 is also valid but more difficult to prove. Here we only treat the special case of relative Galois fields K.

Theorem 115. *Suppose that K is identical with all relative conjugate fields with respect to k (that is, suppose that K is a relative Galois field). Then the only prime ideals of K dividing the relative different $\mathfrak{D}_k$ of K are those which divide a prime ideal of k to a power higher than the first.*

If $\mathfrak{p}$ is a prime ideal in k and $\mathfrak{P}$ is a prime divisor of $\mathfrak{p}$ whose square does not divide $\mathfrak{p}$, then the relatively conjugate prime ideals $\mathfrak{P}^{(i)}$ also divide $\mathfrak{p}$ to exactly the first power. The relative norm $\mathfrak{p}^f$ of $\mathfrak{P}$ is the product of all the $\mathfrak{P}^{(i)}$, and the latter split into sets, each consisting of f primes which coincide with one another; there are exactly m/f distinct ones among the $\mathfrak{P}^{(i)}$. Let $\mathfrak{P}^{(1)}, \ldots, \mathfrak{P}^{(f)}$ be those which are identical with $\mathfrak{P}$.

For the proof of Theorem 115 it suffices, by Theorem 112, to display a number A in K whose different is not divisible by this $\mathfrak{P}$. We choose A to be a primitive root mod $\mathfrak{P}$ which is divisible by $\mathfrak{p}\mathfrak{P}^{-1}$. Now, by the above, $\mathfrak{P}^{(f+1)}, \ldots, \mathfrak{P}^{(m)}$ are different from $\mathfrak{P}$ so

$$\frac{\mathfrak{p}}{\mathfrak{P}^{(i)}} \text{ is divisible by } \mathfrak{P} \text{ for } i = f+1, \ldots, m.$$

Consequently,

$$A^{(i)} \equiv 0 \;(\mathrm{mod}\; \mathfrak{P}) \quad \text{for } i = f+1, \ldots, m.$$

On the other hand, if

$$\Phi(x) = \prod_{r=1}^{m} (x - A^{(r)})$$

is a polynomial over k, then, by (44),

$$(\Phi(x))^{n(\mathfrak{p})} \equiv \Phi(x^{n(\mathfrak{p})}) \;(\mathrm{mod}\; \mathfrak{p});$$

consequently in any case $\Phi(x) \equiv 0 \;(\mathrm{mod}\; \mathfrak{P})$ has the roots 0, A, $A^{n(\mathfrak{p})}, \ldots,$ $A^{n(\mathfrak{p})^{f-1}}$. Since A is a primitive root mod $\mathfrak{P}$, these $f+1$ numbers are surely distinct mod $\mathfrak{P}$. Because of the decomposition of $\Phi(x)$ into factors there must occur at least $f+1$ distinct numbers mod $\mathfrak{P}$ among the numbers

$A^{(1)}, \ldots, A^{(m)}$, and since the last $m - f$ are congruent to zero mod $\mathfrak{P}$, $A^{(1)}, \ldots, A^{(f)}$ are distinct mod $\mathfrak{P}$. The relative different

$$\delta_k(A^{(1)}) = (A^{(1)} - A^{(2)}) \cdots (A^{(1)} - A^{(m)})$$

is thus not divisible by $\mathfrak{P}$ and our theorem is proved.

§39 Decomposition Laws in the Relative Fields $K(\sqrt[l]{\mu})$

We now investigate, as a most important example, the decomposition laws of prime ideals of a ground field k in a relative field K which arise by adjoining an l-th root of some number μ in k. In this case we make the

Hypothesis: The field k contains the lth root of unity $\zeta = e^{2\pi i/l}$, where l is a positive rational prime (possibly 2).

Lemma. *The numbers* $1 - \zeta^a$ *($a \not\equiv 0 \pmod{l}$) are all associated. They satisfy the ideal equation*

$$(l) = (1 - \zeta)^{l-1}. \tag{80}$$

To see this let a and a_1 be rational integers coprime to l. Then we determine a positive rational integer b such that

$$ab \equiv a_1 \pmod{l}, \quad \text{thus } \zeta^{a_1} = \zeta^{ab}.$$

Consequently,

$$\frac{1 - \zeta^{a_1}}{1 - \zeta^a} = \frac{1 - \zeta^{ab}}{1 - \zeta^a} = 1 + \zeta^a + \zeta^{2a} + \cdots + \zeta^{(b-1)a}$$

is an integer and the same follows for the inverse quotient; thus this quotient must be a unit.

Moreover the polynomial

$$1 + x + x^2 + \cdots + x^{l-1} = \frac{x^l - 1}{x - 1} = (x - \zeta)(x - \zeta^2) \cdots (x - \zeta^{l-1}),$$

evaluated at $x = 1$ yields

$$l = (1 - \zeta)(1 - \zeta^2) \cdots (1 - \zeta^{l-1}),$$

from which the ideal equation (80) follows, by what we have just done.

Moreover, from this lemma we infer the fact that the field $k(\zeta)$ has degree exactly $l - 1$. For by §30 the degree of $k(\zeta)$ is at most $\varphi(l) = l - 1$. On the other hand, the prime l becomes the $(l - 1)$st power of an ideal in $k(\zeta)$; consequently, by Theorem 81, the degree is at least $l - 1$, thus exactly $l - 1$. Moreover, $1 - \zeta$ is accordingly also a prime ideal in $k(\zeta)$.

Theorem 116. *If μ is a number in k which is not the lth power of a number in k, then the field $K(\sqrt[l]{\mu}; k)$ has the relative degree l with respect to k. The field $K(\sqrt[l]{\mu}; k)$ is identical with its relative-conjugate fields.*

The number $M = \sqrt[l]{\mu}$ (suppose that the value of the root is somehow fixed) satisfies the equation $x^l - \mu = 0$, whose roots are the l numbers

$$\zeta^a M \qquad (a = 0, 1, \ldots, l - 1).$$

In any case all relative conjugates of M must occur among them. Let these conjugates be the m ($m \leq l$) numbers $\zeta^{a_1} M, \ldots, \zeta^{a_m} M$. As a relative norm of M, their product is a number in k; accordingly M^m belongs to k, but so does $M^l = \mu$. Since l is a prime if $m < l$, then m is relatively prime to l. Consequently M itself can be represented as a product of powers of M^l and M^m and thus is a number in k, in which case the relative degree is $= 1$. Therefore the only possibilities are $m = 1$ or $m = l$; with this the theorem is proved.

From here on we assume that the relative degree of $K(\sqrt[l]{\mu}; k)$ is equal to l. The numbers $M_1 = \sqrt[l]{\mu_1}$ and $M_2 = \sqrt[l]{\mu_2}$ obviously generate the same relative field if an equation

$$\mu_1^a \mu_2^b = \alpha^l$$

holds, where α is a number in k and a and b are rational integers not divisible by l. Each number in K can be put in the form

$$A = \alpha_0 + \alpha_1 M + \cdots + \alpha_{l-1} M^{l-1}$$

in exactly one way, where the $\alpha_0, \ldots, \alpha_{l-1}$ are numbers in k. The relative conjugates of A are obtained by replacing M successively by ζM, $\zeta^2 M, \ldots, \zeta^{l-1} M$. In general sA denotes that number among the relative conjugate numbers which arises if M is replaced by ζM:

$$sA = \alpha_0 + \alpha_1(\zeta M) + \alpha_2(\zeta M)^2 + \cdots + \alpha_{l-1}(\zeta M)^{l-1}$$
$$sM = \zeta M.$$

For each rational integer n ($n \geq 1$)

$$s^1 A = sA, \qquad s^n A = s(s^{n-1} A), \quad \text{thus } s^n M = \zeta^n M$$

so that we always have

$$s^l A = s^{2l} A = \cdots = s^{ml} A = A$$

for each positive rational integer m. These l "substitutions" $s, s^2, \ldots, s^l$ then obviously form a cyclic group of order l, where s^l plays the role of the unit element. The negative powers of s are then defined as in §5:

$$s^0 A = A, \qquad s^{-1} A = s^{l-1} A, \qquad s^{-n} A = s^{n(l-1)} A \qquad (n > 0).$$

From Theorem 55 it follows immediately that *every rational equation between the numbers $A_1, A_2, \ldots$ in K with coefficients in k remains valid if $A_1, A_2, \ldots$ are replaced simultaneously by $sA_1, sA_2, \ldots$ and consequently also if $A_1, A_2, \ldots$ are replaced by $s^m A_1, s^m A_2, \ldots$*

Because of this fact, the cyclic group: $(s, s^2, \ldots, s^{l-1}, s^l)$ is called the Galois group of the field K with respect to k and K is called a *relatively cyclic field* with respect to k.

Since the relative degree l is a prime, by Theorem 54 a number A in K is either distinct from all numbers $sA, s^2A, \ldots, s^{l-1}A$ or it is equal to all these numbers.

We also use the substitution symbol s^m with ideals so that $s^m\mathfrak{A}$ is to denote, among the ideals conjugate to $\mathfrak{A}$, that ideal which arises when all numbers A in $\mathfrak{A}$ are replaced by s^mA.

Theorem 117. *Only the following possibilities exist for the behavior of a prime ideal $\mathfrak{p}$ of k under passage to K:*

$\mathfrak{p}$ *remains a prime ideal in K,*
$\mathfrak{p}$ *becomes the lth power of a prime ideal in k,*
$\mathfrak{p}$ *becomes the product of l distinct prime ideals in K.*

Let $\mathfrak{P}$ be a prime ideal in K which divides $\mathfrak{p}$. Then, by Theorem 107, the relative norm of $\mathfrak{P}$ is

$$\mathfrak{P} \cdot s\mathfrak{P} \cdots s^{l-1}\mathfrak{P} = \mathfrak{p}^{f_1},$$

where f_1 is the relative degree of $\mathfrak{P}$; thus no prime ideal other than the prime ideals $s^m\mathfrak{P}$ divides $\mathfrak{p}$. Now if $\mathfrak{P}$ is equal to one of the $s^m\mathfrak{P}$ ($m \not\equiv 0 \pmod{l}$), and consequently equal to all $s^m\mathfrak{P}$, we then have, for a rational integer a

$$\mathfrak{p} = \mathfrak{P}^a.$$

By taking relative norms it follows that $\mathfrak{p}^l = \mathfrak{p}^{f_1 a}$, $l = f_1 a$; thus $a = 1$ and $\mathfrak{p}$ remains a prime ideal in K or $a = l$ and $\mathfrak{p}$ becomes the lth power of a prime ideal $\mathfrak{P}$. On the other hand if $\mathfrak{P}$ is distinct from all relatively conjugate ideals then a decomposition

$$\mathfrak{p} = \mathfrak{P}^a \cdot (s\mathfrak{P})^{a_1} \cdots (s^{l-1}\mathfrak{P})^{a_{l-1}}$$

holds. If we apply the substitutions $s, s^2, \ldots, s^{l-1}$ to this, we obtain

$$a = a_1 = \cdots = a_{l-1},$$

and

$$\mathfrak{p} = (\mathfrak{P} \cdot s\mathfrak{P} \cdots s^{l-1}\mathfrak{P})^a = \mathfrak{p}^{f_1 a}$$
$$1 = f_1 a, \qquad a = f_1 = 1.$$

In this case $\mathfrak{p}$ is the product of l distinct conjugate ideals $\mathfrak{P}, \ldots, s^{l-1}\mathfrak{P}$, which are all of relative degree 1.

Theorem 118. *Suppose that the prime ideal $\mathfrak{p}$ divides the number μ exactly to the power $\mathfrak{p}^a$. Then, if a is not divisible by l, $\mathfrak{p}$ becomes the lth power of a prime ideal in K: $\mathfrak{p} = \mathfrak{P}^l$. However if $a = 0$ and $\mathfrak{p}$ does not divide l, then $\mathfrak{p}$ becomes the product of l distinct prime ideals in K provided the congruence*

$$\mu \equiv \xi^l \pmod{\mathfrak{p}}$$

can be solved by an integer ξ in k. On the other hand $\mathfrak{p}$ remains a prime ideal in K if this congruence cannot be solved.

PROOF. I. If a is not divisible by l, then we may assume $a = 1$. To see this we choose an integer β in k which is divisible by $\mathfrak{p}$ but not by $\mathfrak{p}^2$. Then since $(a, l) = 1$, we can choose the rational integers x, y in such a way that $\mu^* = \mu^x \beta^{ly}$ is divisible by $\mathfrak{p}$ but not by $\mathfrak{p}^2$, while $\sqrt[l]{\mu^*}$ generates the same relative field as $\sqrt[l]{\mu}$. Thus the new exponent for this μ^* is $a = 1$, which we thus wish to assume already for μ. Taking the lth power of the ideal

$$\mathfrak{P} = (\mathfrak{p}, \sqrt[l]{\mu}),$$

we obtain $\mathfrak{P}^l = (\mathfrak{p}^l, \mu) = \mathfrak{p}$. Thus, by Theorem 108, $\mathfrak{P}$ is a prime ideal in K.

II: Suppose that a is divisible by l. Then we again wish to replace μ by some $\mu^* = \mu\beta^{-a} = \mu(\beta^{-a/l})^l$ which generates the same field $K = K(\sqrt[l]{u^*})$ and where the corresponding exponent $a = 0$.

II(1): Suppose that $\mathfrak{p}$ divides neither l nor μ, and that there exists a ξ in k such that

$$\mu \equiv \xi^l \pmod{\mathfrak{p}}.$$

Accordingly $\mathfrak{p}$ divides the product

$$\mu - \xi^l = (M - \xi)(sM - \xi) \cdots (s^{l-1}M - \xi). \tag{81}$$

However it divides no factor, since, then, as an ideal in k it would have to divide all (relatively conjugate) factors, thus also the difference of two factors, that is,

$$\mathfrak{p} \mid \zeta^a M - \zeta^b M, \qquad \mathfrak{p} \mid (\zeta^a - \zeta^b)M.$$

However, since $\mathfrak{p}$ is relatively prime to M, it would have to divide $\zeta^a - \zeta^b$, that is, by the lemma it would have to divide l, contrary to hypothesis. Thus $\mathfrak{p}$ is not a prime ideal in K and

$$\mathfrak{P} = (\mathfrak{p}, M - \xi)$$

is a factor of $\mathfrak{p}$ which is distinct from 1 and which is distinct from its relative conjugates. Obviously the relative norm is $\mathfrak{p}$.

II(2): Suppose that $\mathfrak{p}$ divides neither l nor μ and that $\mathfrak{p}$ splits into l distinct factors in K,

$$\mathfrak{p} = \mathfrak{P} \cdot s\mathfrak{P} \cdots s^{l-1}\mathfrak{P}.$$

Then $\mathfrak{P}$ has relative degree 1. By Theorem 110, each number in K is thus congruent to a number in k modulo $\mathfrak{P}$, hence there is a ξ such that

$$M \equiv \xi \pmod{\mathfrak{P}};$$

the relative norm of $M - \xi$, that is, $\mu - \xi^l$, is hence divisible by the relative norm of $\mathfrak{P}$, that is,

$$\mu \equiv \xi^l \pmod{\mathfrak{p}}.$$

Thus Theorem 118 is proved.

So just as the decomposition of a prime p in the quadratic field $K(\sqrt{d})$ is connected with quadratic residues in $k(1)$, we see in general the connection between the decomposition of p under passage to $K(\sqrt[l]{\mu}; k)$ with the lth power residues in the field k. The decomposition of factors of l is given by the following theorem:

Theorem 119. *Let* $\mathfrak{l}$ *be a prime factor of* $1 - \zeta$, *which divides* $1 - \zeta$ *exactly to the* ath *power:* $1 - \zeta = \mathfrak{l}^a \mathfrak{l}_1$. *Suppose that* $\mathfrak{l}$ *does not divide* μ. *Then* $\mathfrak{l}$ *splits into* l *factors which are distinct from one another in* $K(\sqrt[l]{\mu}; k)$ *if the congruence*

$$\mu = \xi^l \pmod{\mathfrak{l}^{al+1}} \tag{82}$$

can be solved by a number ξ *in* k. *The ideal* $\mathfrak{l}$ *remains prime in* K *if the congruence*

$$\mu \equiv \xi^l \pmod{\mathfrak{l}^{al}} \tag{83}$$

can indeed be solved, but (82) *cannot be solved. Finally,* $\mathfrak{l}$ *becomes the* lth *power of a prime ideal in* K, *if the congruence* (83) *is also unsolvable.*

I: The solvability of (82) is identical with the decomposition of $\mathfrak{l}$ into distinct factors in K. Namely from $\mathfrak{l} = \mathfrak{L} \cdot s\mathfrak{L} \cdots s^{l-1}\mathfrak{L}$, where the conjugates are distinct from one another, it follows as in the proof of Theorem 110 that every integer in K is congruent to an integer in k modulo every power of $\mathfrak{L}$. Thus to each rational integer b there corresponds ξ in k such that

$$M - \xi \equiv 0 \, (\mathfrak{L}^b);$$

consequently the relative norm of this number $M - \xi$ is divisible by $N_k(\mathfrak{L})^b = \mathfrak{l}^b$ and thus $\mu \equiv \xi^l \pmod{\mathfrak{l}^b}$ is solvable for ξ. Conversely suppose that $\mu \equiv \xi^l \pmod{\mathfrak{l}^{al+1}}$. Let ρ be a nonintegral number in k which can be represented as a quotient

$$\rho = \frac{\mathfrak{r}}{\mathfrak{l}^a}, \qquad (\mathfrak{r}, \mathfrak{l}) = 1,$$

with an integral number ideal $\mathfrak{r}$ which is relatively prime to $\mathfrak{l}$. Then the number $A = \rho(M - \xi)$ is an integer for it is a root of the polynomial

$$f(x) = (x + \rho\xi)^l - \rho^l \mu$$

$$= x^l + \binom{l}{1}\rho\xi x^{l-1} + \binom{l}{2}\rho^2\xi^2 x^{l-2} + \cdots + \binom{l}{l-1}\rho^{l-1}\xi^{l-1}x + \rho^l(\xi^l - \mu).$$

The binomial coefficients are divisible by l, hence by assumption and by (80) they are divisible by $\mathfrak{l}^{a(l-1)}$, so that $\rho^{l-1}\mathfrak{l}^{a(l-1)}$ is an integer, and the constant term is an integer by (82). If we set $\mathfrak{L} = (\mathfrak{l}, A)$, then this ideal is not 1, as $N_k(A) = \rho^l(\xi^l - \mu)$ is divisible by $\mathfrak{l}$. Furthermore $\mathfrak{L}$ is coprime to all conjugates, since the number

$$A - sA = \rho M(1 - \zeta),$$

which is obviously relatively prime to $\mathfrak{l}$, is contained in $(\mathfrak{L}, s\mathfrak{L})$. Hence $\mathfrak{l}$ contains a factor in K which is distinct from all conjugates; thus, by Theorem 117, it splits into l factors which are distinct from each other.

II: If $\mu = \xi^l(\mathfrak{l}^{al})$ is solvable, then we see in the same way that $A = \rho(M - \xi)$ is an integer in K whose relative different is relatively prime to $\mathfrak{l}$. Consequently, by Theorem 114, $\mathfrak{l}$ cannot be the lth power of a prime ideal in K, and hence, if (82) is unsolvable, $\mathfrak{l}$ does not split into l distinct factors, by I. Thus, by Theorem 117, $\mathfrak{l}$ must also be a prime ideal in K.

III: Suppose that $\mu \equiv \xi^l(\mathfrak{l}^{al})$ is unsolvable and let u be the highest exponent for which $\mu \equiv \xi^l \pmod{\mathfrak{l}^u}$ is solvable. In any case $u \geq 1$ since, by Fermat's theorem, every number is congruent to an lth power modulo $\mathfrak{l}$. Moreover u is not divisible by l. For if

$$\mu \equiv \xi^l \pmod{\mathfrak{l}^{bl}}, \qquad 0 < b \leq a - 1,$$

can be solved, then this congruence can also be solved modulo $\mathfrak{l}^{bl+1}$ since if λ is an integer in k such that

$$\lambda \text{ is divisible by } \mathfrak{l}^b \text{ but not by } \mathfrak{l}^{b+1},$$

then for every integer ω

$$(\xi + \lambda\omega)^l \equiv \xi^l + \lambda^l\omega^l \pmod{\mathfrak{l}^{bl+1}}$$

provided $b \leq a - 1$. But since ω^l represents every residue class modulo $\mathfrak{l}$, ω can be determined in such a way that

$$\mu - (\xi + \lambda\omega)^l \equiv 0 \ (\mathfrak{l}^{bl+1}),$$

from which it follows, since $u < al$, that u is not divisible by l. Let $u = bl + v$ $(0 < v \leq l - 1)$, and $u < al$, and let

$$\rho \text{ be a number with ideal denominator } \mathfrak{l}^b.$$

Then we see as above that $A = \rho(M - \xi)$ is an integer which is not divisible by $\mathfrak{l}$, if $\mu \equiv \xi^l \pmod{\mathfrak{l}^u}$, but $N_k(A)$ is divisible by $\mathfrak{l}^v$. Hence $\mathfrak{L} = (\mathfrak{l}, A)$ is an ideal in K which is different from $\mathfrak{l}$ and from (1). Thus $\mathfrak{l}$ is not a prime ideal in K and since case I does not hold, then by Theorem 117, $\mathfrak{l}$ can only be the lth power of a prime ideal in K.

Moreover, we obtain immediately from Theorem 118 and 119

Theorem 120. *The relative discriminant of $K(\sqrt[l]{\mu}; k)$ with respect to k is equal to 1 if and only if μ is the lth power of an ideal in k, and at the same time, provided μ is chosen relatively prime to l, the congruence $\mu \equiv \xi^l (\mathrm{mod}(1 - \zeta)^l)$ can be solved by a number ξ in k.*

As was mentioned previously, the discriminant of a field can never be equal to ± 1. Now it is a fundamental fact for all of arithmetic that the relative discriminant with respect to number fields other than $k(1)$ can very easily be equal to 1. This development originates with *Kronecker*. Hilbert recognized the significance of these fields for general arithmetic and based the theory of the higher reciprocity laws on them. There is, for example, the theorem[6] that a field $K(\sqrt[l]{\mu}); k)$ with relative discriminant 1 exists if and only if the number of ideal classes[7] in k is divisible by l. Such a relative field is called a *Hilbert class field* of k.

[6] For these problems compare §54–58 in Hilbert's Zahlbericht as well as Hilbert's basic paper Über die Theorie der relativ Abelschen Zahlkörper, *Acta mathematica*, Vol. 26 (1902) and *Göttinger Nachrichten* 1898. The contributions of Hilbert have been continued and partly brought to a conclusion by Furtwängler in a long sequence of papers (the two most important are: Allgemeiner Existenzbeweis für den Klassenkörper eines beliebigen algebraischen Zahlkörpers *Math. Ann.* Vol. 63 (1906) and Die Reziprozitätsgesetze für Potenzreste mit Primzahlexponenten in algebraischen Zahlkörpern I,II,III, *Math. Ann.* Vol. 67, 72, 74 (1909 through 1913).

[7] In the case $l = 2$, the foundations must be laid for a more narrow class concept. (Compare with the last section of this book.)

CHAPTER VI

Introduction of Transcendental Methods into the Arithmetic of Number Fields

§40 The Density of the Ideals in a Class

In 1840, *Dirichlet*, in his pioneering paper "Recherches sur diverses applications de l'analyse infinitésimale à la théorie des nombres" (*Crelles Journal*, Vol. 19. *Werke* Vol. 1 p. 411), showed how the powerful methods of the analysis of continuous variables can be used in the solution of purely arithmetic problems. These methods have become of great significance for the arithmetic of number fields. Even today the problem of the class number and the problem of the distribution of prime ideals are still only approachable by these transcendental methods, and at this time they still completely evade a purely arithmetic treatment.

In this chapter we discuss the two problems mentioned and their solutions by Dirichlet's methods.

The basic fact which Dirichlet discovered[1] is that one may speak of a "density" of ideals in a fixed class of ideals of a field K, and that this density is the same for all classes of ideals of K. Indeed, to be more precise the following theorem holds:

Theorem 121. *Let A be an arbitrary class of ideals of K, and let $Z(t; A)$ denote the number of integral ideals in the class A whose norm is $\leq t$. Then the limit*

$$\lim_{t \to \infty} \frac{Z(t; A)}{t} = \varkappa$$

---

[1] *Dirichlet* developed his results only for quadratic fields and not for the ideals discussed here but in the context of quadratic forms (compare §53). The considerations were carried over to general algebraic number fields by *Dedekind*.

139

exists, and is given by the formula

$$\varkappa = \frac{2^{r_1 + r_2} \pi^{r_2} R}{w \, |\sqrt{d}|}$$

which is independent of A and is determined by the field alone.

(The notation is that of §§34, 35).
Proof: Let $\mathfrak{a}$ be an integral ideal of the class A^{-1}, reciprocal to A, so that each ideal of A becomes a principal ideal by multiplication by $\mathfrak{a}$. Accordingly, for each integral ideal $\mathfrak{b}$ in A there exists a single principal ideal (ω) which is divisible by $\mathfrak{b}$ such that

$$\mathfrak{a}\mathfrak{b} = \omega.$$

Consequently $Z(t; A)$ is equal to the number of nonassociated integral numbers ω of the field which are divisible by $\mathfrak{a}$ and whose norm is $\leq t \cdot N(\mathfrak{a})$ in absolute value.

We now attempt to extract a single element from each system of associated numbers by means of inequalities. For this purpose let $\varepsilon_1, \varepsilon_2, \ldots, \varepsilon_r$ be a system of r basic units as in §35. To each number ω of the field different from 0, there is a uniquely determined system of real numbers $c_1, c_2, \ldots, c_r$ such that for the first r conjugates we have:

$$\log \frac{\omega^{(p)}}{\sqrt[n]{N(\omega)}} = c_1 \log|\varepsilon_1^{(p)}| + \cdots + c_r \log|\varepsilon_r^{(p)}| \qquad (p = 1, 2, \ldots, r). \quad (84)$$

Let us call the c_i the exponents of ω. Again $e_p = 1$ if $K^{(p)}$ is real, $e_p = 2$ otherwise. Then since

$$\sum_{p=1}^{r+1} e_p \log \left| \frac{\omega^{(p)}}{\sqrt[n]{N(\omega)}} \right| = 0 \quad \text{and} \quad \sum_{p=1}^{r+1} e_p \log|\varepsilon_k^{(p)}| = 0,$$

Equation (84) also holds for $p = r + 1$ and consequently for all conjugates. Now since by Theorem 100 each unit has the form

$$\zeta \varepsilon_1^{m_1} \varepsilon_2^{m_2} \cdots \varepsilon_r^{m_r},$$

where ζ is one of the existing roots of unity in the field K, while the m_i are rational integers, then the system of associated ω has the exponents

$$c_1 + m_1, \quad c_2 + m_2, \ldots, \quad c_r + m_r.$$

Consequently, to each ω there is an associated number whose exponents satisfy the conditions

$$0 \leq c_i < 1 \qquad (i = 1, 2, \ldots, r).$$

Furthermore among the elements associated to ω there are exactly w distinct elements of this kind. From this it follows that $w \cdot Z(t; A)$ is equal to the number of those integral elements of the field which are divisible by $\mathfrak{a}$ and which satisfy the conditions

$$\left|N(\omega)\right| = \left|\omega^{(1)} \cdot \omega^{(2)} \cdots \omega^{(n)}\right| \le N(\mathfrak{a})t \tag{85}$$

$$\log\left|\frac{\omega^{(p)}}{\sqrt[n]{N(\omega)}}\right| = \sum_{q=1}^{r} c_q \log|\varepsilon_q^{(p)}|;$$

$$0 \le c_q < 1 \qquad (p = 1, \ldots, n). \tag{86}$$

However, in order that ω be divisible by $\mathfrak{a}$ it is necessary and sufficient that

$$\omega^{(p)} = \sum_{k=1}^{n} x_k \alpha_k^{(p)} \qquad (p = 1, 2, \ldots, n) \tag{87}$$

with rational integers $x_1, \ldots, x_n$, where $\alpha_1, \ldots, \alpha_n$ denotes a definite basis of the ideal $\mathfrak{a}$. Consequently, $w \cdot Z(t; A)$ is the number of rational integers $x_1, \ldots, x_n$ which satisfy the three conditions (85), (86), (87), where not all $x_i = 0$.

If we now choose arbitrary real values for the x_i, then to the corresponding $\omega^{(p)}$ there is associated a uniquely determined real number c_q, by Equation (86), provided all $\omega^{(p)} \ne 0$. Now let $x_1, \ldots, x_n$ be the Cartesian rectangular coordinates of a point in n-dimensional space and, to begin with, consider only those points which do not lie on one of the manifolds $\omega^{(p)} = 0$ of lower dimension. Then Inequalities (85) and (86) obviously define, in the complementary space, a domain B_t lying entirely in finite space; for we have

$$\left|\omega^{(p)}\right| = \left|\sqrt[n]{N(\omega)}\right|e^{\sum_{q=1}^{r} c_q \log|\varepsilon_q^{(p)}|} \le \sqrt[n]{tN(\mathfrak{a})}e^{rM}, \qquad (p = 1, 2, \ldots, n)$$

where M denotes the absolute value of the numerically largest of the values $\log|\varepsilon_q^{(p)}|$. We now complete the domain B_t to a closed domain B_t^* which likewise still lies in finite space, by adding on to B_t those finitely many parts of the linear manifold $\omega^{(p)} = 0$ which moreover satisfy the conditions

$$\left|\omega^{(p)}\right| \le e^{rM}\sqrt[n]{tN(\mathfrak{a})}; \qquad (p = 1, 2, \ldots, n)$$

and at least one $\omega^{(p)} = 0$. The number of lattice points $x_1, \ldots, x_n$ (that is, the points with rational integer coordinates) which belong to this closed domain B_t^*, is the number $w \cdot Z(t; A)$ increased by 1 (corresponding to the origin). However, the number of lattice points is asymptotically equal to the volume of this domain. To see this we set $x_k = y_k\sqrt[n]{t}$ and then the domain B_t^* in the x-space goes over into the domain B_1^* in y-space. The lattice points x correspond to those points y whose coordinates have the form

$$\frac{\text{rational integer}}{\sqrt[n]{t}};$$

thus it is the y-space covered with a net of cubes with length of edges $1/\sqrt[n]{t}$, and by the definition of volume, or of the multiple integral, we thus have

$$\lim_{t \to \infty} \frac{w \cdot Z(t; A)}{t} = \int \cdots_{(B_1^*)} \int dy_1 \cdots dy_n = J.$$

At the same time B_1^* is that domain which is described by the following inequalities. Let us set

$$\omega^{(p)} = \sum_{k=1}^{n} y_k \alpha_k^{(p)} \qquad (p = 1, 2, \ldots, n)$$

and now we have

$$0 < |\omega^{(1)} \cdot \omega^{(2)} \cdots \omega^{(n)}| \le N(\mathfrak{a})$$

$$\log\left|\frac{\omega^{(p)}}{\sqrt[n]{N(\omega)}}\right| = \sum_{q=1}^{r} c_q \log|\varepsilon_q^{(p)}|$$

with $0 \le c_q < 1$ $(p, q = 1, 2, \ldots, r)$ or

$$|\omega^{(p)}| \le e^{rM} \sqrt[n]{N(\mathfrak{a})} \text{ and at least one } \omega^{(p)} = 0.$$

Since this last condition defines only manifolds of lower dimension, this part of the domain makes no contribution to the n-fold integral and these conditions can be omitted.

To evaluate the integral J we introduce the real and imaginary parts of the $\omega^{(p)}$ as new variables in place of the y's.

We set

$$z_p = \omega^{(p)} \quad \text{for } p = 1, 2, \ldots, r_1,$$
$$z_p + i z_{p+r_2} = \omega^{(p)} \quad \text{for } p = r_1 + 1, \ldots, r_1 + r_2,$$

so that the functional determinant (as in Theorem 95)

$$\left|\frac{\partial(z_1, \ldots, z_n)}{\partial(y_1, \ldots, y_n)}\right| = 2^{-r_2} N(\mathfrak{a})|\sqrt{d}|.$$

If we then introduce polar coordinates for z_p and z_{p+r_2}:

$$z_p = \rho_p \cos \varphi_{p-r_1} \quad (\rho_p > 0, 0 \le \varphi_{p-r_1} < 2\pi, p = r_1 + 1, \ldots, r_1 + r_2),$$
$$z_{p+r_2} = \rho_p \sin \varphi_{p-r_1}$$

and if we set

$$z_p = \rho_p, \qquad p = 1, 2, \ldots, r_1,$$

for the sake of symmetry, then

$$\left|\frac{\partial(z_1, \ldots, z_n)}{\partial(\rho_1, \ldots, \rho_{r+1}, \varphi_1, \varphi_2, \ldots, \varphi_{r_2})}\right| = \rho_{r_1+1} \cdots \rho_{r_1+r_2}$$

and the domain B_1^* is described in the new variables by

$$0 < \prod_{p=1}^{r+1} |\rho_p|^{e_p} \le N(\mathfrak{a})$$

$$\log|\rho_p| = \frac{1}{n} \log \prod_{k=1}^{r+1} |\rho_k|^{e_k} + \sum_{q=1}^{r} c_q \log|\varepsilon_q^{(p)}|, \qquad 0 \le C_q < 1$$

$$\rho_p > 0 \quad \text{and} \quad 0 \le \varphi_{p-r_1} < 2\pi \quad \text{for } p = r_1 + 1, \ldots, r_1 + r_2.$$

The integration with respect to the φ_i can be carried out; besides we need only integrate over the portion of the domain with $\rho_1 > 0, \ldots, \rho_{r_1} > 0$, if we put the factor 2^{r_1} before the integral. Thus we obtain

$$J = \frac{2^{r_1+r_2}\pi^{r_2}}{N(\mathfrak{a})|\sqrt{d}|} \int \cdots \int \rho_{r_1+1} \cdot \rho_{r_1+2} \cdots \rho_{r_1+r_2} \, d\rho_1 \, d\rho_2 \cdots d\rho_{r+1}$$

$$= \frac{2^{r_1+r_2}\pi^{r_2}}{N(\mathfrak{a})|\sqrt{d}|} \int \cdots \int dv_1 \, dv_2 \cdots dv_{r+1}$$

if we introduce $v = \rho_k^{e_k}$. Then the conditions for the v_i read:

$$0 < v_1 \cdot v_2 \cdots v_{r+1} \leq N(\mathfrak{a}), \qquad v_p > 0$$

$$\log v_p = \frac{e_p}{n} \log(v_1 \cdots v_{r+1}) + e_p \sum_{q=1}^{r} c_q \log|\varepsilon_q^{(p)}|, \qquad 0 \leq c_q < 1.$$

Finally we introduce the $c_1, \ldots, c_r$ as new variables in place of the v_i and set

$$u = v_1 \cdot v_2 \cdots v_{r+1},$$

and we thus obtain

$$\log v_p = \frac{e_p}{n} \log u + e_p \sum_{q=1}^{r} c_p \log|\varepsilon_q^{(p)}|,$$

$$\frac{\partial(v_1, \ldots, v_{r+1})}{\partial(u, c_1, \ldots, c_r)} = \frac{v_1 \cdot v_2 \cdots v_{r+1}}{u} \cdot \begin{vmatrix} e_1 \log|\varepsilon_1^{(1)}| & \cdots & e_1 \log|\varepsilon_r^{(1)}| \\ \vdots & & \vdots \\ e_r \log|\varepsilon_1^{(r)}| & \cdots & e_r \log|\varepsilon_r^{(r)}| \end{vmatrix} = \pm R.$$

Finally we obtain

$$J = \frac{2^{r_1+r_2}\pi^{r_2}}{N(\mathfrak{a})|\sqrt{d}|} R \int_{u=0}^{N(\mathfrak{a})} du \int_{0 \leq c_q < 1} \cdots \int dc_1 \, dc_2 \cdots dc_r = \frac{2^{r_1+r_2}\pi^{r_2}R}{|\sqrt{d}|},$$

and with this Theorem 121 is proved.

§41 The Density of Ideals and the Class Number

If we apply the limit equation just proved for each individual ideal class and then sum over all classes, we obtain at once the connection, found by Dirichlet and Dedekind, between the density of integral ideals of the field and its class number, namely

Theorem 122. *Let $Z(t)$ denote the number of integral ideals of the field whose norm is $\leq t$. Then*

$$\lim_{t \to \infty} \frac{Z(t)}{t} = h\varkappa, \tag{88}$$

where h is the class number of the field.

The number $Z(t)$, in whose definition the concept of class does not occur anymore, can now be calculated by another method, namely with the help of our knowledge of the decomposition of rational primes in the field. In this way the class number is connected with the decomposition laws, and thereby in certain cases a remarkably simple expression for the class number can be derived, of which no other way has led until now.

If $F(m)$ denotes the number of integral ideals of the field whose norm is equal to the positive number m, then obviously

$$Z(t) = \sum_{m=1}^{t} F(m).$$

Here $\sum_{m=1}^{t}$ means that the summation index m runs through all rational integers for which $1 \leq m \leq t$. Now, moreover, for two rational integers a, b

$$F(ab) = F(a) \cdot F(b) \quad \text{if } (a,b) = 1. \tag{89}$$

For, from two integral ideals $\mathfrak{a}$ and $\mathfrak{b}$ with $N(\mathfrak{a}) = a$, $N(\mathfrak{b}) = b$, an ideal $\mathfrak{c} = \mathfrak{a}\mathfrak{b}$ arises with $N(\mathfrak{c}) = ab$. And conversely if $\mathfrak{c}$ is an integral ideal with norm ab, let us set

$$(\mathfrak{c}, a) = \mathfrak{a}_1, \qquad (\mathfrak{c}, b) = \mathfrak{b}_1; \tag{90}$$

from this it follows by multiplication that

$$\mathfrak{a}_1 \mathfrak{b}_1 = (\mathfrak{c}^2, \mathfrak{c}a, \mathfrak{c}b, ab) = \mathfrak{c}\left(\mathfrak{c}, a, b, \frac{ab}{\mathfrak{c}}\right) = \mathfrak{c}.$$

By passage to the conjugate, we obtain from (90) that $N(\mathfrak{a}_1)$, as a divisor of a^n, is thus coprime to b and $N(\mathfrak{b}_1)$ is coprime to a, while the product $N(\mathfrak{a}_1) \cdot N(\mathfrak{b}_1) = ab$. Consequently, $N(\mathfrak{a}_1) = a$, $N(\mathfrak{b}_1) = b$, and $\mathfrak{c}$ is thus decomposed into two factors whose norms are a and b respectively. The assertion (89) follows from this.

Generally, by use of this formula, the calculation of $F(m)$ can be reduced to the calculation of $F(p^k)$ where p^k is a power of a prime.

The calculations for determining $F(p^k)$, and with this $F(m)$, are now simplified considerably by the introduction of a new function, by which the limit process (88) is transformed into a limit process which is more conveniently accessible to calculation. This function is the *zeta-function of Dirichlet-Dedekind*.

§42 The Dedekind Zeta-Function

By a Dirichlet series, we mean a series of the form

$$\sum_{n=1}^{\infty} \frac{a_n}{n^s};$$

where $a_1, a_2, \ldots$ is a given sequence of numbers, s is a variable which assumes only real values in the following discussion, and the symbol n^s denotes the positive value of the power. The a_n are called the coefficients of the series. In case the series converges, it represents a function of s.

Lemma (a). *The series* $\sum_{n=1}^{\infty} 1/n^s$ *converges for* $s > 1$, *and represents a continuous function of* s, *the so-called Riemann zeta-function* $\zeta(s)$. *Moreover*

$$\lim_{s \to 1} (s - 1)\zeta(s) = 1.$$

It follows from the definition of the definite integral that

$$\int_n^{n+1} \frac{dx}{x^s} < \frac{1}{n^s} < \int_{n-1}^n \frac{dx}{x^s} \qquad (n > 1).$$

Hence the series converges for $s > 1$; consequently, as a series with only positive continuous terms it represents a continuous function $\zeta(s)$, and

$$\int_1^{\infty} \frac{dx}{x^s} < \zeta(s) < \int_1^{\infty} \frac{dx}{x^s} + 1,$$

$$1 < (s - 1)\zeta(s) < s$$

from which the limit relation follows.

Lemma (b). *Let us set*

$$S(m) = a_1 + a_2 + \cdots + a_m; \text{ hence } a_n = S(n) - S(n - 1).$$

If there exists a number σ $(\sigma > 0)$ *for which the quotient*

$$\left| \frac{S(m)}{m^\sigma} \right| < A, \qquad (m = 1, 2, \ldots) \tag{91}$$

where A is a constant independent of m, then the series $\sum_{n=1}^{\infty} a_n/n^s$ *converges for* $s > \sigma$ *and represents a continuous function of* s.

Namely, for all positive integers m and h

$$\sum_{n=m}^{m+h} \frac{a_n}{n^s} = \sum_{n=m}^{m+h} \frac{S(n) - S(n - 1)}{n^s}$$

$$= \frac{S(m + h)}{(m + h)^s} - \frac{S(m - 1)}{m^s} + \sum_{n=m}^{m+h-1} S(n) \left(\frac{1}{n^s} - \frac{1}{(n + 1)^s} \right).$$

Since

$$\frac{1}{n^s} - \frac{1}{(n + 1)^s} = s \int_n^{n+1} \frac{dx}{x^{s+1}},$$

it thus follows, if we keep (91) in mind, that for $s > \sigma$

$$\left| \sum_{n=m}^{m+h} \frac{a_n}{n^s} \right| < \frac{2A}{m^{s-\sigma}} + As \int_m^{\infty} \frac{dx}{x^{s-\sigma+1}} = \frac{2A}{m^{s-\sigma}} + \frac{As}{s-\sigma} \frac{1}{m^{s-\sigma}}.$$

Consequently, the series converges for $s > \sigma$, and indeed uniformly in each interval $\sigma + \delta \leq s \leq \sigma + \delta'$ (where $\delta' > \delta > 0$); thus it represents a continuous function of s there.

Lemma (c). *If in the above notation*

$$\lim_{m \to \infty} \frac{S(m)}{m} = c,$$

then, if s approaches 1 (from $s > 1$),

$$\lim_{s \to 1} (s - 1) \sum_{n=1}^{\infty} \frac{a_n}{n^s} = c.$$

By (b) the series converges for $s > 1$. If we set

$$S(n) = cn + \varepsilon_n n,$$

where $\lim_{n \to \infty} \varepsilon_n = 0$ by hypothesis, and

$$\varphi(s) = \sum_{n=1}^{\infty} \frac{a_n}{n^s},$$

then, as above, it follows that for $s > 1$

$$|\varphi(s) - c\zeta(s)| = s \left| \sum_{n=1}^{\infty} n\varepsilon_n \int_n^{n+1} \frac{dx}{x^{s+1}} \right| < s \sum_{n=1}^{\infty} |\varepsilon_n| \int_n^{n+1} \frac{dx}{x^s}.$$

For an arbitrary $\delta > 0$ we now choose an integer N such that $|\varepsilon_n| < \delta$ for $n \geq N$, and we choose C in such a way that $|\varepsilon_n| < C$ for all n. It then follows that

$$|(s-1)\varphi(s) - c(s-1)\zeta(s)| < Cs(s-1) \sum_{n=1}^{N-1} \int_n^{n+1} \frac{dx}{x} + \delta s(s-1) \sum_{N}^{\infty} \int_n^{n+1} \frac{dx}{x^s}$$

$$< Cs(s-1) \log N + \delta s(s-1) \cdot \int_N^{\infty} \frac{dx}{x^s}.$$

Since the last expression tends to δ as s tends to 1, we have

$$\lim_{s \to 1} \{(s-1)\varphi(s) - c(s-1)\zeta(s)\} = 0$$

and, keeping (a) in mind, our Lemma (c) is proved.

We now assign to each algebraic number field k a function of a continuous variable s, the so-called *zeta-function of k*, namely

$$\zeta_k(s) = \sum_{\mathfrak{a}} \frac{1}{N(\mathfrak{a})^s}, \tag{92}$$

which Dirichlet introduced for quadratic fields and which Dedekind introduced for arbitrary k. Here $\mathfrak{a}$ is to run once through all ideals of k which are different from zero. If we use the notation $F(n)$ from the preceding section, then the series can also be written as

$$\zeta_k(s) = \sum_{n=1}^{\infty} \frac{F(n)}{n^s},$$

and from Theorem 122 and Lemmas (b) and (c) we obtain

Theorem 123. $\zeta_k(s)$ *is defined by the convergent series* (92) *for* $s > 1$ *as a continuous function of* s *and as* s *approaches* 1

$$\lim_{s \to 1} (s - 1)\zeta_k(s) = h\varkappa.$$

From this formula we now have a chance of calculating h if we express $\zeta_k(s)$ in an essentially different form with the help of the prime ideals of k.

Theorem 124. *For* $s > 1$, *the equation*

$$\zeta_k(s) = \prod_{\mathfrak{p}} \frac{1}{1 - \dfrac{1}{N(\mathfrak{p})^s}}. \tag{93}$$

holds where $\mathfrak{p}$ *runs through all distinct prime ideals* $\mathfrak{p}$ *of* k.

To begin with, this product converges, since $\sum_{\mathfrak{p}} 1/N(\mathfrak{p})^s$ converges as the constituent of the series for $\zeta_k(s)$. For a single factor we obtain a convergent series of positive terms

$$\frac{1}{1 - N(\mathfrak{p})^{-s}} = 1 + \frac{1}{N(\mathfrak{p})^s} + \frac{1}{N(\mathfrak{p}^2)^s} + \cdots. \tag{94}$$

If we multiply these expressions in a purely formal way for all $\mathfrak{p}$, then we obtain a series with terms

$$\frac{1}{N(\mathfrak{p}_1^{\alpha_1}\mathfrak{p}_2^{\alpha_2} \cdots \mathfrak{p}_r^{\alpha_r})^s},$$

where each product of powers of prime ideals appears exactly once in the norm symbol. However, by the fundamental theorem we obtain each integral ideal of k exactly once in this form, that is, all terms of the convergent series $\zeta_k(s)$ appear exactly once. Since the series converges absolutely for $s > 1$ in each single factor and the product converges for $s > 1$, the equality of the values of the series, that is, the validity of (93), follows from the formal agreement of the terms of the series.

Theorem 125. *Following Dedekind the determination of the class number* h *is reduced to the determination of the prime ideals of the field by the equation*

$$h \cdot \varkappa = \lim_{s \to 1} (s - 1) \prod_{\mathfrak{p}} \frac{1}{1 - \dfrac{1}{N(\mathfrak{p})^s}} \tag{95}$$

This fundamental fact is only another way of writing (88), as has been mentioned already; however, it is more convenient as a starting point for the further calculation than the former equation.

While a useful expression for the class number can now be derived in those fields, where the decomposition of the rational primes p is known (compare §51, where the calculation is carried out for quadratic fields), in the reverse direction we can also derive results about the prime ideals from Theorems 123 and 124 if we make use only of the fact that in any case $h \cdot \varkappa$ is different from zero. This will be discussed in the next sections.

§43 The Distribution of Prime Ideals of Degree 1, in Particular the Rational Primes in Arithmetic Progressions

From Theorem 123, we obtain immediately: The Dedekind zeta-function $\zeta_k(s)$ becomes infinitely large to the first order, as s approaches 1, so that

$$\log \zeta_k(s) = \log \frac{1}{s-1} + g(s), \tag{96}$$

where $g(s)$ is a function which remains bounded as s tends to 1. From the product representation (93) we then have

Theorem 126. *If $\mathfrak{p}_1$ runs through the distinct prime ideals of degree one in k, then, for $s > 1$,*

$$\sum_{\mathfrak{p}_1} \frac{1}{N(\mathfrak{p}_1)^s} = \log \frac{1}{s-1} + g_1(s), \tag{97}$$

where $g_1(s)$ again remains bounded as s tends to 1. Hence there are infinitely many prime ideals of degree one in k.

Proof: Let $\mathfrak{p}_f$ run through the distinct prime ideals of degree f for $f = 1, 2, \ldots, n$. (Of course $\mathfrak{p}_f$ need not exist for each f.) Since at most n distinct prime ideals of k divide a given rational prime p then, in any case, for $s > 1$,

$$1 \leq \prod_{\mathfrak{p}_f} \frac{1}{1 - \frac{1}{N(\mathfrak{p}_f)^s}} \leq \prod_p \frac{1}{\left(1 - \frac{1}{p^{fs}}\right)^n} = \zeta(fs)^n.$$

Thus as s tends to 1, the product over $\mathfrak{p}_f$ remains between two fixed positive bounds for $f \geq 2$. The fact that $\zeta_k(s)$ becomes infinite is thus brought about by the prime ideal $\mathfrak{p}_1$ alone, and indeed by passing to the logarithm,

$$\log \zeta_k(s) = -\sum_{\mathfrak{p}_1} \log\left(1 - \frac{1}{N(\mathfrak{p}_1)^s}\right) + f(s), \tag{98}$$

where again $f(s)$ remains bounded. However, since $N(\mathfrak{p}_1) \geq 2$, we have for $s \geq 1$

$$- \log\left(1 - \frac{1}{N(\mathfrak{p}_1)^s}\right) = \frac{1}{N(\mathfrak{p}_1)^s} + \varphi(\mathfrak{p}_1, s),$$

$$0 \leq \varphi(\mathfrak{p}_1, s) = \frac{1}{2} \frac{1}{N(\mathfrak{p}_1)^{2s}} + \frac{1}{3} \frac{1}{N(\mathfrak{p}_1)^{3s}} + \cdots$$

$$< \frac{1}{N(\mathfrak{p}_1)^{2s}}\left(1 + \frac{1}{2^s} + \frac{1}{4^s} + \cdots\right) < \frac{2}{N(\mathfrak{p}_1)^{2s}},$$

and hence the sum over $\mathfrak{p}_1$ is

$$0 \leq \sum_{\mathfrak{p}_1} \varphi(\mathfrak{p}_1, s) \leq 2 \sum_{\mathfrak{p}_1} \frac{1}{N(\mathfrak{p}_1)^{2s}} \leq 2n \sum_{p} \frac{1}{p^{2s}} \leq 2n \sum_{p} \frac{1}{p^2},$$

that is, bounded for $s \geq 1$. Hence, in combination with (98) it follows that

$$\log \zeta_k(s) - \sum_{\mathfrak{p}_1} \frac{1}{N(\mathfrak{p}_1)^s}$$

remains bounded as s tends to 1, and with this, by (96), we have proved (97). Hence if s approaches 1 the sum over $\mathfrak{p}_1$ becomes arbitrarily large and thus it must consist of infinitely many terms.

This general theorem, valid for every algebraic number field, now permits us to prove very important facts of rational arithmetic, which relate to the distribution of primes.

We choose the field of mth roots of unity for the field k. By Theorem 92, the norms of the prime ideals of degree 1 are precisely the rational primes with the congruence property $p \equiv 1 \pmod{m}$ except for finitely many exceptions. Consequently, from Theorem 126 follows

Theorem 127. *There are infinitely many positive rational primes with the property $p \equiv 1 \pmod{m}$.*

If n_0 is the degree of the field of mth roots of unity (which by §30 is no larger than $\varphi(m)$), then exactly n_0 distinct prime ideals of k divide such a number p, and Equation (97) thus reads

$$n_0 \sum_{p \equiv 1(m)} \frac{1}{p^s} = \log \frac{1}{s-1} + g_1(s). \tag{99}$$

Dirichlet has shown how one can also obtain information about the existence of primes in other residue classes mod m by relatively simple formal considerations. For this purpose we introduce the residue characters modulo m, as they were defined in §15.

Theorem 128. *If $\chi(n)$ denotes a residue character of n mod m, then the Dirichlet series*

$$L(s,\chi) = \sum_{n=1}^{\infty} \frac{\chi(n)}{n^s}$$

is absolutely convergent for $s > 1$ and as long as $s > 1$ we have the product representation

$$L(s,\chi) = \prod_p \frac{1}{1 - \dfrac{\chi(p)}{p^s}}, \tag{100}$$

in which p runs through all positive rational primes. If moreover, χ is not the principal character, then the infinite series for $L(s,\chi)$ is convergent even for $s > 0$.

First of all the absolute convergence of the series and the product representation for $s > 1$ are obtained immediately from the fact that the coefficients $\chi(n)$ are not larger than 1 in absolute value, as $\chi(n)$ is either a root of unity, or, in case $(n, m) > 1$, equal to 0. Since the rule

$$\chi(ab) = \chi(a)\chi(b)$$

holds for all pairs of positive integers a, b, we now obtain

$$\frac{1}{1 - \dfrac{\chi(p)}{p^s}} = 1 + \frac{\chi(p)}{p^s} + \frac{\chi(p^2)}{p^{2s}} + \cdots$$

for each individual factor of the infinite product; by absolute convergence we thereby obtain Equation (100) from this by multiplication, in the same way as above in the proof of Theorem 124.

Finally, if χ is not the principal character χ_1 mod m, then, by the basic property of characters, $\sum_n \chi(n) = 0$, where n runs through any complete system of residues mod m. Thus if the integer $x = y \cdot m + r$, where y and r are integers and $0 \le r < m$, then

$$\left| \sum_{n=1}^{x} \chi(n) \right| = \left| \sum_{n=1}^{ym} \chi(n) + \sum_{n=0}^{r} \chi(n) \right| = \left| \sum_{n=0}^{r} \chi(n) \right| \le m$$

is thus bounded as x grows to infinity and, by Lemma (b) of the preceding paragraph, the Dirichlet series converges for $s > 0$. In particular it follows from this that if χ is not the principal character, the functions $L(s,\chi)$ are also still continuous at the point $s = 1$.

Theorem 129. *For each character χ mod m, if $s > 1$,*

$$\log L(s,\chi) = \sum_p \frac{\chi(p)}{p^s} + g(s,\chi),$$

where $g(s,\chi)$ remains bounded as s approaches 1.

If we define the log function, for $s > 0$, by the convergent series

$$\log \frac{1}{1 - \dfrac{\chi(p)}{p^s}} = \frac{\chi(p)}{p^s} + \frac{1}{2}\frac{\chi(p^2)}{p^{2s}} + \frac{1}{3}\frac{\chi(p^3)}{p^{3s}} + \cdots = \frac{\chi(p)}{p^s} + \frac{f(s,p)}{p^{2s}},$$

where obviously

$$|f(s,p)| \le 1 \quad \text{for } p \ge 2, s \ge 1,$$

then the sum of these expressions, extended over all positive primes, converges for $s > 1$, and thus this sum represents one of the infinitely many values of $\log L(s, \chi)$. Then Theorem 129 holds for this value.

Moreover, for the principal character $\chi = \chi_1$ we have, more precisely,

$$L(s, \chi_1) = \log \frac{1}{s - 1} + H(s), \qquad (101)$$

where $H(s)$ remains finite for $s \ge 1$.

For if we choose the field $k(1)$ for k in (97), then we obtain that

$$\sum_p \frac{1}{p^s} - \log \frac{1}{s - 1}$$

remains finite as $s \to 1$; on the other hand $\chi_1(p)$ is equal to 1 in general and different from 1 (i.e., equal to 0) only for the finitely many primes p which divide m. Thus (101) is, in fact, proved.

In order to go from here to sums which are extended only over the primes of a residue class mod m, let a be an arbitrary rational integer which is coprime to m and let b be a rational integer such that

$$ab \equiv 1 \ (\text{mod } m).$$

Then, as long as $s > 1$, if $\sum_\chi$ denotes a sum to be extended over all characters χ mod m, we have

$$\sum_\chi \chi(b) \log L(s, \chi) = \sum_\chi \chi(b) \sum_p \frac{\chi(p)}{p^s} + \sum_\chi \chi(b) g(s, \chi).$$

The last sum, which we denote by $f(s)$, remains finite in any case as s tends to 1. However, in the double sum

$$\sum_\chi \chi(b)\chi(p) = \sum_\chi \chi(bp) = \begin{cases} 0, & \text{if } bp \not\equiv 1 \ (\text{mod } m), \\ \varphi(m), & \text{if } bp \equiv 1 \ (\text{mod } m), \end{cases}$$

so that we obtain

$$\sum_\chi \chi(b) \log L(s, \chi) = \varphi(m) \sum_{p \equiv a \, (\text{mod } m)} \frac{1}{p^s} + f(s), \qquad (102)$$

where the sum is to be extended only over the positive primes p which are $\equiv a \ (\text{mod } m)$.

Finally, let us now allow s to tend toward the critical value 1. On the left-hand side, by (101), the term which is formed by the principal character $\chi = \chi_1$ becomes infinitely large and positive. *Thus if the remaining summands remain finite, the entire left-hand side of (102) grows beyond all bounds.* Consequently the sum on the right must contain infinitely many terms; hence there are infinitely many p which are $\equiv a \pmod{m}$.

Accordingly, the verification of the following assertion remains the essential point in Dirichlet's train of thought:

If χ is not the principal character, then the quantities $\log L(s, \chi)$ remain finite as s tends to 1.

Since these $L(s, \chi)$ are continuous functions of s, for $s > 0$, by the last part of Theorem 128, the assertion is identical with

Theorem 130. *If χ is not the principal character, then*

$$L(1, \chi) = \lim_{s \to 1} L(s, \chi) \neq 0.$$

The nonvanishing of the L-series is now an immediate consequence of the fact that $\zeta_k(s)$ becomes infinite to the first order at $s = 1$. For by (102) it follows for $a = b = 1$ that

$$\sum_{\chi} \log L(s, \chi) = \varphi(m) \sum_{p \equiv 1 \,(\text{mod } m)} \frac{1}{p^s} + G(s),$$

and, if we use (99), it follows from this that

$$\sum_{\chi} \log L(s, \chi) = \frac{\varphi(m)}{n_0} \log \frac{1}{s - 1} + G_1(s) \tag{103}$$

with $G(s)$ and $G_1(s)$ remaining finite. On the left-hand side only the term corresponding to the principal character $\chi_1(s)$ becomes infinitely large, by (101), and for the remaining part we thus obtain

$$\sum_{\chi \neq \chi_1} \log L(s, \chi) = \left(\frac{\varphi(m)}{n_0} - 1\right) \log \frac{1}{s - 1} + G_2(s),$$

$$\prod_{\chi \neq \chi_1} L(s, \chi) = \left(\frac{1}{s - 1}\right)^{(\varphi(m)/n_0) - 1} e^{G_2(s)}.$$

As has been mentioned already, we have $\varphi(m) \geq n_0$. The right-hand side now becomes infinitely large as s tends to 1, if $\varphi(m) > n_0$, while the product on the left surely remains finite, since this holds for each factor. Thus it follows *first* that $\varphi(m) = n_0$; *secondly*, then, the entire right-hand side is

$$e^{G_2(s)},$$

which as an exponential quantity surely does not tend to 0. Hence this is also the case for the left-hand side, that is, since each factor on the left has a finite limit, Theorem 130 is in fact true.

With this, as shown above, the famous result of Dirichlet is proved.

Theorem 131. *If* $(a, m) = 1$, *then there are infinitely many positive primes p, for which* $p \equiv a \pmod{m}$. *That is* $p = mx + a$ *represents a prime infinitely often for* $x = 1, 2, 3, \ldots$.

As a side result we obtain

$$\varphi(m) = n_0$$

from the proof, that is, the exact degree of the field of mth roots of unity also follows from the decomposition laws. With this it is thus proved that the algebraic equation for $\zeta = e^{2\pi i/m}$ is irreducible over the field of rational numbers.

If we go once more through the chain of conclusions which led us to the proof of Theorem 131, then the verification that $L(1, \chi) \neq 0$ appears as the most difficult point, and this verification was carried out from Equation (103) and the fact that the function $\zeta_k(s)$ becomes infinite to the first order at $s = 1$. This last fact was again based on the theorems in §40 about the density of ideals, in whose proofs the entire theory of units was required. It is now of importance that instead of these number-theoretic methods, more precise knowledge of the function-theoretic properties of the $L(s, \chi)$ can be used with the same success. Several remarks about this follow for purposes of orientation.

To begin with, it can be proved by Lemma (b), §42, that $L(s, \chi)$ is differentiable at $s = 1$ (by termwise differentiation of the series) and that hence, if $L(1, \chi) = 0$, this function would have a zero of at least the first order at $s = 1$, for then

$$\lim_{s \to 1} \frac{L(s, \chi)}{s - 1} = \lim_{s \to 1} \frac{L(s, \chi) - L(1, \chi)}{s - 1} = \left. \frac{dL(s, \chi)}{ds} \right|_{s = 1}$$

exists. On the other hand, the product of all $\varphi(m)$ series is a convergent series with only positive terms for $s > 1$. For if p is an element of order f in the group of residue classes mod m, then, by Theorem 32, the $\varphi(m)$ numbers $\chi(p)$ are all fth roots of unity, each root occurring equally often. Thus

$$\prod \left(1 - \frac{\chi(p)}{p^s} \right) = \left(1 - \frac{1}{p^{fs}} \right)^e, \qquad \left(e = \frac{\varphi(m)}{f} \right),$$

and accordingly $\prod_\chi L(s, \chi)$ is a series with positive coefficients, and indeed > 1 for all $s > 1$. Now since the series $L(s, \chi_1)$ corresponding to the principal character agrees with $\zeta(s)$, except for unimportant factors, this series becomes infinite to the first order at $s = 1$. Furthermore, since the remaining $L(s, \chi)$ either become zero to at least the first order at $s = 1$ or in any case have a finite limit, at most one single factor $L(s, \chi)$ can be equal to zero. And indeed, the χ for which this happens must then be a real character (which takes only the values ± 1, 0, that is, a quadratic character mod m). For if χ is not a real character, then the conjugate imaginary function $\bar{\chi}$ is likewise a character mod m, but different from χ, and the vanishing of $L(1, \chi)$ implies the non-vanishing of the conjugate imaginary quantity $L(1, \bar{\chi})$ which, by the above, cannot occur. Thus we need only verify that $L(1, \chi) \neq 0$ for all quadratic characters χ.

Mertens[2] has proved this assertion by a direct estimate of all the real terms of the series. In this way we obtain a proof of Dirichlet's theorem which is independent of the theory of fields.

Dirichlet himself used the quadratic reciprocity law with which it is seen that the series $L(s, \chi)$ corresponding to real characters appear as factors in the zeta-functions of certain quadratic number fields, and so, for this reason, cannot be zero at $s = 1$. In contrast to the proof given above he does not need the arithmetic of cyclotomic fields, but only that of quadratic fields.

Pure function-theoretic proofs comprise the latter group; they are capable of the farthest generalization. In these proofs the functions $L(s, \chi)$ are investigated as analytic functions of the complex variable s. It is shown that the $L(s, \chi)$ are regular analytic functions of s for all finite values of s with the exception of $L(s, \chi_1)$ which has a pole of the first order only at $s = 1$. Now if one of the L-series were zero at $s = 1$, the product of all these series would have to be a regular function of s everywhere in the finite plane. The contradiction is then obtained, with the help of a general theorem of function theory, from the fact that such a Dirichlet series with positive coefficients must have at least one singular point in the finite plane.[3]

This idea, which is the foundation of Dirichlet's method of introducing group characters, is capable of far-reaching generalization. We can start, instead of from the classification of rational numbers in $k(1)$ by residue classes mod m, with the numbers of any field, which are divided, in another way, into classes which form an Abelian group.[4] Finally Theorem 126 can also be directly applied to other fields instead of $k(e^{2\pi i/m})$, even to relative fields. Moreover, each time we obtain verification of the existence of infinitely many primes (prime ideals) of the ground field with certain properties from a knowledge of the decomposition laws. These contributions will be carried out more precisely in the next chapter (§48) for quadratic fields.

[2] Mertens Über das Nichtverschwinden Dirichletscher Reihen mit reellen Gliedern. *Sitzungsber. d. Akad. d. Wiss. in Wien. math.-naturw. Klasse,* Vol. 104 (1895).

[3] See E. Landau, Handbuch der Lehre von der Verteilung der Primzahlen (Leipzig 1909) Vol. I §121; or Hecke, Über die L-Funktionen und den Dirichletschen Primzahlsatz für einen beliebigen Zahlkörper, *Nachr. v.d. K. Ges. d. Wissensch. zu Göttingen* 1917.

[4] A general contribution in this direction is due to H. Weber, Über Zahlengruppen in algebraischen Körpern I,II,III. *Math. Ann.* 48, 49, 50, (1897–1898).

CHAPTER VII

The Quadratic Number Field

§44 Summary and the System of Ideal Classes

The quadratic number field, which was already treated as an example in §29 is to be discussed in more detail in this chapter. First we recall once more what was proved in §29.

Let D be a positive or negative rational integer, different from 1, and divisible by no rational square except 1. The number $\sqrt{D}$ then generates the most general quadratic field. Its discriminant is

$$d = \begin{cases} D, & \text{if } D \equiv 1 \pmod 4, \\ 4D, & \text{if } D \equiv 2 \text{ or } 3 \pmod 4. \end{cases}$$

In each case 1, $(d + \sqrt{d})/2$ is a basis. Each integer of the field has the form $\alpha = (x + y\sqrt{d})/2$ with rational integers x, y. An odd positive prime p splits into two distinct or equal prime factors or remains irreducible according to whether the quadratic residue symbol $(\frac{d}{p})$ has the value 1, 0, or -1.

We now define the *quadratic residue symbol with denominator* 2, but only for those numbers d which are discriminants of quadratic fields.

If d is even, let $(\frac{d}{2}) = 0$. If d is odd, let $(\frac{d}{2}) = +1$ if d is a quadratic residue mod 8, and $(\frac{d}{2}) = -1$ if d is a quadratic nonresidue mod 8.

Then the decomposition law for the number 2 in $k(\sqrt{d})$ reads exactly the same formally as the law stated above for odd p.

In a real quadratic field the number of fundamental units is equal to 1 by Theorem 100. Since the only real roots of unity are ± 1, the numbers $\pm \varepsilon^n$ ($n = 0, \pm 1, \pm 2, \ldots$), where ε is a fundamental unit, are all the units of the field; the latter is obviously uniquely determined by the additional condition $\varepsilon > 1$. All the units $\eta = (x + y\sqrt{d})/2$ are obviously obtained from the solution

155

of the equation $N(\eta) = \pm 1$, that is,

$$x^2 - dy^2 = \pm 4 \qquad (104)$$

with rational integers x and y. This is the so-called *Pell equation*.

In imaginary quadratic fields every unit η is a root of unity. For $d < 0$ the above equation (where of course the upper sign must hold) has solutions only for $d \geq -4$ except for the trivial solutions, $y = 0$, $x = \pm 2$, that is, $\eta = \pm 1$. Indeed for $d = -4$, the equation has the two additional solutions $x = 0$, $y = \pm 1$, and for $d = -3$ the equation has four additional solutions $x = \pm 1$, $y = \pm 1$. Thus the number w of roots of unity in $k(\sqrt{-3})$, the field of the third root of unity, is equal to 6 and the number of roots of unity in $k(\sqrt{-1})$ is equal to 4. In all other quadratic fields it is equal to 2.

We now try to find a method from the general theory to decide whether or not two ideals $\mathfrak{a}$, $\mathfrak{b}$ of a quadratic field are equivalent and, by this, to give a complete system of nonequivalent ideals, thus also to calculate the class number.

Since $N(\mathfrak{b}) = \mathfrak{b}\mathfrak{b}'$ is a rational principal ideal, the equivalence $\mathfrak{a} \sim \mathfrak{b}$ means the same thing as $\mathfrak{a}\mathfrak{b}' \sim 1$; thus we must decide whether a given ideal is a principal ideal. If the integral ideal $\mathfrak{a}$ is a greatest common divisor of two principal ideals (α, β), then $\mathfrak{a}$ is the content of the form $\alpha u + \beta v$. Consequently $\mathfrak{a}\mathfrak{a}' = N(\mathfrak{a})$ is the content of $(\alpha u + \beta v)(\alpha' u + \beta' v) = \alpha\alpha' u^2 + uv(\alpha'\beta + \alpha\beta') + \beta\beta' v^2$, that is, $N(\mathfrak{a})$ is the greatest common divisor of the rational numbers $\alpha\alpha'$, $\alpha'\beta + \alpha\beta'$, $\beta\beta'$. If the positive rational number n is obtained for this GCD, then the additional question is whether $\pm n$ is the norm of an integer of the field and then moreover if $N(\omega) = \pm n$ whether the equation $(\omega) = (\alpha, \beta)$ is correct. This is again the case if and only if α/ω and β/ω are integers, for in this case the ideal $(\alpha/\omega, \beta/\omega)$ is an integral ideal with norm 1 by construction, thus it is itself equal to (1).

Thus the only difficulty is in finding all different principal ideals (ω) whose norm has a given value. This leads to the problem of finding all rational integers x, y for which (if we set $\omega = (x + y\sqrt{d})/2$)

$$x^2 - dy^2 = \pm 4n. \qquad (105)$$

For imaginary quadratic fields all solutions can be obtained easily in finitely many attempts. Since $d < 0$ we need test only those pairs of rational integers x, y for which

$$|x| \leq 2\sqrt{n}, \qquad |\sqrt{d}|\,|y| \quad 2\sqrt{n},$$

that is, we determine, by calculation, for which rational integers y with $0 \leq y \leq 2|\sqrt{n/d}|$ the expression $\sqrt{4n + dy^2}$ is a rational number.

In order to find the solutions of (105) with $d > 0$, in the real quadratic field, knowledge of a unit different from ± 1 (not necessarily the fundamental unit) is required. If we assume

$$\eta = \frac{u + v\sqrt{d}}{2} \qquad (v > 0)$$

is a unit in $k(\sqrt{d})$ with $\eta > 1$, then among the numbers $\alpha = \omega\eta^n$ ($n = 0$, $\pm 1, \pm 2, \ldots$) associated to a given ω we can surely find one such that

$$1 \leq \left|\frac{\alpha}{\alpha'}\right| < \eta^2$$

(compare Equation (86)). It is thus sufficient, for our problem, to look only for those solutions $\omega = (x + y\sqrt{d})/2$ for which these inequalities are also satisfied. The inequalities can also be written in the form

$$|\omega'| \leq |\omega| < |\omega'|\eta^2 \quad \text{or} \quad |\omega|\eta^{-2} < |\omega'| \leq |\omega|$$

or, since $|\omega\omega'| = n$,

$$\sqrt{n} \leq |\omega| < \eta\sqrt{n}$$
$$\eta^{-1}\sqrt{n} < |\omega'| \leq \sqrt{n}. \tag{106}$$

Moreover if we assume $\omega > 0$, then by Equation (105) with the plus sign, $\omega' > 0$ and it follows from (106), by addition, that

$$(\eta^{-1} + 1)\sqrt{n} < x < (\eta + 1)\sqrt{n}; \tag{107}$$

on the other hand, by Equation (105) with the minus sign,

$$(\eta^{-1} + 1)\sqrt{n} < y\sqrt{d} < (\eta + 1)\sqrt{n}. \tag{108}$$

In any case we need only examine whether a finite number of values x, y satisfy Equation (105). Then we can determine by simple division whether among the numbers $\omega = (x + y\sqrt{d})/2$ found in this way, there are still associated numbers.

Obtaining a unit η can be achieved in various ways. The proof of Dirichlet's unit theorem (Lemma (b) in §35) yields a process immediately. This is essentially a matter of expanding $\sqrt{d}$ as a continued fraction. The result of §52 about the class number will yield another expression for a unit in $k(\sqrt{d})$ which can also be built up from the dth roots of unity.

In any case, in this way a method is given of deciding by finitely many rational operations, whether two given ideals of a quadratic number field are equivalent.

In order to find the class number in this way, we keep in mind that, by Theorem 96, an integral ideal exists in each class whose norm $\leq |\sqrt{d}|$. Hence we first list all integral ideals whose norms satisfy this condition. To begin with, this can be done for prime ideals by the decomposition theorems (§29), and from this we find all ideals of this type by multiplication. Then the class number is the number of nonequivalent ideals among these finitely many ideals. It is useful to clarify the relationships with several numerical examples.

1. $k(\sqrt{-1})$, $k(\sqrt{-3})$, and $k(\sqrt{\pm 2})$ have class number $h = 1$. The next smallest integers to $|\sqrt{d}|$ are 1, 1, 2 respectively. In the first two fields there is an integral ideal with norm ≤ 1 in each class; this ideal is necessarily (1), thus a principal ideal. In $k(\sqrt{\pm 2})$ we moreover have to investigate the ideals

with norm 2. Here 2 becomes the square of a prime ideal $\mathfrak{p}$; this is obviously $= (\sqrt{\pm 2})$, so it is a principal ideal.

2. In $k(\sqrt{7})$ with $d = 28$, the ideals with norms 2, 3, 4, 5 are to be found. Here the primes 2, 3, 5 now split into prime ideals as follows:

$$2 = \mathfrak{p}_2^2, \qquad 3 = \mathfrak{p}_3\mathfrak{p}_3', \qquad 5 \text{ itself is a prime ideal.}$$

Hence there is only one ideal with norm 4, namely $\mathfrak{p}_2^2 = 2$, thus a principal ideal. Thus, except for the principal class, only the classes represented by $\mathfrak{p}_2$, $\mathfrak{p}_3$, $\mathfrak{p}_3'$ occur. We find, by trial $2 = 3^2 - 7 \cdot 1^2$, that is, $\mathfrak{p}_2 = (3 + \sqrt{7})$, thus $\mathfrak{p}_2 \sim 1$. Since $\mathfrak{p}_2 \sim \mathfrak{p}_2'$, $3 + \sqrt{7}$ and $3 - \sqrt{7}$ must be associated, hence the quotient

$$\eta = \frac{3 + \sqrt{7}}{3 - \sqrt{7}} = \frac{(3 + \sqrt{7})^2}{2} = 8 + 3\sqrt{7}$$

is a unit. If $\mathfrak{p}_3$ were a principal ideal $(a + b\sqrt{7})$, then

$$\pm 3 = a^2 - 7b^2, \text{ thus } \pm 3 \equiv a^2 \ (\text{mod } 7)$$

would have to hold. Accordingly, only the lower sign can hold, as $+3$ is a nonresidue mod 7. Thus, for b, by (108) we need only test the values b with

$$(9 - 3\sqrt{7})\sqrt{3} < b\sqrt{28} < (9 + 3\sqrt{7})\sqrt{3},$$

that is,

$$0 < b < (\sqrt{\tfrac{81}{28}} + \tfrac{3}{2})\sqrt{3} < 3 + \sqrt{\tfrac{27}{4}}, \qquad 0 < b \le 5.$$

Already $b = 1$ yields

$$a = \sqrt{-3 + 7 \cdot 1^2} = 2$$

so that $\mathfrak{p}_3 = (2 + \sqrt{7})$ is a principal ideal. So here also $h = 1$.

3. By the calculations in §23 the class number of $k(\sqrt{-5})$ is different from 1, since it was shown there that the ideal $\mathfrak{p}_3 = (3, 4 + \sqrt{-5})$ is not a principal ideal; however $\mathfrak{p}_3^2 = (2 + \sqrt{-5})$ is indeed a principal ideal. By the above, since $d = -20$, the ideals with norms 2, 3, 4 are to be investigated. We obtain $2 = \mathfrak{p}_2^2$; here $\mathfrak{p}_2$ is not a principal ideal, since 2 is not of the form $a^2 + 5b^2$. The only ideal with norm 4 is the principal ideal $\mathfrak{p}_2^2 = 2$; finally since $\mathfrak{p}_3\mathfrak{p}_3' = 3$ and $\mathfrak{p}_3^2 \sim 1$, $\mathfrak{p}_3 \sim \mathfrak{p}_3'$ and, except for the principal class, the ideal classes represented by $\mathfrak{p}_2$, $\mathfrak{p}_3$ occur here. If $\mathfrak{p}_2$ were not equivalent to $\mathfrak{p}_3$, we would have exactly three distinct classes, and because of the group property the third power of $\mathfrak{p}_2$ would have to be a principal ideal; it would follow already that since $\mathfrak{p}_2^2 \sim 1$, $\mathfrak{p}_2 \sim 1$ which is not the case. Thus $\mathfrak{p}_2 \sim \mathfrak{p}_3$ and consequently $h = 2$.

4. In $k(\sqrt{-23})$, $d = -23$; the values 2, 3, 4 are possible for the norm. We have

$$\left(\tfrac{-23}{2}\right) = +1, \qquad 2 = \mathfrak{p}_2\mathfrak{p}_2'$$
$$\left(\tfrac{-23}{3}\right) = +1, \qquad 3 = \mathfrak{p}_3\mathfrak{p}_3'.$$

Hence the ideals with the norms 2, 3, 4 are

$$\mathfrak{p}_2, \ \mathfrak{p}_2', \ \mathfrak{p}_3, \ \mathfrak{p}_3', \ \mathfrak{p}_2^2, \ \mathfrak{p}_2'^2, \ \mathfrak{p}_2\mathfrak{p}_2'. \qquad (109)$$

Obviously the last one is a principal ideal. In order that $\mathfrak{p}_2 \sim \mathfrak{p}_3$, we would have to have $\mathfrak{p}_2\mathfrak{p}_3' \sim 1$. Since $N(\mathfrak{p}_2\mathfrak{p}_3') = 6$ we must see whether 6 is the norm of a number; this is the case

$$6 = \frac{x^2 + 23y^2}{4}$$

which only holds for $x = \pm 1, y = \pm 1$. Hence there are exactly two principal ideals with norm 6, and these are conjugate so that either $\mathfrak{p}_2\mathfrak{p}_3'$ or $\mathfrak{p}_2\mathfrak{p}_3$ are equal to a principal ideal. Let the notation relative to the conjugates be chosen so that $\mathfrak{p}_2\mathfrak{p}_3' \sim 1$. Consequently by (109) at most

$$1, \quad \mathfrak{p}_2, \quad \mathfrak{p}_2', \quad \mathfrak{p}_2^2, \quad \mathfrak{p}_2'^2$$

remain as nonequivalent ideals. The ideal $\mathfrak{p}_2$ is equivalent neither to $\mathfrak{p}_2'$ nor to $\mathfrak{p}_2^2$; indeed, however, $\mathfrak{p}_2 \sim \mathfrak{p}_2'^2$, which means $\mathfrak{p}_2^3 \sim 1$. Then $N(\mathfrak{p}_2^3) = 8$ and 8 is the norm of the integer $(3 + \sqrt{-23})/2$, which is obviously divisible by no rational number except ± 1. The only ideals, however, which are without rational factors and which have norm 8 are $\mathfrak{p}_2^3$ and $\mathfrak{p}_2'^3$ and consequently one of these, and hence also the other, is a principal ideal.

Thus we find $h = 3$ and the three classes

$$\mathfrak{p}_2, \quad \mathfrak{p}_2^2, \quad \mathfrak{p}_2^3 \sim 1$$

as representatives.

§45 The Concept of Strict Equivalence and the Structure of the Class Group

For the investigation of quadratic fields it is useful to introduce a somewhat modified concept of equivalence.

Definition. We call two nonzero ideals $\mathfrak{a}$, $\mathfrak{b}$ of the quadratic field k *equivalent in the strict sense*, if there is a number λ in k such that

$$\mathfrak{a} = \lambda\mathfrak{b} \quad \text{and} \quad N(\lambda) > 0.$$

We write

$$\mathfrak{a} \approx \mathfrak{b}$$

and consider $\mathfrak{a}$ and $\mathfrak{b}$ in the same *ideal class in the strict sense*.

The classes can be combined in the manner familiar to us to form an Abelian group. If $\mathfrak{M}$ is the group of all nonzero ideals, $\mathfrak{H}_0$ is the group of all principal ideals (μ) with $N(\mu) > 0$, and $\mathfrak{H}$ is the group of all nonzero principal ideals where multiplication means multiplication of ideals, then the ideal classes in the strict sense are cosets or residue classes which arise from the decomposition of $\mathfrak{M}$ modulo $\mathfrak{H}_0$; the factor group $\mathfrak{M}/\mathfrak{H}_0$ is the group of ideal classes in the strict sense. Here the unit element is the system of ideals in $\mathfrak{H}_0$. The class group, in the sense used until now, is the factor group $\mathfrak{M}/\mathfrak{H}$.

It follows from $\mathfrak{a} \approx \mathfrak{b}$ that $\mathfrak{a} \sim \mathfrak{b}$. Conversely, if $\mathfrak{a} \sim \mathfrak{b}$, then obviously $\mathfrak{a} \approx \mathfrak{b}$ or $\mathfrak{a} \approx \mathfrak{b}\sqrt{d}$. A class in the wider sense thus splits into at most two classes relative to the strict concept of equivalence. Hence the class number h_0 in the strict sense is also finite and $\leq 2h$.

Since the number μ is determined only up to a unit factor by the ideal equation $\mathfrak{a} = \mu\mathfrak{b}$, the two concepts of equivalence are identical if in each complete sequence of associated numbers some with positive norm occur. That is, if k is imaginary or k is real and the basic unit in k has norm -1, then $h_0 = h$.

In the cases yet remaining in which k is real and each unit in k has norm ± 1, $\mathfrak{a}$ and $\mathfrak{a}\sqrt{d}$ are obviously not equivalent in the strict sense, and then $h_0 = 2h$.

Now the main problem is to investigate the structure of the class group. However, at present, only a very small part of this has been achieved. The result is formulated in the following theorem.

Theorem 132. *The basis number of the strict class group belonging to 2 is $e_0(2) = t - 1$, where t denotes the number of distinct primes which divide the discriminant d of k.*

By Theorem 28, we must show that there exist exactly 2^{t-1} classes in k whose square is the strict principal class. For this purpose we keep in mind that the t distinct prime ideals $\mathfrak{q}_1, \ldots, \mathfrak{q}_t$, which divide d, have the property that their square is a rational principal ideal, thus ≈ 1, by the decomposition laws mentioned above. We first show that each ideal $\mathfrak{a}$ with $\mathfrak{a}^2 \approx 1$ is necessarily equivalent to a product of powers of these $\mathfrak{q}$. From $\mathfrak{a}^2 \approx 1$ and $\mathfrak{a}\mathfrak{a}' \approx 1$, it follows that

$$\frac{\mathfrak{a}}{\mathfrak{a}'} \approx 1, \qquad \frac{\mathfrak{a}}{\mathfrak{a}'} = \alpha,$$

where α is a number with positive norm, which we also take > 0 if it is real. It is a quotient of two conjugate *ideals*, hence $N(\alpha) = 1$. Consequently, this number is also a quotient of two conjugate *numbers*,

$$\alpha = \frac{1 + \omega}{1 + \omega'}.$$

The ideal

$$\frac{\mathfrak{a}}{1 + \omega} = \frac{\mathfrak{a}'}{1 + \omega'}$$

is equal to its conjugate, hence by the decomposition laws

$$\frac{\mathfrak{a}}{1 + \omega} = r \cdot \mathfrak{q}_1^{a_1} \cdots \mathfrak{q}_t^{a_t}$$

where r is a rational number and the a_i are 0 or 1. However, this means

$$\mathfrak{a} \approx \mathfrak{q}_1^{a_1} \cdots \mathfrak{q}_t^{a_t}$$

as claimed since $N(1 + \omega) = \omega(1 + \omega')^2 > 0$.

Such integral ideals in $k(\sqrt{d})$, which are equal to their conjugates but which do not contain a rational factor (except ± 1), are called *ambiguous ideals*. Ideal classes which are equal to their conjugates are called *ambiguous classes*. Furthermore the above proof shows that an ambiguous ideal occurs in each ambiguous class.

Now we must still show that among the t ambiguous classes $Q_1, \ldots, Q_t$ which are defined by $\mathfrak{q}_1, \ldots, \mathfrak{q}_t$ respectively, there are exactly $t - 1$ independent classes (in the sense of group theory). Now if there is a relation

$$Q_1^{a_1} \cdots Q_t^{a_t} = 1, \tag{110}$$

which is not the trivial relation where all a_i are even, then there is a number α such that

$$\alpha = \mathfrak{q}_1^{a_1} \cdots \mathfrak{q}_t^{a_t}, \qquad N(\alpha) > 0. \tag{111}$$

Here we then have $(\alpha) = (\alpha')$, $\alpha = \eta\alpha'$, where η is a unit, $N(\eta) = +1$. We now distinguish three cases:

(a) $d < 0$, where we at once assume $d < -4$, since for $d = -3$ or $d = -4$ our Theorem 132 is already seen to be true because of $h = 1$ and $t = 1$. Then there are only the units ± 1 in k, hence

$$\alpha = \pm\alpha', \qquad \alpha = r(\sqrt{d})^n \qquad (n = 0 \text{ or } 1), \tag{112}$$

where r is a rational number. With $n = 0$ all exponents a_i in (111) are even. With $n = 1$ at least one a_i is odd, since d is not a square.

(b) $d > 0$ and the norm of the fundamental unit is -1. Here $\eta > 0$ because $N(\alpha) > 0$. Hence $\eta = \varepsilon^{2n}$ with n a rational integer. Since

$$\varepsilon^2 = -\frac{\varepsilon}{\varepsilon'} = \frac{\varepsilon\sqrt{d}}{-\varepsilon'\sqrt{d}}$$

we thus obtain

$$\frac{\alpha}{(\varepsilon\sqrt{d})^n} = \frac{\alpha'}{(-\varepsilon'\sqrt{d})^n}, \qquad \alpha = r(\varepsilon\sqrt{d})^n \tag{113}$$

with rational r. Again, to an even n there corresponds a system of exponents a_i consisting only of even numbers. With n odd at least one a_i is odd.

(c) $d > 0$ and the norm of the fundamental unit ε ($\varepsilon > 0$) is $+1$. Here

$$\eta = \varepsilon^n, \qquad \varepsilon = \frac{1 + \varepsilon}{1 + \varepsilon'}, \qquad \eta = \frac{(1 + \varepsilon)^n}{(1 + \varepsilon')^n}, \tag{114}$$

$$\alpha = r(1 + \varepsilon)^n.$$

The ideal $(1 + \varepsilon)$ is equal to its conjugate, but surely not equal to any rational principal ideal. For if

$$1 + \varepsilon = r_1 \varepsilon^k$$

held with rational r_1, then we would have

$$\varepsilon = \frac{1 + \varepsilon}{1 + \varepsilon'} = \varepsilon^{2k}, \qquad \varepsilon^{2k-1} = 1$$

which is not the case. Consequently, $(1 + \varepsilon)$ has a decomposition

$$(1 + \varepsilon) = \text{rational ideal} \times \mathfrak{q}_1^{b_1} \cdots \mathfrak{q}_t^{b_t},$$

where at least one of the exponents b_t is odd.

Thus in each case we obtain that if a decomposition (111) for α holds, where the exponents a_i are not all even, then α must be of one of the three forms (112), (113), (114), where n is odd. Consequently the exponents a_i in (110) are uniquely determined modulo 2. Hence there is *at most one* nontrivial relation between the t classes $Q_1, \ldots, Q_t$. Conversely, however, there is *actually one* such relation as the decomposition of the principal ideals (in the strict sense) $\sqrt{d}$, $\varepsilon\sqrt{d}$, $1 + \varepsilon$ shows in the cases (a), (b), (c) respectively, where at least one of the exponents $a_1, \ldots, a_t$ is odd.

This means that among the classes Q, there are exactly $t - 1$ independent ones; thus Theorem 132 is proved.

We formulate two important consequences of this:

Theorem 133. *If the discriminant d of $k(\sqrt{d})$ is divisible by a single prime $(t = 1)$, then h_0 and hence also h is odd and thus, provided $d > 0$, the norm of the fundamental unit $= -1$.*

Theorem 134. *If d is the product of two positive primes q_1, q_2, which are $\equiv 3 \pmod 4$, then either q_1 or q_2 is the norm of a principal ideal in the strict sense in $k(\sqrt{q_1 q_2})$.*

To begin with the norm of each unit $= +1$ in such a field. For from $N(\alpha) = -1$ for $\alpha = (x + y\sqrt{q_1 q_2})/2$ it would follow that

$$-4 \equiv x^2 \pmod{q_1 q_2};$$

thus -1 would be a quadratic residue mod q_1. However, by Equation (31) in §16, the residue symbol is

$$\left(\frac{-1}{q_1}\right) = (-1)^{(q_1 - 1)/2} = -1.$$

Moreover, it follows from the proof above that an equivalence

$$\mathfrak{q}_1^{a_1} \mathfrak{q}_2^{a_2} \approx 1 \tag{115}$$

holds in $k(\sqrt{q_1 q_2})$, where a_1 and a_2 are not both even. If both were odd, then we would have $q_1 q_2 = \sqrt{q_1 q_2} \approx 1$, thus there would be a unit η such that $N(\eta \sqrt{q_1 q_2}) > 0$, that is, $N(\eta) = -1$, which is impossible as has just been shown. Hence we may take one of the exponents $= 1$, the other $= 0$ in (115); thus Theorem 134 is proved.

Since in the field $h_0 = 2h$ must hold, in view of Theorem 132, there remains the possibility that h is perhaps odd. This is actually the case here, as one can convince oneself without difficulty by a proof analogous to that of Theorem 132.

§46 The Quadratic Reciprocity Law and a New Formulation of the Decomposition Laws in Quadratic Fields

Theorem 135. *If p and q are odd positive primes, then we have the relations*

(I) $\left(\frac{-1}{p}\right) = (-1)^{(p-1)/2}$,

(II) $\left(\frac{p}{q}\right) = \left(\frac{q}{p}\right)(-1)^{((p-1)/2)((q-1)/2)}$,

(III) $\left(\frac{2}{p}\right) = (-1)^{(p^2-1)/8}$,

We obtain the first formula directly from the definition of the residue symbol, Equation (31) in §16. We can also deduce it from field theory, in a somewhat more involved way, but analogous to the subsequent proof of (II) and (III) as follows: If $\left(\frac{-1}{p}\right) = +1$, then p splits in $k(\sqrt{-1})$ and since $h_0 = 1$, p is the norm of a principal ideal x; hence $p = a^2 + b^2$. Since each square is $\equiv 0$ or 1 (mod 4), we hence have $p \equiv 1$ (mod 4). Conversely, if $p \equiv 1$ (mod 4), then, by the second part of Theorem 133, the number -1 is the norm of an integer $\varepsilon = (a + b\sqrt{p})/2$ in $k(\sqrt{p})$, hence $-4 \equiv a^2$ (mod p), that is, -1 is a quadratic residue mod p; with this (I) is proved.

In the proof of (II) we distinguish three cases:

1. Suppose $p \equiv q \equiv 1$ (mod 4). We show that $\left(\frac{p}{q}\right)$ and $\left(\frac{q}{p}\right)$ are simultaneously $+1$ and, consequently, also simultaneously -1. Thus they are equal to one another, as required by the claim in this case.

For if $\left(\frac{q}{p}\right) = +1$, then the prime p splits into two distinct factors $\mathfrak{p}$, $\mathfrak{p}'$ in $k(\sqrt{q})$. Moreover, we can set

$$\mathfrak{p}^{h_0} = \alpha = \frac{x + y\sqrt{q}}{2},$$

where α is a number of positive norm; thus

$$\mathfrak{p}^{h_0} = \frac{x^2 - qy^2}{4}, \qquad 4\mathfrak{p}^{h_0} \equiv x^2 \ (\text{mod } q).$$

Accordingly p^{h_0} is a quadratic residue mod q, and since h_0 is odd by Theorem 133, p itself is a residue mod q, that is, $(\frac{p}{q}) = +1$. Since the hypothesis is symmetric in p and q, Formula (II) is proved in our case.

2. Suppose $q \equiv 1 \pmod 4$, $p \equiv 3 \pmod 4$. It follows from $(\frac{q}{p}) = +1$, as above, that also $(\frac{p}{q}) = +1$; therefore by (I), $(\frac{-p}{q}) = (\frac{-1}{q})(\frac{p}{q}) = +1$.

Conversely, if $(\frac{-p}{q}) = +1$, we conclude in the same way with the help of the field $k(\sqrt{-p})$, that $(\frac{q}{p}) = +1$; thus we always have

$$\left(\frac{-p}{q}\right) = \left(\frac{q}{p}\right), \quad \text{that is,} \quad \left(\frac{p}{q}\right) = \left(\frac{q}{p}\right).$$

in accordance with (II).

3. Finally if $p \equiv q \equiv 3 \pmod 4$, then we can likewise draw the conclusion that

$$\left(\frac{-p}{q}\right) = +1 \quad \text{implies} \quad \left(\frac{-q}{p}\right) = -1,$$

but the converse cannot be proved in this way. For this we go over to the field $k(\sqrt{pq})$ in which, by Theorem 134, p or q is the norm of an integer $(x + y\sqrt{pq})/2$. Suppose

$$4p = x^2 - pqy^2.$$

Here x must be divisible by p, $x = pu$, so $4 = pu^2 - qy^2$. From this equation we obtain

$$\left(\frac{p}{q}\right) = +1 \quad \text{and} \quad \left(\frac{-q}{p}\right) = +1,$$

that is,

$$\left(\frac{q}{p}\right) = -1;$$

hence $(\frac{p}{q})$ and $(\frac{q}{p})$ are different and (II) is also true.

Finally in order to prove Formula (III), we assume that $(\frac{2}{p}) = +1$. Then p splits in $k(\sqrt{2})$ and since $h = h_0 = 1$, p is the norm of an integer,

$$p = x^2 - 2y^2.$$

From this it follows that $p \equiv x^2 \pmod 8$, if y is even and $p \equiv x^2 - 2 \pmod 8$, if y is odd, that is, since x is odd, $p \equiv \pm 1 \pmod 8$.

Conversely, if $p \equiv \pm 1 \pmod 8$, then we go over to the field $k(\sqrt{\pm p})$ in which h_0 is odd by Theorem 133. In this field 2 splits into distinct factors by the decomposition laws; consequently 2 is a quadratic residue mod p.

Thus we have shown

$$\left(\frac{2}{p}\right) = +1 \quad \text{if and only if} \quad p \equiv \pm 1 \pmod 8.$$

However, this is equivalent to (III).

We now generalize the formula to the case where the two numbers p and q are composite positive odd numbers. The symbol introduced by

Legendre whose "denominator" is a prime was also defined for composite denominators at the end of §31. It should now be noted that the same reciprocity formulas also hold for this "Jacobi symbol."

For the proof let a and b be any odd integers. Since

$$(a - 1)(b - 1) \equiv 0 \pmod 4,$$

we have

$$ab - 1 \equiv a - 1 + b - 1 \pmod 4$$

$$\frac{ab - 1}{2} \equiv \frac{a - 1}{2} + \frac{b - 1}{2} \pmod 2. \tag{116}$$

In the same way it follows from

$$(a^2 - 1)(b^2 - 1) \equiv 0 \pmod{16}$$

that

$$\frac{a^2 b^2 - 1}{8} \equiv \frac{a^2 - 1}{8} + \frac{b^2 - 1}{8} \pmod 2. \tag{117}$$

By repeated application of this process we thus obtain for r odd integers $p_1, \ldots, p_r$

$$\frac{p_1 \cdot p_2 \cdots p_r - 1}{2} = \sum_{i=1}^{r} \frac{p_i - 1}{2} \pmod 2$$

$$\frac{(p_1 \cdot p_2 \cdots p_r)^2 - 1}{8} = \sum_{i=1}^{r} \frac{p_i^2 - 1}{2} \pmod 2.$$

Now suppose that the positive odd numbers P and Q are decomposed into their prime factors

$$P = p_1 \cdot p_2 \cdots p_r, \qquad Q = q_1 \cdot q_2 \cdots q_s.$$

Then, by the definition in §31 and by application of (116) and (117),

$$\left(\frac{-1}{P}\right) = \left(\frac{-1}{p_1}\right) \cdot \left(\frac{-1}{p_2}\right) \cdots \left(\frac{-1}{p_r}\right) = (-1)^{\sum_{i=1}^{r} (p_i - 1)/2} = (-1)^{(P-1)/2}, \tag{118}$$

$$\left(\frac{2}{P}\right) = (-1)^{\sum_{i=1}^{r} (p_i^2 - 1)/8} = (-1)^{(P^2 - 1)/8} \tag{119}$$

and finally

$$\left(\frac{P}{Q}\right) \cdot \left(\frac{Q}{P}\right) = \prod_{\substack{i=1,\ldots,r \\ k=1,\ldots,s}} \left(\frac{p_i}{q_k}\right) = (-1)^{\sum_{i=1}^{r}(p_i-1)/2 \sum_{k=1}^{s}(q_k-1)/2} \prod_{\substack{i=1,\ldots,r \\ k=1,\ldots,s}} \left(\frac{q_k}{p_i}\right)$$

$$\left(\frac{P}{Q}\right) = \left(\frac{Q}{P}\right)(-1)^{((P-1)/2) \cdot ((Q-1)/2)}. \tag{120}$$

Finally we further extend the definition to *negative denominators*, by setting

$$\left(\frac{a}{n}\right) = \left(\frac{a}{-n}\right). \tag{121}$$

Then in order to formulate the reciprocity laws for negative numbers we use the symbol sgn a (read signum a):

$$\operatorname{sgn} a = \begin{cases} +1 & \text{if } a > 0, \\ -1 & \text{if } a < 0. \end{cases}$$

Note that $|a| = a \cdot \operatorname{sgn} a$. With the help of this symbol we obtain at once from (116) that for odd P

$$\left(\frac{-1}{P}\right) = (-1)^{(|P|-1)/2} = (-1)^{(P-1)/2 + (\operatorname{sgn} P - 1)/2}.$$

Consequently for odd P, Q

$$\left(\frac{P}{Q}\right) = \left(\frac{\operatorname{sgn} P}{Q}\right)\left(\frac{|P|}{Q}\right)$$

$$= (-1)^{((\operatorname{sgn} P - 1)/2)((Q-1)/2) + ((\operatorname{sgn} P - 1)/2)((\operatorname{sgn} Q - 1)/2)}\left(\frac{|P|}{Q}\right).$$

Moreover, by (120)

$$\left(\frac{|P|}{Q}\right) = \left(\frac{|P|}{|Q|}\right) = \left(\frac{|Q|}{|P|}\right)(-1)^{((|P|-1)/2)((|Q|-1)/2)}$$

$$= \left(\frac{Q}{P}\right)(-1)^{((\operatorname{sgn} Q - 1)/2)((|P|-1)/2) + ((|P|-1)/2)((|Q|-1)/2)}.$$

Finally we obtain from these formulas

Theorem 136. (General Quadratic Reciprocity Law) *If P and Q are odd rational integers, then*

$$\left(\frac{-1}{P}\right) = (-1)^{(P-1)/2 + (\operatorname{sgn} P - 1)/2}, \qquad \left(\frac{2}{P}\right) = (-1)^{(P^2-1)/8}$$

$$\left(\frac{P}{Q}\right) = \left(\frac{Q}{P}\right) \cdot (-1)^{((P-1)/2)((Q-1)/2) + ((\operatorname{sgn} P - 1)/2)((\operatorname{sgn} Q - 1)/2)}.$$

Finally we generalize the definition of the *residue symbol to even denominators*, although we restrict the numerator. By Theorem 45 the residue class group mod 8 and modulo higher powers of 2 is no longer cyclic, but instead it has two basis classes. Each odd number is $\equiv (-1)^a 5^b \pmod{2^k}$ ($k \geq 3$), where the exponent a is uniquely determined mod 2 and the exponent b is uniquely determined mod 2^{k-2}. The numbers with $a \equiv 0 \pmod 2$ form a cyclic subgroup of $\Re(2^k)$; these are the numbers which are $\equiv 1 \pmod 4$. Among the classes of this subgroup, those classes which are squares are to be fixed by a single character. Corresponding to this we define:

Definition. If a is a rational integer $\equiv 0$ or $1 \pmod 4$ we set

$$\left(\frac{a}{2}\right) = \left(\frac{a}{-2}\right) = \begin{cases} 0, & \text{if } a \equiv 0 \pmod 4, \\ +1, & \text{if } a \equiv 1 \pmod 8, \\ -1, & \text{if } a \equiv 5 \pmod 8. \end{cases} \tag{122}$$

By Theorem 136, $\left(\frac{a}{2}\right) = \left(\frac{2}{a}\right)$ if the first symbol has a meaning. Moreover for two such numbers a and a' it follows that

$$\left(\frac{a}{2}\right) = \left(\frac{a'}{2}\right) \quad \text{if } a \equiv a' \pmod 8$$

$$\left(\frac{a \cdot a'}{2}\right) = \left(\frac{a}{2}\right) \cdot \left(\frac{a'}{2}\right).$$

Finally, in general we set

$$\left(\frac{a}{2}\right)^c = \left(\frac{a}{2^c}\right), \qquad \left(\frac{a}{mn}\right) = \left(\frac{a}{m}\right) \cdot \left(\frac{a}{n}\right) \tag{123}$$

for arbitrary denominators, if $a \equiv 0$ or $1 \pmod 4$. This definition remains in agreement with the stipulation in §44 because each field discriminant is $\equiv 0$ or $1 \pmod 4$.

Theorem 137. *If d is the discriminant of a quadratic field and n, m are positive integers, then*

$$\left(\frac{d}{n}\right) = \left(\frac{d}{m}\right) \quad \text{if } n \equiv m \pmod d, \tag{124}$$

$$\left(\frac{d}{n}\right) = \left(\frac{d}{m}\right) \cdot \operatorname{sgn} d \quad \text{if } n \equiv -m \pmod d. \tag{125}$$

Accordingly, $\left(\frac{d}{n}\right)$ thus represents a residue character mod d for positive n. For the proof we must split off the highest power of 2 dividing d, n, m. Let

$$d = 2^a d', \qquad n = 2^b n', \qquad m = 2^c m'$$

with odd d', n', m'.

Case 1: $a > 0$. The case $b > 0$ is trivial here, since then, by hypothesis, we must have $c > 0$, and both symbols in (124) and (125) have the value zero. Thus suppose that $b = c = 0$. Then by Theorem 136

$$\left(\frac{2^a d'}{n}\right) = \left(\frac{2}{n}\right)^a \left(\frac{d'}{n}\right) = (-1)^{a(n^2 - 1)/8} \left(\frac{n}{d'}\right)(-1)^{((n-1)/2)((d'-1)/2)} \tag{126}$$

and the analogous equation holds for m. Since d is divisible at least by 4, the first factors for n and m agree. The same holds true for the other two factors in case $n \equiv m \pmod d$; however if $n \equiv -m \pmod d$, then the factors differ precisely by the factor $\operatorname{sgn} d'$.

Case 2: $a = 0$, thus $d \equiv 1 \pmod 4$.

$$\left(\frac{d}{2^b n'}\right) = \left(\frac{d}{2}\right)^b \left(\frac{d}{n'}\right) = \left(\frac{2}{d}\right)^b \left(\frac{n'}{d}\right) = \left(\frac{n}{d}\right), \tag{127}$$

from which the assertion can be read off immediately.

Now we see from this theorem that the decomposition law for quadratic fields, as it was proved in §29, is indeed formally of a quite different type than that for cyclotomic fields, but on the other hand shows a great similarity relative to content. Theorem 137 shows that *if two positive primes belong to the same residue class* mod d, *then they split in exactly the same manner in* $k(\sqrt{d})$. Thus $k(\sqrt{d})$ *is also a class field* which belongs to a classification of the rational numbers mod d. For if we consider those numbers n for which $(\frac{d}{n})$ has the same nonzero value to be of the same "type," then the positive integers relatively prime to d split into two types. By Theorem 137 all natural numbers congruent to a mod d belong to the same type as a. Consequently, one type consists of certain $\frac{1}{2}\varphi(d)$ residue classes mod d, which are relatively prime to d. If we assume that $a_1, a_2, \ldots, a_m$ $(m = \frac{1}{2}\varphi(d))$ are the numbers which are incongruent mod d which belong to the same type as 1 (all quadratic residues mod d occur among them), then the decomposition law reads:

Let p be a positive prime relatively prime to d and let f be the smallest positive exponent such that p^f is congruent to one of the numbers $a_1, \ldots, a_m$ modulo d. Then p splits into $2/f$ distinct prime ideals in $k(\sqrt{d})$. All of these have degree f.

In particular, if the discriminant d is an odd prime, $d = (-1)^{(q-1)/2}q$, then by Formula (126), $(\frac{d}{n}) = (\frac{n}{q})$ and moreover

$$\left(\frac{n}{q}\right) \equiv n^{(q-1)/2} \pmod q.$$

The exponent f, which has just been discussed, is thus the smallest positive exponent for p for which

$$p^{f(q-1)/2} \equiv 1 \pmod q.$$

§47 Norm Residues and the Group of Norms of Numbers

By means of the quadratic field $k(\sqrt{d})$, a distinguished group of residue classes among the rational numbers is defined for each modulus n. Namely, let n be a rational integer. In the group $\mathfrak{R}(n)$ of residue classes relatively prime to n we then consider those residue classes which can be represented by norms of integers in $k(\sqrt{d})$. These obviously form a subgroup of $\mathfrak{R}(n)$, which we call the group of *norm residues mod n* (for the field $k(\sqrt{d})$), and which we denote by $\mathfrak{N}(n)$. In particular, an integer a which is relatively prime to n is called a norm residue mod n, if there is an integer α in k such that

$$a \equiv N(\alpha) \pmod{n},$$

otherwise a is called a *norm nonresidue* mod n. (Those a not relatively prime to n thus remain quite outside of consideration in this sense.)

It will now be shown that in general $\mathfrak{N}(p)$ and $\mathfrak{R}(p)$ are identical; these two groups are distinct only if the prime p divides the discriminant d.

Theorem 138. *If the odd prime p does not divide the discriminant d, then each rational integer relatively prime to p is a norm residue mod p for $k(\sqrt{d})$.*

We distinguish two cases in the proof.

1. p splits into two distinct factors $\mathfrak{p}$ and $\mathfrak{p}'$, of degree 1, in $k(\sqrt{d})$. Then there is a number π in $k(\sqrt{d})$ which is divisible by $\mathfrak{p}$ but not by $\mathfrak{p}'$, and for each integer α

$$N(\pi'\alpha + \pi) \equiv \pi'^2\alpha \pmod{\mathfrak{p}}.$$

From this it follows that the rational numbers $N(\pi'\alpha + \pi)$ run through a complete system mod $\mathfrak{p}$, hence also mod p, if α runs through a complete system of residues mod $\mathfrak{p}$.

2. p is irreducible in $k(\sqrt{d})$, thus p is a prime of degree 2. Let ρ be a primitive root mod p in $k(\sqrt{d})$. Then

$$\rho^p \equiv \rho' \pmod{p} \quad \text{and hence} \quad N(\rho) \equiv \rho^{p+1} \pmod{p}. \qquad (128)$$

For if the quadratic function $f(x) = x^2 + ax + b$ with integral coefficients has the roots ρ, ρ', then the functional congruence

$$f(x)^p \equiv f(x^p) \pmod{p}$$

implies

$$0 \equiv f(\rho^p) \equiv (\rho^p - \rho)(\rho^p - \rho') \pmod{p},$$

from which (128) follows. Hence the residue classes of

$$N(\rho^a) \equiv \rho^{a(p+1)} \pmod{p}$$

are mutually distinct, for $a = 1, 2, \ldots, p - 1$, since two powers of ρ yield the same residue class only if the exponents are congruent mod $(N(p) - 1)$, that is, mod$(p^2 - 1)$. Hence there are $N(\rho^a)$ rational residue classes modulo p which are relatively prime to p.

Theorem 139. *If the odd prime q divides the discriminant d of $k(\sqrt{d})$, then exactly one half of the classes of $\mathfrak{R}(q)$ are norm residues mod q, and indeed these are the classes of $\mathfrak{R}(q)$ which can be represented as the square of a class.*

If q is the prime ideal of $k(\sqrt{d})$ which divides q, then each number α in k is congruent to a rational number mod q, say r. However, since $q = q'$ it follows from $\alpha \equiv r \pmod{q}$ that $\alpha' \equiv r \pmod{q}$ and

$$N(\alpha) \equiv r^2 \pmod{q}, \quad \text{hence also mod } q,$$

that is, if $(r, q) = 1$,

$$\left(\frac{N(\alpha)}{q}\right) = +1.$$

Conversely if the condition $(\frac{a}{q}) = +1$ is satisfied, then there is a rational integer x with $a \equiv x^2 \pmod{q}$, and since $a \equiv N(x) \pmod{q}$, a is a norm residue. Moreover we see that for arbitrary composite moduli m, n:

Lemma. *Suppose that $(m, n) = 1$. Then if for each a there exist two integers α and β in $k(\sqrt{d})$ such that*

$$a \equiv N(\alpha) \pmod{m} \quad and \quad a \equiv N(\beta) \pmod{n},$$

there is also an integer γ in $k(\sqrt{d})$ for which

$$a \equiv N(\gamma) \pmod{mn}.$$

To see this we choose positive exponents b, c such that

$$m^b \equiv 1 \pmod{n} \quad \text{and} \quad n^c \equiv 1 \pmod{m}$$

(say $b = \varphi(n)$ and $c = \varphi(m)$). Then

$$\gamma = n^c \alpha + m^b \beta$$

has the asserted properties.

As far as the prime 2 is concerned, we consider the group $\Re(2^a)$ for $a = 2$ or 3.

Theorem 140. *If the discriminant d of $k(\sqrt{d})$ is odd, then each odd number is a norm residue mod 8. However, if d is even, then exactly half of all incongruent odd numbers mod 8 are norm residues mod 8.*

For the proof we test the residue classes in $k(\sqrt{d})$ mod 8. We find, with $\alpha = x + y\sqrt{d}$ and d odd, that

$$x = 0, 1, 2, 1$$
$$y = 1, 0, 1, 2$$
$$N(\alpha) \equiv 3, 1, 7, 5 \pmod{8}, \quad \text{if } d \equiv 5 \pmod{8}$$
$$N(\alpha) \equiv 7, 1, 3, 5 \pmod{8}, \quad \text{if } d \equiv 1 \pmod{8}$$

and thus the first assertion is proved.

We deal with the second part of the theorem in the same manner. For even d exactly the following residue classes mod 8 appear as norm residues mod 8:

$$N(\alpha) \equiv 1 \text{ or } 5 \ (\text{mod } 8), \qquad \text{if } \ \frac{d}{4} \equiv 3 \ (\text{mod } 4)$$

$$N(\alpha) \equiv 1 \text{ or } -1 \ (\text{mod } 8), \qquad \text{if } \ \frac{d}{4} \equiv 2 \ (\text{mod } 8) \tag{129}$$

$$N(\alpha) \equiv 1 \text{ or } 3 \ (\text{mod } 8), \qquad \text{if } \ \frac{d}{4} \equiv 6 \ (\text{mod } 8).$$

Note that for $d/4 \equiv 3$ (mod 4) the only norm residues mod 4 lie in the residue class of 1 mod 4, hence also that $\mathfrak{N}(4)$ is different from $\mathfrak{R}(4)$.

We now wish to express this state of affairs somewhat more clearly by using the general concepts of group theory from §10. Only the norm residues modulo the divisors of d will be of interest. Let $q_1, q_2, \ldots, q_t$ be the t distinct positive primes dividing d, with the exception that *when d is even, the number q_t denotes the highest power of 2 dividing d.* Then for each $i = 1, \ldots, t$ the group $\mathfrak{N}(q_i)$ of the norm residues in $k(\sqrt{d})$ is a group of index 2 in $\mathfrak{R}(q_i)$. By Theorem 33, the fact that a class in $\mathfrak{R}(q_i)$ belongs to this subgroup is thus expressed by the fact that a completely determined character of the group $\mathfrak{R}(q_i)$ has the value 1 for this class. This character $\chi_i(n)$ can be given immediately, where we denote the representative of the residue class by the argument n, as is common with residue classes. For by Theorem 139

$$\chi_i(n) = \left(\frac{n}{q_i}\right) \quad \text{if } q_i \text{ is odd.} \tag{130}$$

The group $\mathfrak{R}(8)$ has two basis classes, each of order 2; consequently it has three distinct subgroups of index 2, and as (129) shows, each of these also appears once as $\mathfrak{N}(8)$. The three quadratic characters mod 8 which are different from 1 are

$$(-1)^{(n-1)/2}, \ (-1)^{(n^2-1)/8}, \ (-1)^{(n-1)/2+(n^2-1)/8},$$

and for even d we thus find the last character

$$\chi_t(n) = \begin{cases} (-1)^{(n-1)/2}, & \text{if } \ \dfrac{d}{4} \equiv 3 \ (\text{mod } 4), \\[2mm] (-1)^{(n^2-1)/8}, & \text{if } \ \dfrac{d}{4} \equiv 2 \ (\text{mod } 8), \\[2mm] (-1)^{(n-1)/2+(n^2-1)/8}, & \text{if } \ \dfrac{d}{4} \equiv 6 \ (\text{mod } 8), \end{cases} \tag{131}$$

that is,

$$\chi_t(n) = (-1)^{a(n^2-1)/8+((d'-1)/2)((n-1)/2)} \quad \text{if } d = 2^a d', \ d' \text{ odd.} \tag{132}$$

In view of the lemma, it thus follows immediately that

Theorem 141. *The group* $\mathfrak{N}(d)$ *of norm residues* mod *d for a quadratic field with discriminant d has index* 2^t *in* $\mathfrak{R}(d)$, *where t is the number of distinct prime factors of d. In order that a number n be a norm residue* mod *d, it is necessary and sufficient that the t residue characters*

$$\chi_i(n) \qquad (i = 1, \ldots, t)$$

defined by (130) *and* (132), *have the value* $+1$.

To make a study of the literature easier let us note that *Hilbert* also defined the norm residue concept for those numbers n which are not coprime to p, and that in the remaining cases the definition has a different form:

Definition of the *Hilbert norm residue symbol*: Let n and m be rational integers, m not a square, p a prime (including 2). If the number n is congruent to the norm of an integer in $k(\sqrt{m})$ modulo each power p^e, then let us set

$$\left(\frac{n, m}{p}\right) = +1$$

and call n the norm residue of the field $k(\sqrt{m})$ mod p. In each other case let this symbol be equal to -1, and let n be called a norm nonresidue mod p.

If n is not divisible by p and p divides the discriminant of $k(\sqrt{m})$, then

$$\left(\frac{n, d}{q_i}\right) = \chi_i(n) \qquad (q_i \text{ odd})$$

$$\left(\frac{n, d}{2}\right) = \chi_i(n) \qquad (d \text{ even}).$$

On the other hand, if p does not divide nd, then we have $\left(\frac{n, d}{p}\right) = +1$.

§48 The Group of Ideal Norms, the Group of Genera, and Determination of the Number of Genera

As in the case of norms of numbers, the norms of ideals of k can now also be studied. Those residue classes mod d which can be represented by norms, taken to be positive, of ideals of $k(\sqrt{d})$ relatively prime to d obviously form a subgroup of $\mathfrak{R}(d)$. Let this subgroup be called the *group of ideal norms* mod d and let it be denoted by $\mathfrak{J}(d)$. $\mathfrak{R}(d)$ is obviously a subgroup of $\mathfrak{J}(d)$. For if a class mod d can be represented by the norm of a number $N(\alpha)$, then $N(\alpha + dx)$ belongs to the same class for x a rational integer, and for sufficiently large x $N(\alpha + dx)$ is obviously positive. Thus it is the norm of the principal ideal $(\alpha + dx)$.

Since the structure of $\mathfrak{N}(d)$ is already known to us by Theorem 141, we need only investigate the factor group $\mathfrak{J}/\mathfrak{N}$. Since $\mathfrak{N}$ has order $\varphi(d)/2^t$, the order of $\mathfrak{J}(d)$ is a multiple of this number; on the other hand, the order is a divisor of the order $\varphi(d)$ of $\mathfrak{R}(d)$. Consequently the degree of $\mathfrak{J}/\mathfrak{N}$ is equal to 2^u, where the integer $u \leq t$. The first principal result will be the equation $n = t - 1$; the second important result will be the disclosure of the connection of this group with the group of ideal classes and Theorem 133.

The factor group $\mathfrak{J}/\mathfrak{N}$ arises if we do not regard as distinct norms of ideals which differ mod d only by a factor which is the norm of a number in k. For these ideals we obtain a division into classes which we can define in a useful way as follows:

We consider two integral ideals $\mathfrak{a}$ and $\mathfrak{b}$ in k, coprime to d, to be of the same *genus* if there is a number α in k such that

$$|N(\mathfrak{a})| \equiv N(\alpha)|N(\mathfrak{b})| \pmod{d}.$$

In the manner familiar to us we combine the genera in k to form the Abelian *group of genera* by defining the product of two genera G_1 and G_2 as that genus to which the ideal $\mathfrak{a}_1 \cdot \mathfrak{a}_2$ belongs, where $\mathfrak{a}_1$ and $\mathfrak{a}_2$ are ideals from G_1 and G_2 respectively. The group of genera is obviously isomorphic to the group $\mathfrak{J}/\mathfrak{N}$. The unit element of this group is called the *principal genus*; it is that genus which contains the ideal 1, thus the principal ideal, in the strict sense. Ideals which are equivalent in the strict sense obviously belong to the same genus if they are coprime to d; consequently each genus consists of a certain number of ideal classes in the strict sense. Since the classes which belong to the principal genus—let their number be f—obviously form a subgroup of the class group, f is a divisor of h_0, and each genus contains exactly f classes. If g denotes the number of different genera, then

$$h_0 = g \cdot f.$$

The square of each genus is the principal genus. Namely, if for each $\mathfrak{a}$, we set $a = |N(\mathfrak{a})|$, we have

$$|N(\mathfrak{a}^2)| = N(a).$$

Thus the order g of the group of genera must be a power of 2, $g = 2^u$, as we already found above for the group $\mathfrak{J}/\mathfrak{N}$. However, we obtain at once a more precise statement about u if we keep in mind that the number of distinct classes, which can be represented as squares is, by Theorem 133 exactly $h_0/2^{t-1}$ because of Theorem 129. Consequently,

$$f \geq \frac{h_0}{2^{t-1}}, \qquad g = \frac{h_0}{f} \leq 2^{t-1}, \qquad u \leq t - 1. \tag{133}$$

Now to prove the equation $u = t - 1$, we attempt to construct the group characters for the group of genera. We obtain these at once from the t

functions $\chi_i(n)$ of the preceding section. For each norm residue n mod d, the $\chi_i(n)$ have the value 1. If we now define for each integral ideal a from $k(\sqrt{d})$, relatively prime to d, the t functions

$$\gamma_i(a) = \chi_i(|N(a)|) \qquad (i = 1, \ldots, t), \tag{134}$$

then each $\gamma_i(a)$ has the same value for ideals of the same genus. Moreover, $\gamma_i(ab) = \gamma_i(a) \cdot \gamma_i(b)$, so we have:

Theorem 142. *The t functions $\gamma_i(a)$ are group characters of the genus represented by a.*

Now by §10, the group of characters of an Abelian group is isomorphic to the group. There are u independent elements in the group of genera and no more, because this group is of order 2^u and each element has at most order 2. Consequently, there are also exactly u independent characters. Hence, among the t characters at least $t - u$ relations must hold. That is, since $t - u \geq 1$:

Theorem 143. *At least one relation must hold for all ideals a of the field which are coprime to d, namely,*

$$\prod_{i=1}^{t} \gamma_i^{c_i}(a) = 1,$$

where the rational integers c_i are independent of a and are not all divisible by 2.

Thus for $t = 1$ the equation

$$\gamma_1(a) = \chi_1(|N(a)|) = 1$$

must hold. In fact this is exactly one part of the quadratic reciprocity law, which has not been used until now (in §47 and 48). We see that the proof of this equation is essentially reduced to the fact that h_0 is odd for fields with $t = 1$, as in our proof in §46.

Conversely, we now wish to obtain the equation

$$\prod_{i=1}^{t} \gamma_i(a) = 1 \tag{135}$$

from the quadratic reciprocity law. For this we show that for each positive integer n relatively prime to d the equation

$$\prod_{i=1}^{t} \chi_i(n) = \left(\frac{d}{n}\right) \tag{136}$$

is valid. For odd d we have

$$\prod_{i=1}^{t} \chi_i(n) = \prod_{i=1}^{t} \left(\frac{n}{q_i}\right) = \left(\frac{n}{q_1 \cdots q_t}\right) = \left(\frac{n}{d}\right),$$

and, by (127), this quantity is equal to the reciprocal symbol. However, if $q_t = 2^a$ $(a > 0)$ and $d = 2^a d'$, then

$$\prod_{i=1}^{t-1} \chi_i(n) = \left(\frac{n}{d'}\right), \qquad \chi_t(n) = (-1)^{a(n^2-1)/8 + ((d'-1)/2)((n-1)/2)}$$

and (136) likewise follows from (126).

From this we now obtain at once the character relation (135) for prime ideals of degree 1. Namely, for such $\mathfrak{a} = \mathfrak{p}$, by the decomposition theorems, we have $\left(\frac{d}{N(\mathfrak{p})}\right) = +1$. However if $\mathfrak{a}$ is a prime ideal of degree 2, then $N(\mathfrak{a})$ is a rational square, hence each $\gamma_i(\mathfrak{a}) = 1$. However, if (135) is valid for each prime ideal not dividing d, then it is also valid for each $\mathfrak{a}$ with $(\mathfrak{a}, d) = 1$.

The fact that the number of genera g is exactly 2^{t-1} is now proved most easily with the use of transcendental methods if we show that there is only the one relation (135) between the t characters $\gamma_i(\mathfrak{a})$ and hence that there are $t - 1$ independent characters $\gamma_i(\mathfrak{a})$ of the group of genera, whose degree is at least 2^{t-1}, consequently exactly 2^{t-1}, by (133).

Theorem 144. *Let $e_1, e_2, \ldots, e_t$ be t numbers ± 1 such that $e_1 \cdot e_2 \cdots e_t = 1$. Then there are infinitely many prime ideals $\mathfrak{p}$ of degree 1 in $k(\sqrt{d})$ for which*

$$\gamma_i(\mathfrak{p}) = e_i \qquad (i = 1, 2, \ldots, t).$$

If we set $N(\mathfrak{p}) = p$, then the assertion obviously states that there are infinitely many rational primes p which satisfy the conditions

$$\chi_i(p) = e_i \qquad (i = 1, \ldots, t)$$

and

$$\left(\frac{d}{p}\right) = +1.$$

By (136), the last condition is now a consequence of the first t conditions, since $e_1 \cdot e_2 \cdots e_t = +1$. Thus we need only keep these conditions in mind.

Since each $\chi_i(n)$ is a residue character mod q_i, the single equation

$$\chi_i(n) = e_i$$

thus requires that n belong to certain residue classes mod q_i, and there are always such rational integers n. The fact that the t equations hold simultaneously thus requires that n belong to certain residue classes modulo each of the t moduli q_i. By Theorem 15, this means that n belongs to certain residue classes mod $q_1 \cdot q_2 \cdots q_t$ that is, n belongs to certain residue classes mod d (which are of course relatively prime to d). Now however, by Theorem 131, in each residue class mod d which is relatively prime to d, there are also infinitely many positive rational primes. Thus our theorem is proved.

We proved the existence of these primes by the theory of cyclotomic fields of the $|d|$th roots of unity in §43. It is important that the existence of infinitely many p with $\chi_i(p) = e_i$ can also be deduced from the theory of quadratic fields alone (as well as by transcendental methods) as we still wish to show in §49.

As already shown above, it follows from Theorem 144 that $g = 2^{t-1}$, thus $f = h_0/2^{t-1}$. That is, the number of classes in the principal genus is equal to the number of those ideal classes which can be represented as squares of classes. Thus we have proved:

Theorem 145. (Fundamental Theorem of Genera). *In the quadratic field with discriminant d the number of genera is equal to 2^{t-1}. A complete system of independent characters of the genus group is formed from any $t-1$ of the functions*

$$\gamma_i(\mathfrak{a}) = \chi_i(|N(\mathfrak{a})|) \qquad (i = 1, \ldots, t).$$

In order that an ideal class be a square, it is necessary and sufficient that it belongs to the principal genus.

Gauss first found this theorem and gave a purely number-theoretic proof for it. Such a proof is also presented in Hilbert's *Bericht*.

From the last part of the above theorem we can further conclude that in order for the ideal $\mathfrak{a}$, relatively prime to d, to be equivalent to the square of an ideal, it is necessary and sufficient that $|N(\mathfrak{a})|$ be a norm residue mod d, that is, that the congruence

$$|N(\mathfrak{a})| \equiv x^2 \pmod{d}$$

be solvable with x a rational integer. Then the ideal norm $|N(\mathfrak{a})|$ is also the norm of an integral or fractional *number* of the field. For from $\mathfrak{a} \approx \mathfrak{b}^2$ there follows the existence of a number α, of the field, with

$$\mathfrak{a} = \alpha\mathfrak{b}^2, \qquad N(\alpha) > 0.$$

Hence

$$|N(\mathfrak{a})| = N(\alpha) \cdot |N(\mathfrak{b}^2)| = N(\alpha) \cdot N(\mathfrak{b})^2 = N(\alpha b), \quad \text{where } b = |N(\mathfrak{b})|.$$

§49 The Zeta Function of $k(\sqrt{d})$ and the Existence of Primes with Prescribed Quadratic Residue Characters

In order to express the zeta-function $\zeta(s)$ of $k(\sqrt{d})$ by simpler functions, we consider those factors in the infinite product

$$\zeta_k(s) = \prod_{\mathfrak{p}} \frac{1}{1 - N(\mathfrak{p})^{-s}},$$

which are derived from prime ideals $\mathfrak{p}$ which divide a definite rational prime p. By the decomposition laws we see at once that this partial product

$$\prod_{\mathfrak{p}|p} (1 - N(\mathfrak{p})^{-s}) = (1 - p^{-s})\left(1 - \left(\frac{d}{p}\right)p^{-s}\right).$$

Consequently $\zeta_k(s)$ becomes the product

$$\zeta_k(s) = \prod_p \frac{1}{1 - p^{-s}} \cdot \prod_p \frac{1}{1 - \left(\dfrac{d}{p}\right) p^{-s}},$$

where p runs through all positive primes, which likewise converges for $s > 1$. Hence

$$\zeta_k(s) = \zeta(s) \cdot L(s)$$

$$L(s) = \prod_p \frac{1}{1 - \left(\dfrac{d}{p}\right) p^{-s}}. \tag{137}$$

If we substitute this expression for $\zeta_k(s)$ into Formula (95) for the class number, then we obtain

$$h \cdot \varkappa = \lim_{s \to 1} L(s). \tag{138}$$

From this we conclude that $L(s)$ tends to a finite limit different from 0 as s approaches 1. Now we wish to derive results about the distribution of the symbol $\left(\frac{d}{p}\right)$ from this fact in a manner similar to that used in §43. It follows from (138) that

$$\lim_{s \to 1} \log L(s) \text{ is finite.} \tag{139}$$

As with $L(s, \chi)$ in §43, we find

$$\log L(s) = -\sum_p \log\left(1 - \left(\frac{d}{p}\right) p^{-s}\right),$$

$$= \sum_p \sum_{m=1}^{\infty} \frac{1}{m p^{ms}} \left(\frac{d}{p}\right)^m$$

$$= \sum_p \left(\frac{d}{p}\right) \frac{1}{p^s} + H(s),$$

where $H(s)$ is a convergent Dirichlet series for $s > \frac{1}{2}$, which thus has a limit as $s \to 1$. Hence by (139)

$$\lim_{s \to 1} \sum_p \left(\frac{d}{p}\right) \frac{1}{p^s} \text{ is finite.} \tag{140}$$

This assertion is obviously still true if we omit finitely many p from the sum, and consequently also if we replace d by an integer differing from d by a rational square. That is,

Theorem 146. *If a is an arbitrary positive or negative rational integer, which is not a square, then the function*

$$L(s; a) = \sum_{p > 2} \left(\frac{a}{p}\right) \frac{1}{p^s}$$

has a finite limit as $s \to 1$.

A formalism similar to that of §43 then leads us to

Theorem 147. *Let* $a_1, a_2, \ldots, a_r$ *be any rational integers such that a product of powers*

$$a_1^{u_1} a_2^{u_2} \cdots a_r^{u_r}$$

is a rational square only if all u_i *are even. Moreover, let* $c_1, c_2, \ldots, c_r$ *take arbitrary values* ± 1. *Then there are infinitely many primes* p *which satisfy the conditions*

$$\left(\frac{a_i}{p}\right) = c_i \quad \text{for } i = 1, 2, \ldots, r. \tag{141}$$

For the proof we set, for the sake of symmetry,

$$L(s; 1) = \sum_{p > 2} \frac{1}{p^s}$$

(a function which grows beyond all bounds as $s \to 1$ by §43), and we form the sum

$$\sum_{u_1, \ldots, u_r} c_1^{u_1} c_2^{u_2} \cdots c_r^{u_r} L(s; a_1^{u_1} \cdot a_2^{u_2} \cdots a_r^{u_r}) = \varphi(s) \tag{142}$$

consisting of 2^r terms ($s > 1$), where each u_i runs through the values 0, 1. Thus the definition of L yields

$$\varphi(s) = \sum_{p > 2} \left(1 + c_1\left(\frac{a_1}{p}\right)\right)\left(1 + c_2\left(\frac{a_2}{p}\right)\right) \cdots \left(1 + c_r\left(\frac{a_r}{p}\right)\right) \frac{1}{p^s}. \tag{143}$$

As can be easily seen, in this sum over p, only the terms p^{-s} in which p satisfies Conditions (141) of the assertion have a nonzero factor (and in fact the factor 2^r), except for the finitely many p dividing a. Now

$$\lim_{s \to 1} \varphi(s) = \infty$$

since $L(s, 1)$ grows beyond all bounds in each sum (142), while, by our hypothesis, all remaining $L(s; a)$ remain finite by Theorem 146. Consequently, infinitely many nonzero terms must also appear in (143) and our theorem is proved.

In particular, from this it follows for $r = 1$:

In each quadratic field there are infinitely many prime ideals of the first degree as well as of the second degree.

If, in the notation of the preceding section we choose the $a_i = \pm q_i$ and $r = t$, so that each a_i itself is a discriminant of a field and the product $a_1 \cdot a_2 \cdots a_t$ is exactly d, then by Formula (136) applied to each individual field $k(\sqrt{a_i})$,

$$\chi_i(p) = \left(\frac{a_i}{p}\right) \quad (i = 1, \ldots, t)$$

and thus Theorem 144 of the preceding section has been proved without the Dirichlet theorem on primes, that is, without the theory of cyclotomic fields.

§50 Determination of the Class Number of $k(\sqrt{d})$ without Use of the Zeta-Function

We now turn to a determination of the number h of ideal classes (in the wider sense) by the methods of Chapter 6. At first we wish to undertake this determination, as in §41, just from the density of ideals, without use of $\zeta_k(s)$. Afterwards we wish to apply the formally more elegent method of Theorem 125 with the help of $\zeta_k(s)$.

To use the first method we must determine the function $F(n)$, the number of integral ideals of the field with norm n. By (89), $F(ab) = F(a)F(b)$ for a and b relatively prime.

Lemma. *For each power p^k of the prime p,*

$$F(p^k) = \sum_{i=0}^{k} \left(\frac{d}{p^i}\right) = 1 + \sum_{i=1}^{k} \left(\frac{d}{p}\right)^i \tag{144}$$

Case (a): $\left(\frac{d}{p}\right) = -1$. If $N(a) = p^k$, then we must have $a = p^u$ with positive rational u; hence $2u = k$, i.e.

$$F(p^k) = \begin{cases} 1, & \text{if } k \text{ even} \\ 0, & \text{if } k \text{ odd} \end{cases}$$

in agreement with Equation (144).

Case (b): $\left(\frac{d}{p}\right) = 0$. p is the square of a prime ideal $\mathfrak{p}$, and it follows from $N(a) = p^k$ that $a = \mathfrak{p}^u$, so $u = k$ and $F(p^k) = 1$.

Case (c): $\left(\frac{d}{p}\right) = +1$. p is the product of two distinct prime ideals $\mathfrak{p}$, $\mathfrak{p}'$ and it follows from $N(a) = p^k$ that $a = \mathfrak{p}^u \cdot \mathfrak{p}'^{u'}$ with $u + u' = k$. Then, for $u = 0, 1, \ldots, k$, the $k + 1$ pairs of numbers u, $k - u$ yield exactly $k + 1$ distinct ideals a and we obtain

$$F(p^k) = k + 1$$

as asserted in the lemma.

Theorem 148. *For each natural number n we have*

$$F(n) = \sum_{m|n} \left(\frac{d}{m}\right),$$

where m runs through all distinct positive divisors of n.

If we decompose n into its distinct prime factors

$$n = p_1^{k_1} \cdot p_2^{k_2} \cdots p_r^{k_r},$$

then

$$F(n) = F(p_1^{k_1}) \cdot F(p_2^{k_2}) \cdots F(p_r^{k_r}) = \prod_{i=1}^{r} \sum_{c_i=0}^{k_i} \left(\frac{d_i}{p_i^{c_i}}\right)$$

$$F(n) = \sum_{\substack{c_1=0,\ldots,k_1 \\ c_2=0,\ldots,k_2 \\ \vdots}} \left(\frac{d}{p_1^{c_1} p_2^{c_2} \cdots p_r^{c_r}}\right) = \sum_{m|n}\left(\frac{d}{m}\right).$$

We henceforth set

$$\left(\frac{d}{n}\right) = \chi(n), \qquad n > 0$$

in order to thereby emphasize the fact already proved by Theorem 137, that for positive n, $\left(\frac{d}{n}\right)$ is a residue character mod d.

We now substitute the expression found for $F(n)$ into Formula (88) of §41 and we obtain

$$h \cdot \varkappa = \lim_{x\to\infty} \frac{\sum_{n\leq x} F(n)}{x} = \lim_{x\to\infty} \frac{1}{x} \sum_{n\leq x} \sum_{m|n} \chi(m).$$

In the finite double sum we set (with m' integral)

$$n = m \cdot m', \qquad \sum_{n\leq x} F(n) = \sum_{\substack{m,m'>0 \\ m \cdot m' \leq x}} \chi(m),$$

where m, m' run through all natural numbers whose product is $\leq x$. Thus m' is to run through those integers with the property

$$1 \leq m' \leq \frac{x}{m},$$

whose number is $[x/m]$, where $[u]$ denotes the largest integer $\leq u$. Consequently we obtain

$$\sum_{1\leq n\leq x} F(n) = \sum_{1\leq m\leq x} \chi(m)\left[\frac{x}{m}\right]$$

$$= x \sum_{1\leq m\leq x} \frac{\chi(m)}{m} + \sum_{1\leq m\leq x} \chi(m)\left(\left[\frac{x}{m}\right] - \frac{x}{m}\right).$$

After division by x the first sum has the limit

$$\sum_{m=1}^{\infty} \frac{\chi(m)}{m},$$

as $x \to \infty$, since the series converges for $s = 1$ by Theorem 128 since it is the series $L(s, \chi)$ for $s = 1$. Thus we obtain

$$h \cdot \varkappa = \sum_{n=1}^{\infty} \frac{\chi(n)}{n} + \lim_{x\to\infty} \frac{1}{x} \sum_{1\leq n\leq x} \chi(n)\left(\left[\frac{x}{n}\right] - \frac{x}{n}\right).$$

However, the last limit is equal to 0, by the following general limit theorem:[1]

Let $a_1, a_2, \ldots$ be a sequence of coefficients such that

$$\lim_{x \to \infty} \frac{1}{x} \sum_{n \le x} a_n = 0 \quad \text{and} \quad \sum_{n \le x} |a_n| \le x \quad \text{for all } x > 0.$$

Then

$$\lim_{x \to \infty} \frac{1}{x} \sum_{n \le x} a_n \left(\left[\frac{x}{n} \right] - \frac{x}{n} \right) = 0.$$

By the proof of Theorem 128, these hypotheses agree for $a_n = \chi(n)$. Thus we obtain

$$h \cdot \varkappa = \sum_{n=1}^{\infty} \frac{\chi(n)}{n}. \tag{145}$$

In the next section this equation will be proved in a shorter way using the zeta-function and accordingly the sum will be treated further.

§51 Determination of the Class Number with the Help of the Zeta-Function

In §49 we have already represented $\zeta_k(s)$ as $\zeta(s) \cdot L(s)$ where

$$L(s) = \prod_p \frac{1}{1 - \chi(p)p^{-s}} \tag{137}$$

and concluded from this that

$$h \cdot \varkappa = \lim_{s \to 1} L(s). \tag{138}$$

Now since $\chi(n)$ is a residue character mod d for natural numbers n, the function defined by (137) is identical with some $L(s, \chi)$ from §43 and

$$L(s) = \sum_{n=1}^{\infty} \frac{\chi(n)}{n^s}$$

from which, by Theorem 128, the equation

$$h \cdot \varkappa = L(1) = \sum_{n=1}^{\infty} \frac{\chi(n)}{n^s} \tag{145}$$

follows, which we have just obtained in §50 without using the zeta-function. If we compare the two proofs of this formula, then we see, by the decomposition laws, that the representation of $\zeta_k(s)$ as $\zeta(s)L(s)$ means the same thing as the determination of $F(n)$ by Theorem 148 as far as content is concerned.

---

[1] For this theorem compare E. Landau, Über einige neuere Grenzwertsätze. *Rendiconti del Circolo Matematico di Palermo* 34 (1912).

We now know that for the quadratic field with discriminant d

$$
\varkappa = \begin{cases} \dfrac{2 \log \varepsilon}{|\sqrt{d}|} & \text{if } d > 0 \text{ and } \varepsilon \text{ is the fundamental unit, } \varepsilon > 1, \\[3mm] \dfrac{2\pi}{w|\sqrt{d}|} & \text{if } d < 0; (w = 2 \text{ for } d < -4). \end{cases}
$$

Thus, for positive d we obtain from (145) the remarkable

Theorem 149. *The expression*

$$
\varepsilon^{2h} = e^{\sqrt{d} \sum_{n=1}^{\infty} \left(\frac{d}{n}\right)/n}
$$

represents a unit of infinite degree in $k(\sqrt{d})$ *with* $d > 0$. *And*

$$
\varepsilon^{2h} + \varepsilon'^{2h} = \varepsilon^{2h} + \varepsilon^{-2h} = e^{\sqrt{d}L(1)} + e^{-\sqrt{d}L(1)} = A
$$

is a rational integer such that this unit is the larger of the two roots of the equation

$$
x^2 - Ax + 1 = 0.
$$

The rational integer A can thus also be calculated numerically by estimating the remainder of the convergent series $L(1)$ and with this we have a transcendental method for finding a unit in real quadratic fields.

However, in each case the series $L(1)$ can be summed in a very visible way, and in particular a surprisingly simple expression is obtained for h for imaginary quadratic fields.

Since $\chi(n)$ is a periodic function of the integral argument n with period $|d|$ for $n > 0$, the idea of expanding $\chi(n)$ in a kind of finite Fourier series seems natural. Thus we try to determine the $|d|$ quantities c_n ($n = 0, 1, \ldots,$ $|d| - 1$) in such a way that

$$
\chi(a) := \sum_{n=0}^{|d|-1} c_n \zeta^{an} \qquad (\zeta = e^{2\pi i/|d|}) \tag{146}
$$

for

$$
a = 0, 1, \ldots, |d| - 1.
$$

These $|d|$ linear equations for the c_n can certainly be solved uniquely, as the determinant of the coefficients ζ^{an} is surely different from 0. For the calculation it is useful to define $\chi(n)$ and c_n for arbitrary n, therefore for negative rational integers n, by setting

$$
\chi(n) = \chi(m) \quad \text{and} \quad c_m = c_n, \quad \text{if } n \equiv m \pmod{d}.
$$

With this equation (146) is true for each rational integer a.

(For $\chi(n)$ with negative n, this does not always correspond to the condition $\chi(n) = \left(\frac{d}{n}\right)$, since by our earlier agreement $\left(\frac{d}{n}\right) = \left(\frac{d}{-n}\right)$.)

By Theorem 137, we obtain

$$\chi(n) = \chi(-n) \cdot \operatorname{sgn} d, \tag{147}$$

and from this follows an analogous property of c_n. For if we set

$$\chi(-a) = \sum_n c_n \zeta^{-an},$$

where n runs through an arbitrary complete system of residues mod d, then since the same holds for $-n$, we obtain

$$\chi(-a) = \sum_n c_{-n} \zeta^{an}, \qquad \chi(a) = \sum_n c_{-n} \operatorname{sgn} d \; \zeta^{an}.$$

Since the c_n are uniquely determined by (146)

$$c_{-n} = c_n \operatorname{sgn} d. \tag{147a}$$

We will take up the determination of the c_n later; however, even now we can put $L(1)$ into an essentially different form:

$$L(1) = \sum_{n=1}^{\infty} \frac{\chi(n)}{n} = \sum_{n=1}^{\infty} \frac{1}{n} \sum_{q=0}^{|d|-1} c_q \zeta^{qn}.$$

Now as is known from $\zeta^q \neq 1, |\zeta| = 1$,

$$-\log(1 - \zeta^q) = \sum_{n=1}^{\infty} \frac{\zeta^{qn}}{n},$$

in particular, this series converges for $q \not\equiv 0 \pmod d$. Thus we must have $c_0 = 0$ since $\sum_{n=1}^{\infty} 1/n$ diverges but the whole series $L(1)$ converges. Hence we write

$$L(1) = \sum_{q=1}^{|d|-1} c_q \sum_{n=1}^{\infty} \frac{\zeta^{qn}}{n}.$$

If we take the terms with q and with $|d| - q$ together and we consider (147a), we obtain

$$L(1) = \frac{1}{2} \sum_{q=1}^{|d|-1} c_q \sum_{n=1}^{\infty} \frac{\zeta^{qn} + \operatorname{sgn} d \cdot \zeta^{-qn}}{n}$$

and thus obtain two essentially different expressions for $d > 0$ and for $d < 0$:

1. $d < 0$

$$L(1) = i \sum_{q=1}^{|d|-1} c_q \sum_{n=1}^{\infty} \frac{\sin \dfrac{2\pi q n}{|d|}}{n}.$$

However, it is known that

$$\sum_{n=1}^{\infty} \frac{\sin 2\pi n x}{n} = \pi(\tfrac{1}{2} - x) \quad \text{for } 0 < x < 1.$$

Hence
$$L(1) = \frac{\pi i}{2} \sum_q c_q - \pi i \sum_{q=1}^{|d|-1} c_q \frac{q}{|d|}.$$

By (146), if we set $a = 0$ the first sum is equal to 0, hence
$$L(1) = -\frac{\pi i}{|d|} \sum_{n=1}^{|d|-1} n c_n$$

$$h = -\frac{wi}{2|\sqrt{d}|} \sum_{n=1}^{|d|-1} n c_n. \tag{148}$$

2. $d > 0$

$$L(1) = \frac{1}{2} \sum_{q=1}^{d-1} c_q \sum_{n=1}^{\infty} \frac{\zeta^{qn} + \zeta^{-qn}}{n} = \sum_{q=1}^{d-1} c_q \operatorname{Re}(\log(1 - \zeta^q))$$

$$= -\sum_{q=1}^{d-1} c_q \log|1 - \zeta^q| = -\sum_{q=1}^{d-1} c_q \log|e^{\pi i q/d} - e^{-\pi i q/d}|,$$

where $\operatorname{Re}(u)$ denotes the real part of u and the last symbol log denotes the real value. Thus
$$h = \frac{-|\sqrt{d}|}{2 \log \varepsilon} \sum_{n=1}^{d-1} c_n \log \sin \frac{\pi n}{d}. \tag{149}$$

In the two final formulas for h, the c_n must still be calculated from Equations (146), which will now be done.

§52 Gauss Sums and the Final Formula for the Class Number

To derive c_n we obtain immediately from the defining equations, by multiplication by ζ^{-am} and summation over a mod d,

$$\sum_{a=0}^{|d|-1} \chi(a)\zeta^{-am} = \sum_{n=0}^{|d|-1} c_n \sum_{a=0}^{|d|-1} \zeta^{a(n-m)} = c_m \cdot |d|$$

$$c_n = \frac{1}{|d|} \sum_{a=0}^{|d|-1} \chi(a)\zeta^{-an} = \frac{\chi(-1)}{|d|} \sum_a \chi(-a)\zeta^{-an}$$

$$= \frac{1}{d} \sum_{a=0}^{|d|-1} \chi(a)\zeta^{an}.$$

This last sum is called a Gauss sum. Gauss first investigated it and obtained its value, where the chief difficulty is the determination of the sign. In this section we wish to establish only its simplest properties, and to postpone closer investigation to the next chapter, where we treat the analogue of Gauss sums for arbitrary algebraic number fields.

In this section we set, for an arbitrary discriminant d of a quadratic field and an arbitrary rational integer n,

$$G(n,d) = \sum_{a \bmod d} \chi(a)e^{2\pi ian/|d|} \qquad (150)$$

where

$$\chi(-a) = \chi(a) \text{ sgn } d$$

and for a positive a

$$\chi(a) = \left(\frac{d}{a}\right).$$

It follows from the definition that

$$G(n_1, d) = G(n_2, d) \quad \text{if } n_1 \equiv n_2 \text{ (mod } d).$$

We show moreover that calculating $G(n, d)$ can be reduced to calculating $G(n, q)$, where q is a discriminant which is divisible only by a single prime. For this purpose, we set, in the notation of §47, in case $t > 1$,

$$d = (\pm q_1) \cdot (\pm q_2) \cdots (\pm q_t),$$

where the signs are chosen so that each $\pm q_i$ itself is a discriminant. Moreover, we define the residue character

$$\left.\begin{array}{c} \chi_r(n) = \left(\dfrac{\pm q_r}{n}\right) \\[2mm] \chi_r(-n) = \chi_r(n) \text{ sgn } (\pm q_r) \end{array}\right\} (r = 1, \ldots, t), n > 0, \qquad (151)$$

so that the Gauss sum $G(n, \pm q_r)$ can also be formed from the $\chi_r(n)$. Finally, we choose a special system of residues a mod d, namely,

$$a = a_1 \frac{|d|}{q_1} + \cdots + a_t \frac{|d|}{q_t},$$

where each a_r is to run through a complete system of residues mod q_r. Here

$$\chi(n) = \chi_1(n) \cdot \chi_2(n) \cdots \chi_t(n)$$

$$\chi_r(a) = \chi_r(a_r) \cdot \chi_r\left(\frac{|d|}{q_r}\right)$$

$$G(n, d) = \sum_{a_1, \ldots, a_t} \chi_r(a_1) \cdots \chi_t(a_t)e^{2\pi in(a_1/q_1 + \ldots + a_t/q_t)}C$$

with

$$C = \prod_{r=1}^{t} \chi_r\left(\frac{|d|}{q_r}\right). \qquad (152)$$

Hence we have

$$G(n, d) = C \prod_{r=1}^{t} G(n, \pm q_r), \qquad C = \pm 1. \qquad (153)$$

From this equation we obtain

$$G(n, d) = 0 \quad \text{if } (n, d) \neq 1, \qquad (154)$$

for if n has one odd prime q_r as a common factor with d, then for this q_r

$$G(n, \pm q_r) = G(0, \pm q_r) = \sum_{a \bmod q_r} \chi_r(a) = 0$$

by Theorem 31, since χ_r is a character mod q_r. However, if n and d have 2 as a common factor, then $G(n, -4)$ or $G(n, \pm 8)$ appears as the last factor in the product (153). But for even n this product is 0, as one can convince oneself by a calculation.

As the third property for G we find that for rational integers c, n

$$G(cn, d) = \chi(c)G(n, d), \quad \text{if } (c, d) = 1. \tag{155}$$

For

$$\chi(c)G(cn, d) = \sum_{a \bmod d} \chi(ac)e^{2\pi inac/|d|} = G(n, d),$$

since, along with a, ac also runs through a complete system of residues mod d. Since $\chi^2(c) = 1$, the assertion then follows.

Theorem 150. *For each rational integer n*

$$G(n, d) = \chi(n)G(1, d), \qquad c_n = \chi(n)\frac{G(1, d)}{d}.$$

For if n and d are not coprime, then by (154) both sides of the first equation are equal to 0. However, if $(n, d) = 1$, then we choose c in (155) in such way that $cn \equiv 1 \pmod d$; thus $\chi(c) = \chi(n)$.

For the complete determination of c_n we are only lacking the value of $G(1, d)$, which is independent of n.

Theorem 151. $G^2(1, d) = d$.

By Equation (153) we need only prove the theorem for those d which are divisible by only a single prime. For $d = -4$ or $d \pm 8$ the truth follows by a direct calculation. However for $|d|$ an odd prime we find

$$G^2(1, \pm q) = \sum_{a,b} \chi(a)\chi(b)\zeta^{a+b} = \sum_{a=1}^{q-1} \chi(a) \sum_{b=1}^{q-1} \chi(ab)\zeta^{a+ab}$$

$$= \sum_{b=1}^{q-1} \chi(b) \sum_{a=1}^{q-1} \zeta^{(b+1)a}.$$

Now

$$1 + \zeta^n + \cdots + \zeta^{(q-1)n} = \begin{cases} 0, & \text{if } (n, q) = 1, \\ q, & \text{if } n \equiv 0 \pmod q. \end{cases}$$

Thus

$$G^2(1, \pm q) = -\sum_{b \equiv -1 \,(\bmod q)} \chi(b) + (q - 1)\chi(-1)$$

$$= q\chi(-1) - \sum_{b \bmod d} \chi(b) = \pm q.$$

Now the problem arises of finding which of the two values of $\sqrt{d}$ defines the number $G(1,d)$, which was given in a transcendental way by means of the exponential function. This is the famous problem of determining the sign of the Gauss sums, which we will settle in the next chapter.

Theorem 152. *The class number h of the quadratic field with discriminant d has the value*

$$h = -\frac{\rho}{|d|} \sum_{n=1}^{|d|-1} n\left(\frac{d}{n}\right), \qquad \rho = \frac{-iG(1,d)}{|\sqrt{d}|} = \pm 1 \quad \text{for } d < -4.$$

$$h = \frac{\rho}{2 \log \varepsilon} \log \frac{\prod\limits_a \sin \frac{\pi a}{d}}{\prod\limits_b \sin \frac{\pi b}{d}}, \qquad \rho = \frac{G(1,d)}{|\sqrt{d}|} = \pm 1 \quad \text{for } d > 0.$$

In the second expression a and b run through those numbers $1, 2, \ldots, d-1$ for which

$$\left(\frac{d}{a}\right) = -1, \qquad \left(\frac{d}{b}\right) = +1$$

respectively. The final result will be that we always have $\rho = +1$ (§58). The formula for the class number of an imaginary field then becomes remarkably simple, and from its structure it appears to belong completely to elementary number theory. In spite of this, until now no one has succeeded in proving this formula by purely number-theoretic methods without the transcendental techniques of Dirichlet. Up to now we have not been able to even show that the expression for h is always positive by other methods. At present we can only take this formula, which is still completely incomprehensible to us, as a fact for calculations.

The second formula behaves likewise. In particular, we obtain from it the fact that the quotient $\prod_a / \prod_b$ is a unit of the field $k(\sqrt{d})$. This latter formula can also be proved rather easily from the theory of the $2d$th roots of unity, to which this number obviously belongs. However, until now it also has not been proved by purely number-theoretic methods that this unit is > 1 and that it is connected with the class number in the manner described above.

§53 Connection between Ideals in $k(\sqrt{d})$ and Binary Quadratic Forms

To conclude this chapter we present the connection between the modern theory of quadratic fields and the classical theory of binary quadratic forms, for which Gauss laid the foundations.

By a binary *quadratic form* in the variables x, y we mean an expression of the form

$$F(x, y) = Ax^2 + Bxy + Cy^2,$$

where the coefficients of the form, A, B, and C, are quantities independent of x and y and not all 0. Obviously such a form can always be represented as a product of two homogeneous linear functions of x, y:

$$F(x, y) = (\alpha a + \beta y)(\alpha' x + \beta' y). \tag{156}$$

The four quantities α, β, α', β' are, of course, not uniquely determined by A, B, C. For example, if $A \neq 0$

$$F(x, y) = \left(\sqrt{A}x + \frac{B + \sqrt{B^2 - 4AC}}{2\sqrt{A}} \, y \right)\left(\sqrt{A}x + \frac{B - \sqrt{B^2 - 4AC}}{2\sqrt{A}} \, y \right).$$

By comparing coefficients we confirm at once that

$$D = B^2 - 4AC = (\alpha\beta' - \alpha'\beta)^2 = \begin{vmatrix} \alpha & \beta \\ \alpha' & \beta' \end{vmatrix}^2. \tag{157}$$

This expression is called the *discriminant* (or also the *determinant*) of the form.

If we apply a homogeneous linear transformation

$$x = ax_1' + by_1', \qquad y = cx_1' + dy_1', \tag{158}$$

to the variables x, y, then $F(x, y)$ is transformed into a quadratic form in x', y'. If we choose the form (156), then

$$\begin{aligned}
F(ax_1 + by_1, cx_1 + dy_1) \\
= ((\alpha a + \beta c)x_1 + (\alpha b + \beta d)y_1)((\alpha' a + \beta' c)x_1 + (\alpha' b + \beta' d)y_1) \\
= A_1 x_1^2 + B_1 x_1 y_1 + C_1 y_1^2 = F_1(x_1, y_1).
\end{aligned}$$

The connection between the A, B, C, and the A_1, B_1, C_1 does not matter for us. However, we note that for the discriminant,

$$\begin{aligned}
D_1 = B_1^2 - 4A_1 C_1 &= \begin{vmatrix} \alpha a + \beta c & \alpha b + \beta d \\ \alpha' a + \beta' c & \alpha' b + \beta' d \end{vmatrix}^2 \\
&= \begin{vmatrix} \alpha & \beta \\ \alpha' & \beta' \end{vmatrix}^2 \begin{vmatrix} a & b \\ c & d \end{vmatrix}^2, \tag{159}
\end{aligned}$$

$$D_1 = D(ad - bc)^2.$$

If the discriminant of the transformation, $ad - bc$, is different from 0, then conversely the form $F_1(x_1, y_1)$ transforms into the original form $F(x, y)$ by a suitable transformation of x_1, y_1. For it follows from (158) that

$$x_1 = \frac{dx - by}{ad - bc}, \qquad y_1 = \frac{-cx + ay}{ad - bc}.$$

This transformation is said to be *reciprocal* (inverse) to the transformation (158). Its determinant is $1/(ad - bc)$.

We now consider exclusively those transformations where the coefficients a, b, c, d are rational integers with determinant $ad - bc = +1$, the so called *integral unimodular transformations*. The reciprocal of such a transformation likewise has this property, as the above formulas show.

Definition. If a form $F(x, y)$ is transformed into the form $F_1(x_1, y_1)$ by an integral unimodular transformation, then we say F is *equivalent* to F_1, in symbols

$$F \sim F_1.$$

By what has just been demonstrated, we also have $F_1 \sim F$, as F_1 is transformed to F by the reciprocal transformation. Thus the equivalence is symmetric in F and F_1. Furthermore $F \sim F$ always holds.

Lemma (a). *If*

$$F_1 \sim F \quad and \quad F_1 \sim F_2$$

holds for the three quadratic forms F, F_1, F_2, *then*

$$F \sim F_2.$$

In fact if there are two unimodular transformations with integral coefficients a, b, c, d and a_1, b_1, c_1, d_1 respectively, for which

$$F(ax + by, cx + dy) = F_1(x, y) \quad and \quad F_1(a_1 x + b_1 y, c_1 x + d_1 y) = F_2(x, y),$$

then, in the first equation, we set

$$x = a_1 x_1 + b_1 y_1, \qquad y = c_1 x_1 + d_1 y_1.$$

From now on we omit the index 1 from the variables x_1, y_1 whose designation does not really matter. By combination with the second equation it then follows that

$$F((aa_1 + bc_1)x + (ab_1 + bd_1)y, (ca_1 + dc_1)x + (cb_1 + dd_1)y) = F_2(x, y).$$

The arguments of F are obtained from x, y by an integral homogeneous linear transformation, and the determinant of its coefficients is

$$\begin{vmatrix} aa_1 + bc_1 & ab_1 + bd_1 \\ ca_1 + dc_1 & cb_1 + dd_1 \end{vmatrix} = (ad - bc)(a_1 d_1 - b_1 c_1) = 1.$$

That is, $F \sim F_2$. Thus the equivalence is transitive.

By a class of equivalent forms we mean the collection of all forms which are equivalent to a given form, say F, and we call F a representative of the class. By (159), all forms of a class have the same discriminant.

We restrict ourselves mainly to real forms, that is, to those forms with real coefficients. If F is a real form, then the same holds for all forms equivalent to F.

Theorem 153. *If D is the discriminant of F and $D > 0$, then $F(x, y)$ may take positive as well as negative values for appropriate real values of x, y. If $D < 0$, then either the value of F is ≥ 0 for all real x, y or $F(x, y) \leq 0$ for all real x, y; $F(x, y) = 0$ is possible only for $x = y = 0$.*

For the proof we consider the decomposition

$$A \cdot F(x, y) = \left(Ax + \frac{B}{2} y \right)^2 - \frac{D}{4} y^2.$$

Now if $D = B^2 - 4AC < 0$, then we must have $A \neq 0$ and it follows from the equation that

$$AF(x, y) \geq 0,$$

where the equality sign holds only if $y = 0$ and $Ax + \frac{B}{2}y = 0$, that is, $x = y = 0$. Consequently, $F(x, y)$ always has the sign of A if $x^2 + y^2 \neq 0$.

On the other hand, if $D > 0$, then, to begin with, suppose that $A \neq 0$. Then

$$A \cdot F(1, 0) = A^2 > 0$$
$$A \cdot F(B, -2A) = -DA^2 < 0;$$

hence both signs are possible for F and F can obviously also be zero for real x, y without x and y both vanishing.

If $D > 0$ and $A = 0$, then the equation

$$F(x, y) = y(Bx + Cy)$$

shows the truth of the assertion.

The form F is called *indefinite* if $D > 0$; on the other hand it is called *definite* if $D < 0$, and in the latter case it is called *positive definite* (respectively *negative definite*) if $F(x, y) \geq 0$ (respectively $F(x, y) \leq 0$).

From now on we consider integral forms exclusively, that is, forms with integral coefficients. The discriminant D is obviously congruent to 0 or 1 (mod 4).

Now let e be the discriminant of the quadratic field $k(\sqrt{e})$. We wish to develop a method *by which we can assign to each ideal class of $k(\sqrt{e})$ (in the strict sense) a class of equivalent forms with discriminant e.*

For this purpose let $\mathfrak{a}$ be an arbitrary integral ideal of a given class of $k(\sqrt{e})$.

We let

$$\alpha_1, \alpha_2 \text{ be a basis for } \mathfrak{a}, \text{ for which}$$
$$\alpha_1\alpha_2' - \alpha_2\alpha_1' = N(\mathfrak{a})\sqrt{e} \text{ is positive or} \tag{161}$$
$$\text{positive imaginary.}$$

To each ideal $\mathfrak{a}$ we assign the form

$$F(x, y) = \frac{(\alpha_1 x + \alpha_2 y)(\alpha_1' x + \alpha_2' y)}{|N(\mathfrak{a})|}.$$

This form obviously has rational integral coefficients, as the divisor of the coefficients of the numerator is equal to the product $\mathfrak{a}\mathfrak{a}' = N(\mathfrak{a})$ by Theorem 87. Furthermore, by Equation (157) the discriminant is

$$D = \frac{(\alpha_1\alpha_2' - \alpha_2\alpha_1')^2}{N(\mathfrak{a})^2} = e.$$

If a form F is derived from the ideal $\mathfrak{a}$ in this way, then we say: F *belongs to* $\mathfrak{a}$ and we write $F \to \mathfrak{a}$.

With $e < 0$ we obviously obtain only positive definite forms, for the first coefficient is

$$A = \frac{\alpha_1\alpha_1'}{|N(\mathfrak{a})|} = \frac{|N(\alpha_1)|}{|N(\mathfrak{a})|} > 0.$$

Lemma (b). *For each indefinite $(e > 0)$ or positive definite $(e < 0)$ integral form F with discriminant e there is an ideal $\mathfrak{a}$ such that $F \to \mathfrak{a}$.*

To begin with, the form $F(x, y) = Ax^2 + Bxy + Cy^2$, where $B^2 - 4AC = e$, is a primitive polynomial since if p divides A, B, C, then e/p^2 must also be a discriminant, which is possible only for discriminants of fields when $p = \pm 1$. We now consider the ideal

$$\mathfrak{m} = \left(A, \frac{B - \sqrt{e}}{2} \right),$$

where $\sqrt{e}$ denotes the positive (respectively positive imaginary) value. By Theorem 87, $N(\mathfrak{m}) = \mathfrak{m} \cdot \mathfrak{m}'$ is the content of the form

$$\left(Ax + \frac{B - \sqrt{e}}{2} y \right) \left(Ax + \frac{B + \sqrt{e}}{2} y \right) = AF(x, y)$$

$$N(\mathfrak{m}) = |A|.$$

Consequently the two numbers A and $(B - \sqrt{e})/2$ in $\mathfrak{m}$ are a basis for $\mathfrak{m}$, since the square of their determinant has the value $N^2(\mathfrak{m})e$. Hence in exactly the same way $\alpha_1 = \lambda A$ and $\alpha_2 = \lambda(B - \sqrt{e})/2$ is a basis for $\lambda\mathfrak{m}$ if λ is a number in k ($\lambda \neq 0$). Since

$$\alpha_1\alpha_2' - \alpha_2\alpha_1' = \lambda\lambda'A\sqrt{e},$$

this basis still has Property (161) if

$$\lambda\lambda'A > 0.$$

Hence we choose

(1) if $e < 0$, $\lambda = 1$ (since by hypothesis A is then > 0),
(2) if $e > 0$ and $A > 0$, again $\lambda = 1$,
(3) if $e > 0$ and $A < 0$, $\lambda = \sqrt{e}$.

Then in each case

$$\lambda\lambda'A = |N(\lambda m)|$$

and consequently $F \to \lambda m$.

Theorem 154. *Equivalent forms belong to equivalent ideals (in the strict sense) and conversely.*

We obtain

$$F(x, y) = \frac{(\alpha_1 x + \alpha_2 y)(\alpha'_1 x + \alpha'_2 y)}{|N(\mathfrak{a})|}$$

$$G(x, y) = \frac{(\beta_1 x + \beta_2 y)(\beta'_1 x + \beta'_2 y)}{|N(\mathfrak{b})|},$$ (162)

respectively from the basis α_1, α_2 for $\mathfrak{a}$ and the basis β_1, β_2 for $\mathfrak{b}$. Thus the two bases have Property (161).

Now if $F \sim G$, then there are rational integers a, b, c, d, with $ad - bc = 1$ such that

$$F(ax + by, cx + dy) = G(x, y),$$ (163)

$$\frac{((a\alpha_1 + c\alpha_2)x + (b\alpha_1 + d\alpha_2)y) \cdot ((a\alpha'_1 + c\alpha'_2)x + (b\alpha'_1 + d\alpha'_2)y)}{|N(\mathfrak{a})|}$$

$$= \frac{(\beta_1 x + \beta_2 y)(\beta'_1 x + \beta'_2 y)}{N(\mathfrak{b})}.$$

Since the quotients $-\beta_2/\beta_1$ and $-\beta'_2/\beta'_1$ are defined uniquely (except for order) as the zeros of $G(x, 1)$, we have

$$\frac{a\alpha_1 + c\alpha_2}{b\alpha_1 + d\alpha_2} = \frac{\beta_1}{\beta_2} \text{ or } \frac{\beta'_1}{\beta'_2}.$$

Thus there is a λ such that

$$a\alpha_1 + c\alpha_2 = \lambda\beta_1 \text{ or } \lambda\beta'_1$$
$$b\alpha_1 + d\alpha_2 = \lambda\beta_2 \text{ or } \lambda\beta'_2.$$

In both cases we have by (163)

$$\lambda\lambda' = \left|\frac{N(\mathfrak{a})}{N(\mathfrak{b})}\right| > 0.$$

Consequently only the first of these two cases can hold, as in the second we would have

$$(ad - bc)(\alpha_1\alpha'_2 - \alpha_2\alpha'_1) = -\lambda\lambda'(\beta_1\beta'_2 - \beta_2\beta'_1)$$

contrary to the assumption (161).

Since $ad - bc = 1$, $\lambda\beta_1$, $\lambda\beta_2$ is now also a basis for $\mathfrak{a}$; thus

$$\mathfrak{a} = \lambda(\beta_1, \beta_2) = \lambda \cdot \mathfrak{b}$$
$$\mathfrak{a} \approx \mathfrak{b}.$$

Conversely, suppose that $\mathfrak{a} \approx \mathfrak{b}$ and λ is a number with positive norm such that $\mathfrak{a} = \lambda\mathfrak{b}$. Then $\lambda\beta_1$, $\lambda\beta_2$ must be a basis for $\mathfrak{a}$ which thus arises from α_1, α_2 by an integral transformation with determinant ± 1. Hence there are rational integers a, b, c, d, such that

$$a\alpha_1 + c\alpha_2 = \lambda\beta_1, \qquad b\alpha_1 + d\alpha_2 = \lambda\beta_2.$$

It follows from Property (161) for both pairs and $N(\lambda) > 0$ that $ad - bc = +1$ and

$$\lambda\lambda' = \left| \frac{N(\mathfrak{a})}{N(\mathfrak{b})} \right|$$

and from this Equation (163) follows, that is, $F \sim G$.

By virtue of the fact stated in Theorem 154, the h_0 ideal classes of $k(\sqrt{d})$ are assigned in a well-defined and invertible manner to the classes of forms of discriminant e (when $e < 0$ only to the positive definite forms). *The number of nonequivalent integral forms with discriminant d is hence finite, and indeed equal to h_0, or, with $e < 0$ equal to $2h_0$, if we include positive definite and negative definite forms.* For example, each positive form with discriminant -4 is equivalent to $x^2 + y^2$, since $k(\sqrt{-4})$ has class number 1.

A large part of ideal theory can then be translated into the language of the theory of forms and vice versa. The latter is of particular interest for the classical theory of reduced forms, with the help of which it is possible to set up, by inequalities, a complete system of nonequivalent forms and to give, with this system, a far more convenient process for setting up all ideal classes than in §44.[2]

The theory of units (with norm $+1$) again appears in the theory of forms in the following way. All integral unimodular transformations which take a given form into itself can be listed. In fact for each unit ε with $N(\varepsilon) = +1$, along with α_1 and α_2, $\varepsilon\alpha_1$ and $\varepsilon\alpha_2$ is also always a basis for $\mathfrak{a}$, and thus there is a relation

$$\varepsilon\alpha_1 = a\alpha_1 + c\alpha_2, \qquad \varepsilon\alpha_2 = b\alpha_1 + d\alpha_2$$

where a, b, c, d are rational integers with determinant ± 1. If F again is taken as in (162), then obviously

$$F(ax + by, cx + dy) = F(x, y).$$

[2] This reduction theory likewise appears in the theory of elliptic modular functions which has a close relationship to quadratic number fields. Compare, for example, Klein-Fricke *Vorl. üb. d. Theorie d. Ellipt. Modulfunktionen*, Leipzig 1890–1892, Vol. I, 243–269, Vol. II, 161–203, as well as H. Weber, *Elliptische Funktionen and algebraische Zahlen* (= *Lehrbuch d. Algebra* Vol. III) 2nd edition, Braunschweig 1908.

To a large extent the theory of forms is concerned with the problem of which numbers can be represented by $F(x, y)$ if x, y run through all pairs of rational integers. Obviously this goes back to the problem of which numbers appear as norms of integral ideals in a given ideal class.

The difficult composition theory of classes of forms can be expressed very simply in the language of ideal theory if composition of forms is defined by that of the ideal classes.

The investigation of those forms whose discriminant is $Q^2 e$, where Q denotes a rational integer, is reduced to the number ring in $k(\sqrt{e})$ with conductor Q (§36). Of the numbers which occur in an ideal, only those which belong to this ring will be considered. In this way the concept of ideals of a ring and the concept of ideal classes arise. These concepts are then applied to a class of forms of discriminant $Q^2 e$.

CHAPTER VIII

The Law of Quadractic Reciprocity in Arbitrary Number Fields

§54 Quadratic Residue Characters and Gauss Sums in Arbitrary Number Fields

We first encountered Gauss sums when determining the class number of quadratic fields. Expressions of this type occur in many other problems and Gauss was the first to recognize the great importance which these sums have in number theory. His attention was directed to the connection between these sums and the quadratic reciprocity law and he showed how a proof for the reciprocity law is obtained by determining the value of these sums. Today we know a number of methods of evaluating these sums. Among them there is a transcendental method, due to Cauchy, which is of particular interest since it is capable of generalization.

The concept of the Gauss sum for an arbitrary algebraic number field was formulated by the author in 1919.[1] The Cauchy method of determining the value can in fact be carried over, thereby yielding a transcendental proof of the quadratic reciprocity law in each algebraic number field. This proof is to be presented in the following.

We lay the foundations for the investigation of an algebraic number field k, of degree n. First we will extend the concepts and theorems of §16, about quadratic residues, to the field k. We can be very brief here, as we have become sufficiently acquainted with the basic general group-theoretic concepts.

An integer or an integral ideal in k is called *odd* if it is relatively prime to 2.

---

[1] The so-called Lagrange resolvents in the theory of cyclotomic fields are generalizations in another direction.

Definition. Let $\mathfrak{p}$ be an odd prime ideal in k, α an arbitrary integer in k which is not divisible by $\mathfrak{p}$. We call α a *quadratic residue mod* $\mathfrak{p}$ and set

$$\left(\frac{\alpha}{\mathfrak{p}}\right) = +1,$$

if there is an integer ξ in k such that $\alpha \equiv \xi^2$ (mod $\mathfrak{p}$). In the other case we call α a *quadratic nonresidue* mod $\mathfrak{p}$ and we set

$$\left(\frac{\alpha}{\mathfrak{p}}\right) = -1.$$

Finally we set

$$\left(\frac{\alpha}{\mathfrak{p}}\right) = 0 \quad \text{if } \alpha \equiv 0 \ (\text{mod } \mathfrak{p}).$$

By Theorem 84 we see, as in §16, that for each integer α the symbol $\left(\frac{\alpha}{\mathfrak{p}}\right)$ denotes that one of the three numbers $0, 1, -1$ for which

$$\alpha^{(N(\mathfrak{p})-1)/2} \equiv \left(\frac{\alpha}{\mathfrak{p}}\right) (\text{mod } \mathfrak{p}). \tag{164}$$

If we have to deal with residue symbols in different number fields, we will distinguish them by attaching an index.

For integers α, β we have, once again,

$$\left(\frac{\alpha}{\mathfrak{p}}\right) = \left(\frac{\beta}{\mathfrak{p}}\right), \quad \text{if } \alpha \equiv \beta \ (\text{mod } \mathfrak{p})$$

$$\left(\frac{\alpha \cdot \beta}{\mathfrak{p}}\right) = \left(\frac{\alpha}{\mathfrak{p}}\right)\left(\frac{\beta}{\mathfrak{p}}\right).$$

Now let an odd integral ideal $\mathfrak{n}$ be decomposed into its prime ideal factors

$$\mathfrak{n} = \mathfrak{p}_1 \cdot \mathfrak{p}_2 \cdots \mathfrak{p}_r.$$

We then define for an arbitrary integer α (in k)

$$\left(\frac{\alpha}{\mathfrak{n}}\right) = \left(\frac{\alpha}{\mathfrak{p}_1}\right) \cdot \left(\frac{\alpha}{\mathfrak{p}_2}\right) \cdots \left(\frac{\alpha}{\mathfrak{p}_r}\right). \tag{165}$$

Thus this symbol is zero if α is not relatively prime to $\mathfrak{n}$, otherwise it is ± 1. We again have the rules of calculation

$$\left(\frac{\alpha}{\mathfrak{n}}\right) = \left(\frac{\beta}{\mathfrak{n}}\right), \quad \text{if } \alpha \equiv \beta \ (\text{mod } \mathfrak{n})$$

$$\left(\frac{\alpha\beta}{\mathfrak{n}}\right) = \left(\frac{\alpha}{\mathfrak{n}}\right) \cdot \left(\frac{\beta}{\mathfrak{n}}\right)$$

for any integers α and β. If k is the field of rational numbers, then the two definitions (164) and (165) agree with the earlier ones in §16.

We now assign a sum to each nonzero integer or fraction ω of k in the following manner:

Let $\mathfrak{d}$ denote the different of k and let $\mathfrak{d}\omega$ be represented as the quotient of relatively prime integral ideals $\mathfrak{a}$ and $\mathfrak{b}$:

$$\omega = \frac{\mathfrak{b}}{\mathfrak{a}\mathfrak{d}}; \qquad (\mathfrak{a}, \mathfrak{b}) = 1.$$

By Theorem 101 the trace $S(v\omega)$ is a rational integer for each integer v which is divisible by $\mathfrak{a}$. Consequently for integral v the number

$$e^{2\pi i S(v\omega)}$$

depends only on the residue class to which v belongs mod $\mathfrak{a}$. If we now form the sum

$$C(\omega) = \sum_{\mu \bmod \mathfrak{a}} e^{2\pi i S(\mu^2 \omega)}, \tag{166}$$

where μ runs through some complete system of residues mod $\mathfrak{a}$, then we obtain a number which depends only on ω, and which is independent of the special choice of the system of residues. We call such a sum a *Gauss sum in k* which belongs to the denominator $\mathfrak{a}$. We agree here that an addition to the $\sum$ sign, like "μ mod $\mathfrak{a}$" is to mean that the summation letter μ is to run through a complete system of residues mod $\mathfrak{a}$ possibly with further side conditions which will be stated.

In the rational number field these $C(\omega)$ are formally different from the Gauss sums defined in §52; however, as we will see at once, the latter can be reduced to the $C(\omega)$. If the denominator $\mathfrak{a}$ is $= 1$, then obviously $C(\omega) = 1$. We write

$$e^x = \exp x$$

whenever the formulas become more easily visible in this way.

Lemma (a). *Let $\mathfrak{d}\omega$ have denominator $\mathfrak{a}$. Then, if $\mathfrak{a} \neq 1$,*

$$\sum_{\mu \bmod \mathfrak{a}} e^{2\pi i S(\mu\omega)} = 0.$$

If α is an integer, then $\mu + \alpha$ runs through a complete system of residues mod $\mathfrak{a}$ whenever μ does. If we denote the value of the above sum by A, then we have

$$A = A \cdot e^{2\pi i S(\alpha\omega)}. \tag{167}$$

The exponential factor cannot be equal to 1 for each integer α, since then $S(\alpha\omega)$ would always be a rational integer and so, by Theorem 101, $\mathfrak{d}\omega$ would have to be an integer contrary to the hypothesis. Hence it follows from (167) that $A = 0$.

If $\varkappa_1, \varkappa_2, \alpha$ are integers, relatively prime to the denominator $\mathfrak{a}$ of $\mathfrak{d}\omega$, then

$$C(\varkappa_1\omega) = C(\varkappa_2\omega) \quad \text{if } \varkappa_1 \equiv \varkappa_2\alpha^2 \pmod{\mathfrak{a}}, \tag{168}$$

since $\mu\alpha$ runs through a complete system of residues mod $\mathfrak{a}$ simultaneously with μ; thus $C(\varkappa_2\omega) = C(\varkappa_2\alpha^2\omega)$. However, for μ an integer,

$$S(\mu^2\varkappa_1\omega) - S(\mu^2\varkappa_2\alpha^2\omega) = S(\mu^2(\varkappa_1 - \varkappa_2\alpha^2)\omega)$$

is then a rational integer because of the hypothesis; hence

$$C(\varkappa_2\alpha^2\omega) = C(\varkappa_1\omega).$$

We show moreover that the Gauss sums belonging to the denominator $\mathfrak{a}$ can be reduced to those with denominator $\mathfrak{a}_1$ and $\mathfrak{a}_2$, if $\mathfrak{a} = \mathfrak{a}_1\mathfrak{a}_2$ and the integral ideals $\mathfrak{a}_1$ and $\mathfrak{a}_2$ are relatively prime.

To prove this let $\mathfrak{c}_1$, $\mathfrak{c}_2$ be auxiliary integral ideals such that

$$\mathfrak{a}_1\mathfrak{c}_1 = \alpha_1, \qquad \mathfrak{a}_2\mathfrak{c}_2 = \alpha_2$$

are integers and $(\mathfrak{a}, \mathfrak{c}_1\mathfrak{c}_2) = 1$. In (166) we set

$$\omega = \frac{\beta}{\alpha_1\alpha_2}, \quad \text{where } \beta = \frac{b\mathfrak{c}_1\mathfrak{c}_2}{\mathfrak{d}}.$$

We obtain a complete system of residues mod $\mathfrak{a}_1$ in the form

$$\mu = \rho_1\alpha_2 + \rho_2\alpha_1,$$

where each of ρ_1 and ρ_2 run through a complete system of residues mod $\mathfrak{a}_1$ and mod $\mathfrak{a}_2$ respectively. Then

$$e^{2\pi i S(\mu^2\omega)} = e^{2\pi i S(\rho_1^2\alpha_2\beta/\alpha_1) + 2\pi i S(\rho_2^2\alpha_1\beta/\alpha_2)};$$

hence

$$C(\omega) = C\left(\frac{\beta}{\alpha_1\alpha_2}\right) = C\left(\frac{\alpha_2\beta}{\alpha_1}\right) C\left(\frac{\alpha_1\beta}{\alpha_2}\right). \tag{169}$$

Using Equation (169), the calculation of $C(\omega)$ can be reduced to the calculation of a Gauss sum, whose denominator is a power of a prime ideal.

For odd denominators the reduction can be carried still further, namely until the denominators are prime ideals.

For let the denominator $\mathfrak{a} = \mathfrak{p}^a$ be the power of an odd prime ideal $\mathfrak{p}$, with $a \geq 2$. If $\mathfrak{c}$ again denotes an integral auxiliary ideal which is not divisible by $\mathfrak{p}$, such that $\mathfrak{p}\mathfrak{c} = \alpha$ is a number, then

$$\omega = \frac{\beta}{\alpha^a}, \quad \text{where } \beta = \frac{b\mathfrak{c}^a}{\mathfrak{d}},$$

and we have the recursion formula

$$C\left(\frac{\beta}{\alpha^a}\right) = N(\mathfrak{p}) \, C\left(\frac{\beta}{\alpha^{a-2}}\right) \tag{170}$$

The sum on the right obviously belongs to the denominator $\mathfrak{p}^{a-2}$.

To prove our formula, we choose a complete system of residues mod $\mathfrak{p}^a$ in the form

$$\mu + \rho\alpha^{a-1},$$

where
$$\mu \bmod \mathfrak{p}^{a-1}, \quad \rho \bmod \mathfrak{p}$$

each run through a complete system of residues. Then

$$C\left(\frac{\beta}{\alpha^a}\right) = \sum_{\mu \bmod \mathfrak{p}^{a-1}} \sum_{\rho \bmod \mathfrak{p}} \exp\left\{2\pi i S\left(\frac{(\mu + \rho\alpha^{a-1})^2\beta}{\alpha^a}\right)\right\}$$

$$= \sum_{\mu \bmod \mathfrak{p}^{a-1}} \left\{\exp\left\{2\pi i S\left(\frac{\mu^2\beta}{\alpha^a}\right)\right\} \sum_{\rho \bmod \mathfrak{p}} \exp\left\{2\pi i S\left(\frac{2\mu\rho}{\alpha}\beta\right)\right\}\right\}.$$

By Lemma (a) the sum over ρ is equal to zero if 2μ is not divisible by $\mathfrak{p}$, that is, (since $\mathfrak{p}$ is odd) if μ is not divisible by $\mathfrak{p}$. In the other case this sum is equal to $N(\mathfrak{p})$, since each of its terms is equal to 1. Consequently

$$C\left(\frac{\beta}{\alpha^a}\right) = N(\mathfrak{p}) \sum_{\substack{\mu \bmod \mathfrak{p}^{a-1} \\ \mu \equiv 0 \,(\bmod \mathfrak{p})}} \exp\left\{2\pi i S\left(\frac{\mu^2\beta}{\alpha^a}\right)\right\}.$$

Thus μ is assumed to run through all numbers $\nu\alpha$ where ν is a complete system of residues mod $\mathfrak{p}^{a-2}$, that is, the asserted Equation (170) is true.

By repeated application of this formula, for even a, we arrive at the sum $C(\beta)$, which belongs to the denominator 1 and thus is equal to 1. Thus we obtain

Lemma (b). *If the denominator of $\mathfrak{d}\beta/\alpha^a$ is equal to $\mathfrak{p}^a$, where $\mathfrak{p}$ is an odd prime ideal which divides α precisely to the first power, then*

$$C\left(\frac{\beta}{\alpha^a}\right) = \begin{cases} N(\mathfrak{p})^{a/2}, & \text{if } a \text{ is even,} \\ N(\mathfrak{p})^{(a-1)/2} C\left(\dfrac{\beta}{\alpha}\right), & \text{if } a \text{ is odd.} \end{cases}$$

A similar reduction is possible for prime ideals $\mathfrak{p}$ which divide 2. However, we do not need to use this in later applications.

Theorem 155. *Suppose that the denominator $\mathfrak{a}$ of $\mathfrak{d}\omega$ is an odd ideal. Then for every integer $\varkappa$ which is relatively prime to $\mathfrak{a}$*

$$C(\varkappa\omega) = \left(\frac{\varkappa}{\mathfrak{a}}\right) C(\omega).$$

To begin with, the theorem is true if $\mathfrak{a}$ is a prime ideal $\mathfrak{p}$, since if we apply Lemma (a), we have

$$\sum_{\mu \bmod \mathfrak{p}} \left(\frac{\mu}{\mathfrak{p}}\right) e^{2\pi i S(\mu\omega)} = \sum_{\mu \bmod \mathfrak{p}} \left(\left(\frac{\mu}{\mathfrak{p}}\right) + 1\right) e^{2\pi i S(\mu\omega)}.$$

In this second sum only those terms where μ is a quadratic residue mod $\mathfrak{p}$ have a value different from zero, except for the term which corresponds to the residue class $\mu = 0$. Hence the value of this sum is

$$1 + 2 \sum_{\mu^2} e^{2\pi i S(\mu^2 \omega)}$$

where μ^2 now only runs through the distinct quadratic residues, excluding 0. This is precisely the sum $C(\omega)$, since each square except 0 occurs in this sum exactly twice. Hence

$$C(\omega) = \sum_{\mu \bmod \mathfrak{p}} \left(\frac{\mu}{\mathfrak{p}}\right) e^{2\pi i S(\mu\omega)}. \tag{171}$$

If we replace μ by $\mu\varkappa$, a process which does not change the value of the sum, we obtain the equation stated in the theorem.

By Lemma (b) the assertion is also valid if the denominator $\mathfrak{a}$ is a power $\mathfrak{p}^a$ of a prime ideal since, for even a, $\left(\frac{\varkappa}{\mathfrak{p}^a}\right) = \left(\frac{\varkappa}{\mathfrak{p}}\right)^a = 1$ and the Gauss sum actually has the same value for ω and for $\varkappa\omega$. However, by what has just been proved, the additional factor $\left(\frac{\varkappa}{\mathfrak{p}}\right) = \left(\frac{\varkappa}{\mathfrak{p}^a}\right)$ appears for odd a.

Finally, Formula (169) then immediately implies the truth of our theorem for an arbitrary odd denominator.

We deduce from (171) that in fact the sums $G(1, d)$, defined in §52 for the rational number field, are closely connected to the Gauss sums $C(\omega)$; and if $C(\omega)$ is determined, then $G(1, d)$ is also determined.

Finally we further conclude from (169) and Lemma (b):

Theorem 156. *If the Gauss sum $C(\omega)$ belongs to the denominator $\mathfrak{a}$, and if $\mathfrak{a}$ is the square of an odd ideal then*

$$C(\omega) = \left|\sqrt{N(\mathfrak{a})}\right|.$$

§55 Theta-Functions and Their Fourier Expansions

The analytic tool which will lead to the determination of the Gauss sums is the theta-function of n variables. The two concepts are connected in the following way.

Let us take the ground field $k = k(1)$ as the simplest case. We then investigate the function of τ defined by the following series:

$$\theta(\tau) = \sum_{m=-\infty}^{\infty} e^{-\pi\tau m^2}.$$

This series (a so-called simple theta-series) converges as long as the real part of τ is positive. The imaginary axis is seen to be the natural boundary ("singular line") of the analytic function $\theta(\tau)$. Now we investigate the behavior of $\theta(\tau)$ as τ approaches the singular point $\tau = 2ir$, where r is a rational number. It is seen that $\theta(\tau)$ becomes infinite and that

$$\lim_{\tau \to 0} \sqrt{\tau}\theta(\tau + 2ir)$$

exists. This limit is the Gauss sum $C(-r)$ defined in the preceding section, except for unimportant numerical factors. Moreover, the behavior of $\theta(\tau)$ can be determined in yet a second way. There exists a "transformation formula" for $\theta(\tau)$:

$$\theta\left(\frac{1}{\tau}\right) = \sqrt{\tau}\theta(\tau).$$

By this formula the behavior of $\theta(\tau)$ at the point $\tau = 2ir$ is related to the behavior of $\theta(\tau')$ at the point

$$\tau' = \frac{1}{2ir} = -\frac{2i}{4r}.$$

As stated above, the behavior of the latter is related to the Gauss sum $C(1/4r)$. By comparing the two results we obtain a relation between $C(r)$ and $C(-1/4r)$ from which $C(r)$ can be determined and from which, with the help of the formulas of the preceding section, the reciprocity law follows.

Suppose that the field k has degree n, and that k as well as all of its conjugate fields $k^{(p)}$ are real. Then in place of the simple theta-series, the n-tuple theta-series

$$\sum_{\mu} e^{-\pi(t_1\mu^{(1)2} + t_2\mu^{(2)2} + \cdots + t_n\mu^{(n)2})}$$

arises, where $t_1, \ldots, t_n$ are variables with positive real part, and the summation is to be extended over all integers μ of the field k. In this series, we set $t_p = w + 2i\omega^{(p)}$ where ω is a number from k and let the positive quantity w tend to zero.

Finally, if k is a general algebraic number field, among whose conjugates $k^{(1)}, \ldots, k^{(r)}$ are real and the remaining conjugates are not real, then we again have to investigate an n-tuple series. But then we do not get by with one and the same function of $t_1, \ldots, t_n$ to obtain all sums $C(\omega)$, but rather we need the functions

$$\sum_{\mu} \exp\left\{-\pi \sum_{p=1}^{n} t_p|\mu^{(p)}|^2 + 2\pi i \sum_{p=1}^{n} \omega^{(p)}\mu^{(p)2}\right\},$$

which depend on ω, in the neighborhood of the point $t_1 = t_2 = \cdots = t_n = 0$. Here μ again runs through all integers of k.

Even in this sketch of the proof, we should note what these arguments have in common with the transcendental methods of Chapter VI. The fact is that *precise knowledge of the behavior of an analytic function in the neighborhood of its singular points is a source of number-theoretic theorems.*

Because the absolute values of the $\mu^{(p)}$ appear in the individual terms, the derivation of the necessary formulas in the most general case becomes more complicated. In order to make the main ideas of the proof more easily understandable, we first discuss in the next section the formally simpler case in which all conjugate fields of k are real.

To begin with, we develop the train of thought which leads to the definition and exposition of the theta-series and their transformation formulas.

By a *quadratic form in the n variables* $x_1, x_2, \ldots, x_n$ we mean an expression of the form

$$Q(x_1, \ldots, x_n) = \sum_{i,k=1}^{n} a_{ik}x_i x_k = a_{11}x_1^2 + 2a_{12}x_1 x_2 + \cdots,$$

where the coefficients a_{ik} are real or complex quantities independent of $x_1, \ldots, x_n$ with the symmetry property $a_{ik} = a_{ki}$.

A quadratic form with real coefficients is called *positive definite*, if for all real $x_1, \ldots, x_n$

$$Q(x_1, \ldots, x_n) \geq 0,$$

and the equality sign holds only for $x_1 = x_2 = \cdots = 0$. For example $x_1^2 + x_2^2 + \cdots + x_n^2$ is a positive definite form in $x_1, \ldots, x_n$.

Lemma (a). *For each positive definite form $Q(x_1, \ldots, x_n)$ there is a positive quantity c, such that for all real $x_1, \ldots, x_n$*

$$Q(x_1, \ldots, x_n) \geq c(x_1^2 + x_2^2 + \cdots + x_n^2). \tag{172}$$

By hypothesis $Q(y_1, \ldots, y_n) > 0$ for all points of the n-dimensional sphere $y_1^2 + y_2^2 + \cdots + y_n^2 = 1$. Consequently the continuous function Q has a positive minimum c on the surface of the sphere, that is,

$$Q(y_1, \ldots, y_n) \geq c \quad \text{if } y_1^2 + \cdots + y_n^2 = 1.$$

Thus if we set

$$y_i = \frac{x_i}{\sqrt{x_1^2 + \cdots + x_n^2}} \qquad (i = 1, 2, \ldots, n),$$

for arbitrary real x_i not all 0, Formula (172) then follows.

Theorem 157. *Let $Q(x_1, \ldots, x_n) = \sum_{i,k=1}^{n} a_{ik}x_i x_k$ be a quadratic form with real or complex coefficients such that the real part of Q is positive definite. Moreover, let $u_1, \ldots, u_n$ be real variables. Then*

$$\sum_{m_1, \ldots, m_n = -\infty}^{\infty} e^{-\pi Q(m_1 + u_1, \ldots, m_n + u_n)} \tag{173}$$

is an absolutely convergent series and thus represents a function $T(u_1, \ldots, u_n)$. This function, together with all its derivatives with respect to the u_i, is continuous and moreover has period 1 in each of the variables.

The series (173) is called an *n-tuple theta-series*.

To prove this let the real part of Q be denoted by Q_0. By Lemma (a) there is a positive c such that

$$Q_0(m_1 + u_1, \ldots, m_n + u_n) \geq c((m_1 + u_1)^2 + \cdots + (m_n + u_n)^2).$$

Furthermore

$$\left|e^{-\pi Q}\right| = e^{-\pi Q_0} \leq e^{-\pi c \sum_{i=1}^{n} (m_i + u_i)^2}.$$

If we now restrict the real numbers u_i to a domain $|u_i| \leq C/2$, then we obtain

$$\left|e^{-\pi Q}\right| \leq e^{-\pi c \sum_{i=1}^{n} (m_i^2 - C|m_i|) + K},$$

where K is an appropriate constant.

However, since the inequality

$$|m_1| + \cdots + |m_n| \leq \sqrt{n(m_1^2 + \cdots + m_n^2)} \leq \varepsilon\sqrt{n}(m_1^2 + \cdots + m_n^2)$$

is true for every $\varepsilon > 0$, provided that

$$m_1^2 + \cdots + m_n^2 > \frac{1}{\varepsilon^2}, \tag{174}$$

we obtain the estimate

$$\left|e^{-\pi Q}\right| \leq \exp\{-\pi c(1 - \varepsilon C\sqrt{n})(m_1^2 + \cdots + m_n^2) + K\}.$$

If we take ε sufficiently small, then $a = c(1 - \varepsilon C\sqrt{n}) > 0$ and the terms of the given series are thus smaller in absolute value than the corresponding terms of the obviously convergent series with constant terms

$$\sum_{m_1, \ldots, m_n} e^{-\pi a(m_1^2 + \cdots + m_n^2) + K}$$

(with at most finitely many exceptions which do not satisfy (174)). Thus the series of absolute values of (173) is uniformly convergent and the sum is a continuous function of $u_1, \ldots, u_n$. This function $T(u_1, \ldots, u_n)$ has period 1 in each of the variables. For example, if we replace the summation index m_1 by $m_1 - 1$, $T(u_1 + 1, u_2, \ldots, u_n)$ is transformed to $T(u_1, \ldots, u_n)$.

In the same way we can see that the series which arise from T by differentiating termwise, one or more times, converge uniformly. Since

$$Q(m_1 + u_1, \ldots, m_n + u_n) = Q(m_1, \ldots, m_n) + 2\sum_{i,k=1}^{n} a_{ik}m_i u_k + Q(u_1, \ldots, u_n),$$

it is sufficient to investigate the termwise differentiation of

$$\sum_{m_1, \ldots, m_n} \exp\{-\pi Q(m_1, \ldots, m_n) - 2\pi \sum_{i,k=1}^{n} a_{ik}m_i u_k\}.$$

Under differentiation, products of powers of $m_1, \ldots, m_n$ and linear combinations of such expressions are adjoined to the individual terms as factors.

Since $|m| < e^{|m|}$ we have

$$|m_1^{c_1} \cdots m_n^{c_n}| < e^{c_1|m_1| + \cdots + c_n|m_n|} \qquad (c_i \geq 0)$$

and, reasoning as above, we then have the uniform convergence of the differentiated series. Therefore the theorem is completely proved.

We now obtain the transformation formula for the theta-function, which we discussed at the beginning of this section, by expressing the periodicity of the function T in terms of its development in a Fourier series and indeed by using the following fact which we quote from analysis.

Let $\varphi(u_1, \ldots, u_n)$ be a (real or complex) function of n real variables, which is periodic with period 1 in each argument. Moreover, suppose that all partial derivatives of φ up to order 2n are continuous. Then φ can be expanded in an absolutely convergent Fourier series

$$\varphi(u_1, \ldots, u_n) = \sum_{m_1, \ldots, m_n} a(m_1, \ldots, m_n)e^{-2\pi i(m_1 u_1 + \cdots + m_n u_n)},$$

in which the coefficients have the following values:

$$a(m_1, \ldots, m_n) = \int_0^1 \cdots \int_0^1 e^{2\pi i(m_1 u_1 + \cdots + m_n u_n)}\varphi(u_1, \ldots, u_n)\, du_1\, du_2 \cdots du_n.$$

For $n = 1$ this theorem is usually proved in textbooks in analysis. Then it can easily be proved in general by induction on n.

If we set φ equal to the theta-function, which indeed does satisfy our hypotheses, then we obtain

$$a(m_1, \ldots, m_n)$$
$$= \int_0^1 \cdots \int_0^1 e^{2\pi i(m_1 u_1 + \cdots + m_n u_n)} \sum_{k_1, \ldots, k_n = -\infty}^{+\infty} e^{-\pi Q(k_1 + u_1, \ldots, k_n + u_n)}\, dU,$$

for the coefficients, where we set $dU = du_1\, du_2 \cdots du_n$. Now we interchange summation and integration, which is permissible because of the uniform convergence and then we introduce $u_1 - k_1, \ldots, u_n - k_n$ as new variables of integration. As a result the $k_1, \ldots, k_n$ disappear from the integrand; instead they appear in the limits of integration and we obtain

$$a(m_1, \ldots, m_n)$$
$$= \sum_{k_1, \ldots, k_n} \int_{-k_1}^{-k_1 + 1} \cdots \int_{-k_n}^{-k_n + 1} e^{2\pi i(m_1 u_1 + \cdots + m_n u_n) - \pi Q(u_1, \ldots, u_n)}\, dU.$$

The sum of all these integrals can be written as a single integral over the entire infinite space and with this we have proved:

Theorem 158. *The n-tuple theta-function*

$$T(u_1, \ldots, u_n) = \sum_{m_1, \ldots, m_n = -\infty}^{+\infty} e^{-\pi Q(m_1 + u_1, \ldots, m_n + u_n)}$$

admits the representation

$$T(u_1, \ldots, u_n) = \sum_{m_1, \ldots, m_n = -\infty}^{+\infty} a((m))e^{-2\pi i(m_1 u_1 + \cdots + m_n u_n)}$$

where

$$a((m)) = a(m_1, \ldots, m_n)$$

$$= \int_{-\infty}^{+\infty} \cdots \int e^{-\pi Q(u_1, \ldots, u_n) + 2\pi i(m_1 u_1 + \cdots + m_n u_n)} \, du_1 du_2 \cdots du_n.$$

We now wish to substitute specially chosen forms for Q and then evaluate the integrals.

§56 Reciprocity between Gauss Sums in Totally Real Fields

In this section we assume that the algebraic number field k in which we investigate the Gauss sums of §54 is *totally real*, that is, all conjugate fields $k^{(p)}$ are real. Moreover let $\mathfrak{a}$ denote a nonzero ideal in k ($\neq 0$) with basis $\alpha_1, \ldots, \alpha_n$. Then by $t_1, \ldots, t_n$ we understand, for the time being, n positive real variables and we choose for the quadratic form Q of Theorem 158

$$Q(x_1, \ldots, x_n) = \sum_{p=1}^{n} t_p(\alpha_1^{(p)} x_1 + \cdots + \alpha_n^{(p)} x_n)^2,$$

which is obviously positive. The corresponding theta-function is

$$\theta(t, z; \mathfrak{a}) = \sum_{\mu \text{ in } \mathfrak{a}} \exp\left\{-\pi \sum_{p=1}^{n} t_p(\mu^{(p)} + z_p)^2\right\}, \tag{175}$$

where

$$z_p = \sum_{q=1}^{n} \alpha_q^{(p)} u_q \qquad (p = 1, \ldots, n). \tag{176}$$

In the series (175) μ runs through all numbers in $\mathfrak{a}$ exactly once. The Fourier coefficients $a(m_1, \ldots, m_n)$ from Theorem 158 have the values

$$a(m_1, \ldots, m_n) = \int_{-\infty}^{+\infty} \cdots \int \exp\left\{-\pi \sum_{p=1}^{n} t_p z_p^2 + 2\pi i \sum_{p=1}^{n} m_p u_p\right\} dU,$$

where the z_p are again connected with the variables of integration u_p by (176).
We now introduce the z_p as variables of integration in this integral. The inverse of Equations (176) is

$$u_k = \sum_{p=1}^{n} \beta_k^{(p)} z_p \qquad (k = 1, \ldots, n),$$

where by Theorem 102, the numbers $\beta_1, \ldots, \beta_n$ form a basis for the ideal $1/\mathfrak{a}\mathfrak{d}$ in k. We then have

$$\sum_{k=1}^{n} m_k u_k = \sum_{p=1}^{n} \lambda^{(p)} z_p \tag{177}$$

where $\lambda = \sum_{k=1}^{n} \beta_k m_k$ is a number from $1/\mathfrak{a}\mathfrak{d}$.

$$a((m)) = \frac{1}{|N(\mathfrak{a})\sqrt{d}|} \int_{-\infty}^{+\infty} \cdots \int \exp\left\{-\pi \sum_{p=1}^{n} t_p z_p^2 + 2\pi i \sum_{p=1}^{n} \lambda^{(p)} z_p\right\} dz_1 \cdots dz_n.$$

Now for positive t and real λ we have

$$\int_{-\infty}^{+\infty} e^{-\pi t z^2 + 2\pi i \lambda z}\, dz = e^{-\pi \lambda^2/t} \int_{-\infty}^{+\infty} e^{-\pi t(z - i\lambda/t)^2}\, dz = \frac{e^{-\pi \lambda^2/t}}{\sqrt{t}}, \tag{178}$$

where $\sqrt{t}$ denotes the positive value. Thus the coefficient a is the product of n such integrals, and with this we finally obtain from the theorem of the preceding section:

Theorem 159. *The theta-series defined by (175) also admits the representation*

$$\theta(t, z; \mathfrak{a})$$
$$= \frac{1}{N(\mathfrak{a})|\sqrt{d}|\sqrt{t_1 \cdot t_2 \cdots t_n}} \sum_{\lambda \text{ in } 1/\mathfrak{a}\mathfrak{b}} \exp\left\{-\pi \sum_{p=1}^{n} \frac{\lambda^{(p)2}}{t_p} - 2\pi i \sum_{p=1}^{n} \lambda^{(p)} z_p\right\}. \tag{179}$$

On the right side λ runs through all numbers of the ideal $1/\mathfrak{a}\mathfrak{d}$ in k.

We now see at once that this equation also holds for nonreal t, provided only that the real part of each t_p is positive. For then the real part of $1/t_p$ is also positive and the series on both sides of the formula represent analytic functions of $t_1, \ldots, t_n$, which are regular for $\mathfrak{R}(t_p) > 0$ $(p = 1, \ldots, n)$ by the uniform convergence in t. Thus the above formula also holds for arbitrary t which belong to the right half-plane, if by $\sqrt{t_p}$ we understand those single-valued analytic functions which are positive for positive t and thus whose argument lies between $-\pi/4$ and $+\pi/4$, where we set

$$\sqrt{t_1 \cdots t_n} = \sqrt{t_1} \cdot \sqrt{t_2} \cdots \sqrt{t_n}.$$

If we take $z_1 = \cdots = z_n = 0$ and write $\mathfrak{f}$ instead of $\mathfrak{a}$, then we conclude from Theorem 159:

Theorem 160. *The transformation formula*

$$\theta(t; \mathfrak{f}) = \frac{1}{N(\mathfrak{f})|\sqrt{d}|\sqrt{t_1 \cdots t_n}} \theta\left(\frac{1}{t}; \frac{1}{\mathfrak{f}\mathfrak{d}}\right) \tag{180}$$

holds for the functions of $t_1, \ldots, t_n$

$$\theta(t; \mathfrak{f}) = \theta(t, 0; \mathfrak{f}) = \sum_{\mu \text{ in } \mathfrak{f}} \exp\left\{-\pi \sum_{p=1}^{n} t_p \mu^{(p)2}\right\}.$$

Moreover, we deduce from Theorem 159

Lemma (a). *If the complex variables* $t_1, \ldots, t_n$ *simultaneously tend to zero in such a way that the real parts of* $1/t_p$ *tend to plus infinity, then*

$$\lim_{t \to 0} \sqrt{t_1 \cdots t_n}\, \theta(t, z; \mathfrak{a}) = \frac{1}{N(\mathfrak{a})\sqrt{d}},$$

independently of z.

For if we denote the smallest of the n numbers $\mathfrak{R}(1/t_p)$ by r, then

$$\left|\exp\left\{-\pi \sum_{p=1}^{n} \frac{1}{t_p} \lambda^{(p)2}\right\}\right| \le \exp\left\{-\pi r \sum_{p=1}^{n} \lambda^{(p)2}\right\} \le e^{-\pi r c(m_1^2 + \cdots + m_n^2)},$$

where, according to (172), c is a suitably chosen positive constant independent of the t_p. The sum on the right side of (179) with the term $m_1 = \cdots = m_n = 0$ excluded is thus numerically

$$\le \left(\sum_{m=-\infty}^{\infty} e^{-\pi r c m^2}\right)^n - 1 < \left(1 + 2\sum_{m=1}^{\infty} e^{-\pi r c m}\right)^n - 1 = \left(1 + \frac{2e^{-\pi r c}}{1 - e^{-\pi r c}}\right)^n - 1$$

from which Lemma (a) follows if we take the limit as $r \to \infty$.

Formula (180) will now yield the relation we sought between two Gauss sums in k if we take $\mathfrak{f} = 1$. Let ω be a number different from zero in k and let $\mathfrak{d}\omega$ have denominator $\mathfrak{a}$ and numerator $\mathfrak{b}$:

$$\omega = \frac{\mathfrak{b}}{\mathfrak{a}\mathfrak{d}}, \qquad (\mathfrak{a}, \mathfrak{b}) = 1.$$

In (180) we set

$$t_p = x - 2i\omega^{(p)}, \qquad \mathfrak{f} = 1,$$

where x is a positive quantity.

Now, by Lemma (a), we determine how both sides of (180) behave as x approaches 0.

To begin with,

$$\theta(x - 2i\omega; 1) = \sum_{\mu} \exp\left\{-\pi \sum_{p=1}^{n} (x - 2i\omega^{(p)})\mu^{(p)2}\right\}$$

$$= \sum_{\rho \bmod \mathfrak{a}} e^{2\pi i S(\omega \rho^2)} \sum_{\nu \text{ in } \mathfrak{a}} \exp\left\{-\pi \sum_{p=1}^{n} x(\nu^{(p)} + \rho^{(p)})^2\right\}$$

since $\mu = \nu + \rho$ runs through all integers of the field if ρ runs through a complete system of residues mod $\mathfrak{a}$ and ν runs through all numbers from $\mathfrak{a}$.

Here the inner sum over v is again a theta-series, hence

$$\theta(x - 2i\omega; 1) = \sum_{\rho \bmod \mathfrak{a}} e^{2\pi i S(\rho^2 \omega)} \theta(x, \rho; \mathfrak{a}).$$

Finally by Lemma (a) we have

$$\lim_{x \to 0} x^{n/2} \theta(x - 2i\omega; 1) = \frac{C(\omega)}{N(\mathfrak{a})|\sqrt{d}|}, \tag{181}$$

where $C(\omega)$ denotes the Gauss sum of §54.

In exactly the same way we deduce the behavior of the right side in (180) at $x = 0$. We set

$$\frac{1}{t_p} = \frac{i}{2\omega^{(p)}} + \tau_p, \quad \text{where } \tau_p = \frac{-ix}{2\omega^{(p)}(x - 2i\omega^{(p)})},$$

and thus the real part of $1/\tau_p$ is

$$\Re\left(\frac{1}{\tau_p}\right) = \frac{4\omega^{(p)^2}}{x};$$

as $x \to 0$ it grows beyond all bounds. Furthermore, by $\mathfrak{c}$ we mean an integral auxiliary ideal so that $\mathfrak{c}\mathfrak{d}$ is a principal ideal $\mathfrak{c}\mathfrak{d} = \delta$, and $(\mathfrak{c}, 2\mathfrak{b}) = 1$. The numbers of $1/\mathfrak{d}$ are then of the form μ/δ where μ runs through all numbers in $\mathfrak{c}$. In this way we obtain

$$\theta\left(\frac{1}{t}; \frac{1}{\mathfrak{d}}\right) = \sum_{\mu \text{ in } \mathfrak{c}} \exp\left\{-\pi \sum_{p=1}^{n}\left(\tau_p + \frac{i}{2\omega^{(p)}}\right)\frac{\mu^{(p)^2}}{\delta^{(p)^2}}\right\}.$$

Now let

$$\mathfrak{b}_1 \text{ be the denominator of } \frac{\mathfrak{d}\mathfrak{c}^2}{4\omega\delta^2} = \frac{\mathfrak{a}}{4\mathfrak{b}}. \tag{182}$$

Then let us set $\mu = v + \rho$ in this sum, where ρ runs through a complete residue system mod $\mathfrak{b}_1$, in which each element is divisible by $\mathfrak{c}$ and v runs through all numbers of $\mathfrak{b}_1\mathfrak{c}$. We obtain

$$\theta\left(\frac{1}{t}; \frac{1}{\mathfrak{d}}\right) = \sum_{\substack{\rho \bmod \mathfrak{b} \\ \rho \equiv 0(\mathfrak{c})}} e^{-2\pi i S(\rho^2/4\omega\delta^2)} \sum_{v \text{ in } \mathfrak{b}_1\mathfrak{c}} \exp\left\{-\pi \sum_{p=1}^{n} \frac{\tau_p}{\delta^{(p)^2}}(v^{(p)} + \rho^{(p)})^2\right\}$$

$$= \sum_{\substack{\rho \bmod \mathfrak{b}_1 \\ \rho \equiv 0(\mathfrak{c})}} e^{-2\pi i S(\rho^2/4\omega\delta^2)} \theta\left(\frac{\tau}{\delta^2}, \rho; \mathfrak{b}_1\mathfrak{c}\right).$$

Thus by Lemma (a) we know that if x, that is, τ_p, tends to zero

$$\lim_{x \to 0} \sqrt{\frac{\tau_1 \cdots \tau_n}{N(\delta)^2}} \, \theta\left(\frac{1}{t}; \frac{1}{\mathfrak{d}}\right) = \frac{A}{N(\mathfrak{b}_1, \mathfrak{c})|\sqrt{d}|},$$

where we set

$$A = \sum_{\substack{\rho \bmod \mathfrak{b}_1 \\ \rho \equiv 0(\mathfrak{c})}} e^{-2\pi i S(\rho^2/4\omega\delta^2)}. \tag{183}$$

From our convention for the meaning of the square root signs we have

$$\lim_{x \to 0} \frac{1}{|x^{n/2}|} \sqrt{\frac{\tau_1 \cdots \tau_n}{N(\delta)^2}} = \frac{1}{|N(2\omega\delta)|},$$

so that we can also write

$$\lim_{x \to 0} x^{n/2} \theta\left(\frac{1}{t}; \frac{1}{\mathfrak{d}}\right) = \frac{|N(2\omega\delta)|}{|N(\mathfrak{b}_1 \mathfrak{c})\sqrt{d}|} \cdot A. \qquad (184)$$

Finally, after multiplication by $x^{n/2}$, if we then let the quantity x tend to zero in Formula (180) with $\mathfrak{f} = 1$ and keep in mind that in the denominator

$$\lim_{x \to 0} \sqrt{(x - 2i\omega^{(1)}) \cdots (x - 2i\omega^{(n)})}$$

$$= \left|\sqrt{N(2\omega)}\right| e^{-(\pi i/4)(\operatorname{sgn}\omega^{(1)} + \operatorname{sgn}\omega^{(2)} + \cdots + \operatorname{sgn}\omega^{(n)})},$$

it follows from (181), (184), and $|d| = N(\mathfrak{d})$ that

$$\frac{C(\omega)}{N(\mathfrak{a})} = |\sqrt{d}| \cdot \frac{\sqrt{N(2\omega)}}{N(\mathfrak{b}_1)} A \cdot e^{(\pi i/4)S(\operatorname{sgn}\omega)}$$

$$\frac{C(\omega)}{|\sqrt{N(\mathfrak{a})}|} = \left|\frac{\sqrt{N(2\mathfrak{b})}}{N(\mathfrak{b}_1)}\right| e^{(\pi i/4)S(\operatorname{sgn}\omega)} A \qquad \left(S(\operatorname{sgn}\omega) = \sum_{p=1}^{n} \operatorname{sgn}\omega^{(p)}\right).$$

Now the quantity A is likewise a Gauss sum and it indeed belongs to the denominator $\mathfrak{b}_1$. For if α denotes an integer divisible by $\mathfrak{c}$ such that $\alpha/\mathfrak{c}$ is coprime to $\mathfrak{b}_1$, then we can replace ρ by $\rho\alpha$ in (183), and let ρ run through a complete system of residues mod $\mathfrak{b}_1$; from this we see that

$$A = C\left(-\frac{1}{4\omega} \frac{\alpha^2}{\delta^2}\right).$$

If we set $\alpha/\delta = \gamma$, we finally obtain

Theorem 161. *The reciprocity*

$$\frac{C(\omega)}{|\sqrt{N(\mathfrak{a})}|} = \left|\frac{\sqrt{N(2\mathfrak{b})}}{N(\mathfrak{b}_1)}\right| e^{(\pi i/4)S(\operatorname{sgn}\omega)} C\left(-\frac{1}{4\omega} \gamma^2\right).$$

holds between the Gauss sums where $\mathfrak{a}$ denotes the denominator of $\mathfrak{d}\omega$, and $\mathfrak{b}$ denotes the numerator of $\mathfrak{d}\omega$. Moreover, $\mathfrak{b}_1$ is the denominator of $\mathfrak{a}/4\mathfrak{b}$ and γ is an arbitrary number of the field such that $\mathfrak{d}\gamma$ is integral and relatively prime to $\mathfrak{b}_1$.

The method of proof, with which we have just become acquainted, will become more transparent if it is first carried out for the special case where the different $\mathfrak{d}$ of the field is a principal ideal, because the introduction of an auxiliary ideal $\mathfrak{c}$ becomes superfluous.

§57 Reciprocity between Gauss Sums in Arbitrary Algebraic Number Fields

Now let the number field k of degree n be arbitrary and let the conjugates be numbered as in §34, such that for all numbers μ in k, $\mu^{(p)}$ is real for $p = 1, 2, \ldots, r_1$ while $\mu^{(p)}$ is the complex conjugate of $\mu^{(p+r_2)}$ for $p = r_1 + 1, \ldots, r_1 + r_2$. We now consider the function

$$\theta(t, z, \omega; \mathfrak{a}) = \sum_{\mu \text{ in } \mathfrak{a}} \exp\left\{-\pi \sum_{p=1}^{n} [t_p|\mu^{(p)} + z_p|^2 - 2i\omega^{(p)}(\mu^{(p)} + z_p)^2]\right\}, \quad (185)$$

belonging to an arbitrary nonzero ideal $\mathfrak{a}$ of k, where μ runs through all numbers of $\mathfrak{a}$ and the symbols have the following meaning:

$t_p > 0$ for all $p = 1, \ldots, n$,

$t_{p+r_2} = t_p$ for $p = r_1 + 1, \ldots, r_1 + r_2$,

$z_p, \omega^{(p)}$ real for $p = 1, 2, \ldots, r_1$,

$\left.\begin{array}{c} z_{p+r_2} \\ \omega^{(p+r_2)} \end{array}\right\}$ are complex conjugates to $\left\{\begin{array}{c} z_p \\ \omega^{(p)} \end{array}\right.$ for $p = r_1 + 1, \ldots, r_1 + r_2$.

If $\alpha_1, \ldots, \alpha_n$ again denotes a basis for $\mathfrak{a}$, and we set

$$z_p = \sum_{k=1}^{n} \alpha_k^{(p)} u_k, \qquad \mu^{(p)} = \sum_{k=1}^{n} \alpha_k^{(p)} m_k, \quad (186)$$

where the $u_1, \ldots, u_n$ are real and $m_1, \ldots, m_n$ are rational integers, then we see that the exponent appearing in (185) is a quadratic form in $m_1 + u_1, \ldots, m_n + u_n$ whose real part is positive definite. Consequently the series converges and Theorem 158 can be applied ot it.

The Fourier coefficient here has the following value:

$$a((m)) = \int_{-\infty}^{+\infty} \cdots \int \exp\left\{-\pi \sum_{p=1}^{n} [t_p|z_p|^2 - 2i\omega^{(p)}z_p^2 - 2im_p u_p]\right\} dU, \quad (187)$$

where the $z_1, \ldots, z_n$ are again related to the variables of integration $u_1, \ldots, u_n$ by (186). If we express the u's in terms of the z's then, by Theorem 102, the exponent takes the following form as does the analogous formula in the preceeding paragraph:

$$-\pi \sum_{p=1}^{n} [t_p|z_p|^2 - 2i\omega^{(p)}z_p^2 - 2i\lambda^{(p)}z_p],$$

where

$$\lambda = \sum_{k=1}^{n} \beta_k m_k$$

is a number in $1/\mathfrak{a}\mathfrak{d}$ and the β form a basis for $1/\mathfrak{a}\mathfrak{d}$ defined by

$$\sum_{p=1}^{n} \beta_q^{(p)} \alpha_k^{(p)} = \begin{cases} 0 & \text{for } q \neq k, \\ 1 & \text{for } q = k. \end{cases}$$

We now introduce the real and imaginary parts of the z_p as real variables of integration in place of the u_p. We set

$$\left.\begin{matrix} z_p = x_p + iy_p \\ z_{p+r_2} = x_p - iy_p \end{matrix}\right\} \quad p = r_1 + 1, \ldots, r_1 + r_2$$

and

$$z_p = x_p, \qquad p = 1, \ldots, r_1.$$

The functional determinant of the $u_1, \ldots, u_n$ with respect to the x, y has absolute value

$$\frac{2^{r_2}}{|N(\mathfrak{a})\sqrt{d}|}, \tag{188}$$

as was used already in §40 and the exponent becomes

$$-\pi \sum_{p=1}^{r_1} (t_p - 2i\omega^{(p)})x_p^2$$

$$-\pi \sum_{p=r_1+1}^{r_1-r_2} \left[2t_p(x_p^2 + y_p^2) - 2i(\omega^{(p)}(x_p + iy_p)^2 + \omega^{(p)}(x_p - iy_p)^2)\right]$$

$$+2\pi i \sum_{p=1}^{r_1} \lambda^{(p)}x_p + 2\pi i \sum_{p=r_1+1}^{r_1+r_2} \left[\lambda^{(p)}(x_p + iy_p) + \overline{\lambda}^{(p)}(x_p - iy_p)\right].$$

(The bar again denotes the complex conjugate.) By this substitution the integral in (187) becomes a product of r single integrals, each with respect to one of the variables $x_1, \ldots, x_{r_1}$, and a product of r_2 double integrals with respect to the r_2 pairs x_p, y_p.

For $p = 1, \ldots, r_1$ we obtain

$$\int_{-\infty}^{\infty} \exp\{-\pi(t_p - 2i\omega^{(p)})x^2 + 2\pi i\lambda^{(p)}x\} \, dx$$

$$= \frac{1}{\sqrt{t_p - 2i\omega^{(p)}}} e^{-\pi\lambda^{(p)2}/(t_p - 2i\omega^{(p)})}. \tag{189}$$

Here the square root should be taken with real part positive.

The double integrals are of the following form

$$J = \int\!\!\!\int_{-\infty}^{+\infty} \exp\{-2\pi t(x^2 + y^2) + 2\pi i(\omega(x + iy)^2 + \overline{\omega}(x - iy)^2$$

$$+ \lambda(x + iy) + \overline{\lambda}(x - iy))\} \, dx \, dy.$$

Now if $\omega = 0$, we obtain, just as before, the value of the integral:

$$J = \frac{e^{(-2\pi/t)|\lambda|^2}}{2t} \quad \text{if } \omega = 0.$$

On the other hand if $\omega \neq 0$, then we bring the quadratic form in x, y into the form of a sum of squares by introducing the real variables u, v:

$$\sqrt{\omega}(x + iy) = u + iv$$
$$\sqrt{\bar{\omega}}(x - iy) = u - iv.$$

Here we choose some fixed $\sqrt{\omega}$, and we choose $\sqrt{\bar{\omega}}$ as its complex conjugate. For the functional determinant we obtain

$$\frac{\partial(x, y)}{\partial(u, v)} = \frac{1}{\sqrt{\omega}} \frac{1}{\sqrt{\bar{\omega}}} = \frac{1}{|\omega|},$$

and the exponent in the integrand now reads

$$-2\pi t \frac{u^2 + v^2}{|\omega|} + 4\pi i(u^2 - v^2) + 2\pi i\left(\frac{\lambda}{\sqrt{\omega}}(u + iv) + \frac{\bar{\lambda}}{\sqrt{\bar{\omega}}}(u - iv)\right)$$

$$= \left(-\frac{2\pi t}{|\omega|} + 4\pi i\right)u^2 + 2\pi i\left(\frac{\lambda}{\sqrt{\omega}} + \frac{\bar{\lambda}}{\sqrt{\bar{\omega}}}\right)u$$

$$+ \left(-\frac{2\pi t}{|\omega|} - 4\pi i\right)v^2 + 2\pi i\left(\frac{i\lambda}{\sqrt{\omega}} - \frac{i\bar{\lambda}}{\sqrt{\bar{\omega}}}\right)v.$$

Thus J is represented as a product of two simple integrals and indeed we find

$$J = \frac{1}{2\sqrt{t^2 + 4|\omega|^2}} \exp\left\{\frac{-2\pi t}{t^2 + 4|\omega|^2}|\lambda|^2 - \frac{2\pi i}{t^2 + 4|\omega|^2}(\lambda^2\bar{\omega} + \bar{\lambda}^2\omega)\right\}, \quad (190)$$

a formula which is clearly still valid for $\omega = 0$.

If we choose λ and ω real in this expression, then the exponent is exactly twice the exponent appearing on the right-hand side of (189).

Finally for $a(m_1, \ldots, m_n)$ we obtain the value

$$a(m_1, \ldots, m_n) = \frac{1}{N(\mathfrak{a})|\sqrt{d}|W(t, \omega)} \exp\left\{-\pi \sum_{p=1}^{n} \tau_p|\lambda^{(p)}|^2 + 2\pi i \sum_{p=1}^{n} \lambda^{(p)2}\varkappa^{(p)}\right\}$$

$$\left.\begin{array}{l}
\tau_p = \dfrac{t_p}{t_p^2 + 4|\omega^{(p)}|^2} \\[3mm]
\varkappa^{(p)} = \dfrac{-\bar{\omega}^{(p)}}{t_p^2 + 4|\omega^{(p)}|^2}, \\[3mm]
W(t, \omega) = \displaystyle\prod_{p=1}^{r_1} \sqrt{t_p - 2i\omega^{(p)}} \cdot \prod_{p=r_1+1}^{r_1+r_2} \sqrt{t_p^2 + 4|\omega^{(p)}|^2} \\[3mm]
\lambda^{(p)} = \displaystyle\sum_{q=1}^{n} \beta_q^{(p)}m_q.
\end{array}\right\} \quad (191)$$

Here the square roots are to be taken with positive real parts.

If, in (185), we choose $z_1, \ldots, z_n$ as well as $u_1, \ldots, u_n$ equal to zero, then we obtain the following transformation formula:

Theorem 162. *The transformation formula*

$$\theta(t, 0, \omega; \mathfrak{f}) = \frac{1}{N(\mathfrak{f}) | \sqrt{d} | W(t, \omega)} \, \theta\left(\tau, 0, \varkappa; \frac{1}{\mathfrak{f}\mathfrak{d}}\right); \tag{192}$$

is true for the function defined by (185), where the relation between t, ω and $\tau, \varkappa$ is given by (191).

In order to find the behavior of the two theta-series appearing here as we approach $t_1 = t_2 = \cdots = t_n = 0$, we must know the behavior of $\theta(t, z, \omega; \mathfrak{f})$ at this point. This is determined by

Lemma (a). *Let $\sigma_1(t_1), \sigma_2(t_2), \ldots, \sigma_n(t_n)$ be functions of $t_1, \ldots, t_n$ respectively such that $\sigma_{p+r_2} = \bar{\sigma}_p$ for $p = r_1 + 1, \ldots, r_1 + r_2$ and σ_p is real for $p = 1, 2, \ldots, r_1$. Then if $t_1, \ldots, t_n$ tend to 0 simultaneously,*

$$\lim_{t \to 0} \sqrt{t_1 t_2 \cdots t_n}\, \theta(t, z, t \cdot \sigma; \mathfrak{f}) = \frac{1}{N(\mathfrak{f}) | \sqrt{d} |},$$

independently of z, provided

$$\lim_{t_p \to 0} \sigma_p(t_p) = 0.$$

To prove this we need only apply Theorem 158 to the series and substitute the value of the coefficient a found above. Specifically if we choose the number λ to denote a single term in place of $m_1, \ldots, m_n$, we have

$$\theta(t, z, t \cdot \sigma; \mathfrak{f}) = M \sum_{\lambda \text{ in } 1/\mathfrak{f}\mathfrak{d}} b(\lambda) e^{2\pi i \sum\limits_{p=1}^{n} \lambda^{(p)} z_p} \tag{193}$$

with the values

$$M = \frac{1}{N(\mathfrak{f}) | \sqrt{d} | W(t, t\sigma)}$$

$$b(\lambda) = \exp\left\{ -\pi \sum_{p=1}^{n} \frac{|\lambda^{(p)}|^2}{t_p(1 + 4|\sigma_p|^2)} - 2\pi i \sum_{p=1}^{n} \frac{\lambda^{(p)2} \bar{\sigma}_p}{t_p(1 + 4|\sigma_p|^2)} \right\}.$$

Now since

$$\lim_{t \to 0} \sqrt{t_1 \cdots t_n} \cdot M = \frac{1}{N(\mathfrak{f}) | \sqrt{d} |} \lim_{t \to 0} \frac{\sqrt{t_1 \cdots t_n}}{W(t, t\sigma)} = \frac{1}{N(\mathfrak{f}) | \sqrt{d} |}$$

and if we move the term with $\lambda = 0$ to the other side, in the series in (193), then we obtain the inequality

$$|\theta(t, \ldots) - M| \leq M \sum_{\substack{\lambda \text{ in } 1/\mathfrak{f}\mathfrak{d} \\ \lambda \neq 0}} \exp\left\{ -\pi \sum_{p=1}^{n} \frac{|\lambda^{(p)}|^2}{t_p(1 + 4|\sigma_p|^2)} \right\},$$

from which we can read off the statement of Lemma (a) in the same way as in the preceding section.

We now obtain a Gauss sum if, in (192), we take ω equal to a number in k different from 0 and $\mathfrak{f} = 1$:

$$\omega = \frac{b}{\mathfrak{a}\mathfrak{d}}, \qquad (\mathfrak{a}, b) = 1.$$

We have

$$\theta(t, 0, \omega; 1) = \sum_{\rho \bmod \mathfrak{a}} e^{2\pi i S(\rho^2 \omega)} \theta(t, \rho, 0; \mathfrak{a}).$$

Thus by Lemma (a)

$$\lim_{t \to 0} \sqrt{t_1 \cdots t_n} \theta(t, 0, \omega; 1) = \frac{C(\omega)}{N(\mathfrak{a}) |\sqrt{d}|}. \tag{194}$$

To investigate the right-hand side of (192) we introduce an auxiliary integral ideal $\mathfrak{c}$ such that

$$\mathfrak{c}\mathfrak{d} = \delta \text{ is a number in } k \text{ and } (\mathfrak{c}, 4b) = 1.$$

Moreover let

$$b_1 \text{ be the denominator of } \frac{\mathfrak{a}}{4b}.$$

Again it follows directly from the definition of the theta-series that

$$\theta\left(\tau, 0, \varkappa; \frac{1}{\mathfrak{d}}\right) = \theta\left(\frac{\tau}{|\delta|^2}, 0, \frac{\varkappa}{\delta^2}; \mathfrak{c}\right) = \sum_{\substack{\rho \bmod b_1 \\ \rho \equiv 0(\mathfrak{c})}} \theta\left(\frac{\tau}{|\delta|^2}, \rho, \frac{\varkappa}{\delta^2}; b_1 \mathfrak{c}\right).$$

By (191) we now have

$$\varkappa^{(p)} = \frac{-\bar{\omega}^{(p)}}{t_p^2 + 4|\omega^{(p)}|^2} = -\frac{1}{4\omega^{(p)}} + \frac{t_p^2}{4\omega^{(p)}(t_p^2 + 4|\omega^{(p)}|^2)}$$

$$\varkappa^{(p)} = -\frac{1}{4\omega^{(p)}} + \tau_p \sigma_p, \quad \text{where } \sigma_p = \frac{t_p}{4\omega^{(p)}},$$

$$\theta\left(\frac{\tau}{|\delta|^2}, \rho, \frac{\varkappa}{\delta^2}; b_1 \mathfrak{c}\right) = \theta\left(\frac{\tau}{|\delta|^2}, \rho, \frac{-1}{4\omega\delta^2} + \frac{\tau\sigma}{\delta^2}; b_1 \mathfrak{c}\right)$$

$$= e^{2\pi i S(-\rho^2/4\omega\delta^2)} \theta\left(\frac{\tau}{|\delta|^2}, \rho, \frac{\tau\sigma}{\delta^2}; b_1 \mathfrak{c}\right).$$

Lemma (a) can again be applied to the last theta-series if we let the t, that is, the τ, tend to zero, so we obtain

$$\lim_{t \to 0} \sqrt{\frac{\tau_1 \cdots \tau_n}{N(\delta)^2}} \, \theta\left(\tau, 0, \varkappa; \frac{1}{\mathfrak{d}}\right) = \frac{1}{N(b_1\mathfrak{c})|\sqrt{d}|} \sum_{\substack{\rho \bmod b_1 \\ \rho \equiv 0(\mathfrak{c})}} e^{-2\pi i S(\rho^2/4\omega\delta^2)}. \tag{195}$$

As was proved at the end of the preceding section, the last sum is again $C(-\gamma^2/4\omega)$, where γ is an arbitrary number in k for which

$$\mathfrak{d}\gamma \text{ is integral and relatively prime to } \mathfrak{b}_1. \tag{196}$$

Then Equation (195) can be written

$$\lim_{t \to 0} \sqrt{t_1 \cdots t_n}\, \theta\left(\tau, 0, \varkappa; \frac{1}{\mathfrak{d}}\right) = \left|\frac{N(2\omega)}{N(\mathfrak{b}_1)}\sqrt{d}\right| C\left(\frac{-\gamma^2}{4\omega}\right). \tag{197}$$

Finally if, after multiplication by $\sqrt{t_1 \cdots t_n}$, we let all the t tend to zero in Formula (192), and keep in mind that

$$\lim_{t \to 0} W(t, \omega) = \left|\sqrt{N(2\omega)}\right| e^{-(\pi i/4)S(\text{sgn } \omega)},$$

where

$$S(\text{sgn } \omega) = \text{sgn } \omega^{(1)} + \cdots + \text{sgn } \omega^{(r_1)} \qquad (= 0 \text{ if } r_1 = 0), \tag{198}$$

then we obtain from (194) and (197):

Theorem 163. *The reciprocity*

$$\frac{C(\omega)}{\left|\sqrt{N(\mathfrak{a})}\right|} = \left|\frac{\sqrt{N(2\mathfrak{b})}}{N(\mathfrak{b}_1)}\right| e^{(\pi i/4)S(\text{sgn } \omega)} C\left(\frac{-\gamma^2}{4\omega}\right) \tag{199}$$

holds for Gauss sums in k. Here $\mathfrak{a}$, $\mathfrak{b}$ are relatively prime integral ideals, $\omega = \mathfrak{b}/\mathfrak{a}\mathfrak{d}$, $\mathfrak{b}_1$ is the denominator of $\mathfrak{a}/4\mathfrak{b}$ and γ and $S(\text{sgn } \omega)$ are defined by (196) and (198).

This equation agrees formally with the conclusion of the preceding section, where, however, it was proved only for totally real fields.[2]

§58 The Determination of the Sign of Gauss Sums in the Rational Number Field

Formula (199) makes it possible for us to determine the value of Gauss sums. In this section we wish to undertake this determination for the field of rational numbers and to settle the question raised at the conclusion of §52, Theorem 152.

The different of $k(1)$ is 1.

---

[2] L. J. *Mordell* (1920), proved this reciprocity formula for quadratic fields without the theta-function, by using only the Cauchy integral theorem: On the reciprocity formula for the Gauss's sums in the quadratic field, *Proc. of the London Math Soc.*, Ser. 2, Vol. 20 (4). A related formula can already be found in A. *Krazer*, Zur Theorie der mehrfachen Gaussschen Summen, Weber Festschrift (1912).

Thus if a, b are relatively prime rational integers, then

$$C\left(\frac{b}{a}\right) = \sum_{n \bmod a} e^{2\pi i (n^2 b/a)}.$$

For odd a the reciprocity formula, Theorem 163, asserts

$$\frac{C\left(\frac{1}{a}\right)}{|\sqrt{a}|} = \frac{e^{(\pi i/4) \operatorname{sgn} a}}{2\sqrt{2}} C\left(\frac{-a}{4}\right) = \frac{e^{(\pi i/4) \operatorname{sgn} a}}{2\sqrt{2}} \sum_{n \bmod 4} e^{-2\pi i (n^2 a/4)}$$

$$C\left(\frac{-a}{4}\right) = 2(1 + e^{-(\pi i/2)a}) = 2(1 + (-i)^a) = 2(1 - i^a)$$

$$e^{(\pi i/4) \operatorname{sgn} a} = \frac{\sqrt{2}}{2}(1 + i \operatorname{sgn} a)$$

$$\frac{C\left(\frac{1}{a}\right)}{\sqrt{a}} = \tfrac{1}{2}(1 + i \operatorname{sgn} a)(1 - i^a) = \begin{cases} 1 & \text{if } a > 0, \ a \equiv 1(4) \\ i & \text{if } a > 0, \ a \equiv 3(4) \end{cases}$$

$$C\left(\frac{1}{a}\right) = \sqrt{(-1)^{(a-1)/2} a}, \quad \text{for } a > 0,$$

where the root is to be taken positive (respectively positive imaginary). On the other hand, for primes a we have

$$C\left(\frac{1}{|a|}\right) = \sum_{n \bmod a} \left(\frac{n}{a}\right) e^{2\pi i (n/|a|)}$$

by (171). However, we have for an odd discriminant $a = d$, by (127),

$$\left(\frac{n}{a}\right) = \left(\frac{a}{n}\right) \quad \text{for } n > 0.$$

Hence we have for odd prime discriminant a

$$\sum_{n=1}^{|a|-1} \left(\frac{a}{n}\right) e^{2\pi i (n/|a|)} = \sqrt{(-1)^{(|a|-1)/2} |a|} = \sqrt{a},$$

where the root is to be taken positive (positive imaginary).

For odd field discriminant d, by Equation (150), the Gauss sum $G(1, d)$ is thus

$$G(1, d) = \sqrt{d} \quad \text{if } d \text{ is an odd prime,} \tag{200}$$

with $\sqrt{d}$ equal to a positive or positive imaginary quantity.

Now if d_1 and d_2 are two odd relatively prime discriminants, then by §52

$$G(1, d_1 d_2) = \sum_{n \bmod d_1 d_2} \left(\frac{n}{d_1}\right)\left(\frac{n}{d_2}\right) e^{2\pi i (n/|d_1 d_2|)}$$

$$= \left(\frac{|d_2|}{d_1}\right)\left(\frac{|d_1|}{d_2}\right) G(1, d_1) G(1, d_2)$$

$$= (-1)^{((\operatorname{sgn} d_1 - 1)/2)((\operatorname{sgn} d_2 - 1)/2)} G(1, d_1) G(1, d_2).$$

From this it follows that if (200) holds for two relatively prime discriminants d_1, d_2, then it is also valid for the product. Consequently, (200) is also true for each odd discriminant.

Finally $G(1, -4)$ and $G(1, \pm 8)$ must still be calculated. We find

$$G(1, -4) = 2i, \qquad G(1, 8) = 2|\sqrt{2}|, \qquad G(1, -8) = 2i|\sqrt{2}|. \qquad (201)$$

Finally, if u is an odd discriminant and q is a discriminant without odd prime factors, then by (152) and (153) in §52 we again have

$$G(1, qu) = \left(\frac{q}{u}\right)\left(\frac{u}{q}\right) G(1, q) G(1, u) = (-1)^{((\operatorname{sgn} q - 1)/2)((\operatorname{sgn} u - 1)/2)} G(1, q) G(1, u),$$

from which, with the values (201), it finally follows that

Theorem 164. *The Gauss sums $G(1, d)$ for the discriminant d of a quadratic number field have the value*

$$G(1, d) = \sqrt{d}$$

with positive (respectively positive imaginary) root.

The numerical factor ρ in the class number formula of Theorem 152 thus has the value $+1$ as was already stated there.

§59 The Quadratic Reciprocity Law and the First Part of the Supplementary Theorem

We will now derive the quadratic reciprocity law for an arbitrary algebraic number field from Formula (199). First we define:

An integer in k is said to be *primary* if it is odd and congruent to the square of a number in k modulo 4.

A number α in k is said to be *totally positive*, if among its conjugates, the r_1 numbers $\alpha^{(1)}, \ldots, \alpha^{(r_1)}$ are positive.

If all the fields conjugate to k are not real ($r_1 = 0$), then each number in k is said to be totally positive. Even then we may not overlook the fact that the statement "α is totally positive" has a meaning only with respect to a given field which contains α. For example -1 is not totally positive in $k(1)$, however it is indeed totally positive in the field $k(i)$.

In order to make the simple ideas underlying our proof clear, we first make the simplifying assumption that the different $\mathfrak{d}$ of the field k is a principal ideal (in the broadest sense), that is, there is a number δ in k such that

$$(\delta) = \mathfrak{d}.$$

Now let α and β be two relatively prime odd integers. If in (199) we set

$$\omega = \frac{1}{\alpha\beta\delta}, \qquad \gamma = \frac{1}{\delta}, \qquad \mathfrak{a} = \alpha\beta, \qquad \mathfrak{b} = 1, \qquad \mathfrak{b}_1 = 4,$$

then

$$\frac{C\left(\dfrac{1}{\alpha\beta\delta}\right)}{|\sqrt{N(\alpha\beta)}|} = \frac{e^{(\pi i/4)S(\mathrm{sgn}\ \alpha\beta\delta)}}{|\sqrt{N(8)}|} C\left(\frac{-\alpha\beta}{4\delta}\right).$$

Moreover by (169) and Theorem 155

$$C\left(\frac{1}{\alpha\beta\delta}\right) = \left(\frac{\alpha}{\beta}\right)\left(\frac{\beta}{\alpha}\right) C\left(\frac{1}{\alpha\delta}\right) C\left(\frac{1}{\beta\delta}\right).$$

If we now assume that all Gauss sums with odd denominator and also all those with denominator 4 are different from 0 (which will be proved generally afterwards) then we can apply the reciprocity formula to the three sums which occur:

$$\left(\frac{\alpha}{\beta}\right) \cdot \left(\frac{\beta}{\alpha}\right) = e^{(\pi i/4)S(\mathrm{sgn}\ \alpha\beta\delta - \mathrm{sgn}\ \alpha\delta - \mathrm{sgn}\ \beta\delta)} \cdot \frac{C\left(-\dfrac{\alpha\beta}{4\delta}\right)\sqrt{N(8)}}{C\left(\dfrac{-\alpha}{4\delta}\right) C\left(\dfrac{-\beta}{4\delta}\right)}. \qquad (202)$$

Now if at least one of the numbers α, β, say α, is primary, then we have by (168)

$$C\left(\frac{-\alpha}{4\delta}\right) = C\left(\frac{-1}{4\delta}\right), \qquad C\left(\frac{-\alpha\beta}{4\delta}\right) = C\left(\frac{-\beta}{4\delta}\right),$$

and it follows from (202) for $\alpha\beta = 1$ that

$$\frac{C\left(\dfrac{-1}{4\delta}\right)}{\sqrt{N(8)}} = e^{-(\pi i/4)S(\mathrm{sgn}\ \delta)}.$$

In this way we obtain

$$\left(\frac{\alpha}{\beta}\right) \cdot \left(\frac{\beta}{\alpha}\right) = e^{(\pi i/4)S(\mathrm{sgn}\ \alpha\beta\delta - \mathrm{sgn}\ \alpha\delta - \mathrm{sgn}\ \beta\delta + \mathrm{sgn}\ \delta)}.$$

However, for real α, β, δ

$$\text{sgn } \alpha\beta\delta - \text{sgn } \alpha\delta - \text{sgn } \beta\delta + \text{sgn } \delta = (\text{sgn } \alpha - 1)(\text{sgn } \beta - 1) \text{ sgn } \delta$$
$$\equiv 0 \text{ (mod 4)}.$$

Hence

$$\left(\frac{\alpha}{\beta}\right) \cdot \left(\frac{\beta}{\alpha}\right) = (-1)^{\sum\limits_{p=1}^{r_1} ((\text{sgn } \alpha^{(p)} - 1)/2)((\text{sgn } \beta^{(p)} - 1)/2)}.$$

And this is the quadratic reciprocity law for two odd relatively prime numbers of which at least one is primary.

We now omit any special assumptions about the field k. The general case in which the different of k is not a principal ideal is made formally more complicated by the fact that we must still introduce accessory auxiliary ideals into the proof.

Lemma (a). *All Gauss sums which belong to odd denominators are nonzero.*

If $C(\omega)$ is a sum belonging to the odd denominator $\mathfrak{a}$, then we obtain all sums with denominator $\mathfrak{a}$ in the form $C(\varkappa\omega)$, where $\varkappa$ runs through a reduced residue system mod $\mathfrak{a}$. For if $C(\omega_1)$ also belongs to the denominator $\mathfrak{a}$, then the integer $\varkappa$ can be determined so that that $\mathfrak{d}(\varkappa\omega - \omega_1)$ is an integral ideal and for this ideal $C(\varkappa\omega) = C(\omega_1)$ by (168). However, by Theorem 155 $C(\varkappa\omega)$ only differs from $C(\omega)$ by the factor ± 1. Thus it is enough to verify the non-vanishing of a single Gauss sum which belongs to the denominator $\mathfrak{a}$.

Let us choose, corresponding to $\mathfrak{a}$, an integral odd ideal $\mathfrak{c}$ relatively prime to $\mathfrak{a}$ such that

$$\mathfrak{a}\mathfrak{c}\mathfrak{d} = \varkappa \text{ is an integer in } k.$$

By (169) the sum $C(1/4\varkappa)$ can be represented as a product of three Gauss sums, belonging respectively to the denominators 4, $\mathfrak{a}$, $\mathfrak{c}$. Consequently for the proof of our lemma it is sufficient to show that $C(1/4\varkappa) \neq 0$. However this follows from (199), for $\omega = 1/4\varkappa$, because the sum on the right-hand side of that equation belongs to the denominator 1 and hence $= 1$.

Lemma (b). *Each Gauss sum which belongs to the denominator 4 is $\neq 0$.*

To prove this let $\mathfrak{a}$ be an odd ideal such that $\mathfrak{a}\mathfrak{d}$ is some number $\varkappa$. Then, for each odd integer μ, $C(1/\mu\varkappa) \neq 0$ by Lemma (a). Thus, by (199), we also have

$$C\left(\frac{-\gamma^2\varkappa\mu}{4}\right) \neq 0.$$

However, if φ is any number of the field such that $\mathfrak{d}\varphi$ has denominator 4, then there is an odd integer μ for which

$$\mathfrak{d}\left(\varphi + \frac{\gamma^2\varkappa\mu}{4}\right) \text{ is an integral ideal.}$$

Since

$$C(\varphi) = C\left(\frac{-\gamma^2 \varkappa \mu}{4}\right),$$

we have $C(\varphi)$ is also $\neq 0$.

Now let α and β be odd relatively prime integers in k. Let

$$\omega = \frac{b}{\mathfrak{d}}$$

where b is integral and relatively prime to $\alpha\beta$. By (169) and Theorem 155 we have

$$C\left(\frac{\omega}{\alpha \cdot \beta}\right) = C\left(\frac{\beta\omega}{\alpha}\right) C\left(\frac{\alpha\omega}{\beta}\right) = \left(\frac{\alpha}{\beta}\right)\left(\frac{\beta}{\alpha}\right) C\left(\frac{\omega}{\alpha}\right) C\left(\frac{\omega}{\beta}\right)$$

$$\left(\frac{\alpha}{\beta}\right)\left(\frac{\beta}{\alpha}\right) = \frac{C\left(\frac{\omega}{\alpha}\right) C\left(\frac{\omega}{\beta}\right)}{C\left(\frac{\omega}{\alpha\beta}\right)}. \tag{203}$$

We now apply Theorem 163 to each of these three sums. In this case we have $b_1 = 4b$ and

$$\frac{C\left(\frac{\omega}{\alpha}\right) \cdot C\left(\frac{\omega}{\beta}\right)}{C\left(\frac{\omega}{\alpha\beta}\right)} = \frac{1}{|\sqrt{N(8b)}|} \frac{C\left(-\frac{\gamma^2\alpha}{4\omega}\right) C\left(-\frac{\gamma^2\beta}{4\omega}\right)}{C\left(-\frac{\gamma^2\alpha\beta}{4\omega}\right)} \cdot e^{(\pi i/4)S(\mathrm{sgn}\,\omega\alpha + \mathrm{sgn}\,\omega\beta - \mathrm{sgn}\,\omega\alpha\beta)}.$$

Hence we again express $\sqrt{N(8b)}$ as a Gauss sum by taking $\alpha = \beta = 1$ in this equation, by which the left-hand side becomes 1. By substitution we obtain

$$\frac{C\left(\frac{\omega}{\alpha}\right) C\left(\frac{\omega}{\beta}\right)}{C\left(\frac{\omega}{\alpha\beta}\right)} = v(\alpha, \beta) \frac{C\left(\frac{-\gamma^2\alpha}{4\omega}\right) C\left(\frac{-\gamma^2\beta}{4\omega}\right)}{C\left(\frac{-\gamma^2\alpha\beta}{4\omega}\right) C\left(\frac{-\gamma^2}{4\omega}\right)}, \tag{204}$$

where

$$v(\alpha, \beta) = e^{(\pi i/4)S(\mathrm{sgn}\,\omega\alpha + \mathrm{sgn}\,\omega\beta - \mathrm{sgn}\,\omega\alpha\beta - \mathrm{sgn}\,\omega)}$$

is independent of ω since for real ω, α, β, $\mathrm{sgn}\,\omega\alpha + \mathrm{sgn}\,\omega\beta - \mathrm{sgn}\,\omega\alpha\beta -$ $\mathrm{sgn}\,\omega = -\mathrm{sgn}\,\omega(\mathrm{sgn}\,\alpha - 1)(\mathrm{sgn}\,\beta - 1)$ is divisible by 4 and consequently

$$v(\alpha, \beta) = (-1)^{\sum\limits_{p=1}^{r_1} ((\mathrm{sgn}\,\alpha^{(p)} - 1)/2)((\mathrm{sgn}\,\beta^{(p)} - 1)/2)} \tag{205}$$

We make the dependence of the right-hand side of (204) on the accessory ω clearer by splitting the Gauss sum with denominator $4b$ into two such sums with denominators 4 and b. Namely, we represent γ as the quotient of two integral ideals, say

$$\gamma = \frac{\mathfrak{c}}{\mathfrak{b}}, \quad \text{where } (\mathfrak{c}, 4\mathfrak{b}) = 1,$$

and choose an auxiliary integral ideal $\mathfrak{m}$ such that

$$\mathfrak{b}\mathfrak{m} = \mu \text{ odd}; \quad \mu \frac{\gamma^2}{\omega} = \frac{m c^2}{\mathfrak{b}} \text{ is set equal to } \varkappa.$$

Then by (169) and Theorem 155 we have

$$C\left(\frac{-\gamma^2 \alpha}{4\omega}\right) = C\left(\frac{-\varkappa\alpha}{4\mu}\right) = C\left(\frac{-\varkappa\mu\alpha}{4}\right) C\left(\frac{-4\varkappa\alpha}{\mu}\right) = \left(\frac{\alpha}{\mathfrak{b}}\right) C\left(\frac{-\varkappa\mu\alpha}{4}\right) C\left(\frac{-4\varkappa}{\mu}\right),$$

and we obtain three more equations if we replace α by 1, β, $\alpha\beta$. Furthermore $\varkappa\mu = \omega\sigma^2$ where $\sigma = mc$ is an integer. In this way we finally obtain from (203) and (204)

$$\left(\frac{\alpha}{\beta}\right) \cdot \left(\frac{\beta}{\alpha}\right) = v(\alpha, \beta) \frac{C\left(\frac{-\omega\alpha}{4}\right) C\left(\frac{-\omega\beta}{4}\right)}{C\left(\frac{-\omega}{4}\right) C\left(\frac{-\omega\alpha\beta}{4}\right)}, \tag{206}$$

where ω is now an arbitrary number of the field for which $\mathfrak{b}\omega$ is integral and odd.

If we assume that at least one of the numbers α, β, say α, is primary, then by (168)

$$C\left(\frac{-\omega\alpha\beta}{4}\right) = C\left(\frac{-\omega\beta}{4}\right); \quad C\left(\frac{-\omega\alpha}{4}\right) = C\left(\frac{-\omega}{4}\right),$$

and from this follows

Theorem 165 (Law of Quadratic Reciprocity). *For two odd relatively prime integers α, β of which at least one is primary*

$$\left(\frac{\alpha}{\beta}\right) \cdot \left(\frac{\beta}{\alpha}\right) = (-1)^{\sum\limits_{p=1}^{r_1} ((\text{sgn } \alpha^{(p)} - 1)/2)((\text{sgn } \beta^{(p)} - 1)/2)}$$

The unit on the right is surely $+1$ if at least one of the two numbers α, β is totally positive.

From this we deduce the following fact for the residue characters of certain distinguished numbers. Let β be a unit or the square of an odd ideal so that in any case $\left(\frac{\alpha}{\beta}\right) = +1$ for each odd relatively prime α, by definition. If we now choose α such that

$$\alpha = \mathfrak{a}c^2 \text{ and } \alpha \text{ is totally positive and primary,}$$

then by Theorem 165

$$\left(\frac{\beta}{\mathfrak{a}}\right) = \left(\frac{\beta}{\alpha}\right) = \left(\frac{\alpha}{\beta}\right) = +1,$$

that is,

Theorem 166. *Each odd ideal* $\mathfrak{a}$, *which can be made into a totally positive primary number by multiplication by a square ideal, has the property*

$$\left(\frac{\varepsilon}{\mathfrak{a}}\right) = +1$$

for all units and squares of ideals ε *as long as they are relatively prime to* $\mathfrak{a}$.

We will prove in the next section that the converse of this theorem also holds.

In addition, Equation (206) yields the value $\left(\frac{\alpha}{\beta}\right)\left(\frac{\beta}{\alpha}\right)$ in every case where α and β are odd nonprimary numbers. If we set

$$r(\alpha) = \frac{C\left(\dfrac{\omega}{4}\alpha\right)}{C\left(\dfrac{\omega}{4}\right)}$$

with fixed ω, then

$$r(\alpha_1) = r(\alpha_2) \quad \text{if } \alpha_1 \equiv \alpha_2 \xi^2 \ (\text{mod } 4)$$

for some odd ξ; and (206) becomes

$$\left(\frac{\alpha}{\beta}\right) \cdot \left(\frac{\beta}{\alpha}\right) = v(\alpha, \beta) \frac{r(\alpha)r(\beta)}{r(\alpha\beta)}; \tag{207}$$

valid for all odd relatively prime α and β.

The second supplementary theorem concerns the case where one of the numbers α, β is no longer odd.

Suppose that the integer λ splits into two ideal factors $\mathfrak{l}\mathfrak{r}$, such that $\mathfrak{r}$ is odd, while $\mathfrak{l}$ contains no odd prime factor:

$$\lambda = \mathfrak{l}\mathfrak{r}, \qquad (2, \mathfrak{r}) = 1.$$

Let α be an odd number relatively prime to λ, $\omega = b/\mathfrak{d}$, $(\mathfrak{b}, 2\alpha\lambda) = 1$. From the equation

$$C\left(\frac{\lambda\omega}{\alpha}\right) = \left(\frac{\lambda}{\alpha}\right)C\left(\frac{\omega}{\alpha}\right),$$

which is true by Theorem 155, we conclude, by applying the reciprocity formula (199), that

$$\left(\frac{\lambda}{\alpha}\right) = \frac{C\left(\dfrac{\lambda\omega}{\alpha}\right)}{C\left(\dfrac{\omega}{\alpha}\right)} = \frac{C\left(\dfrac{-\gamma^2\alpha}{4\omega\lambda}\right)}{C\left(\dfrac{-\gamma^2\alpha}{4\omega}\right)} \frac{e^{(\pi i/4)S(\text{sgn }\lambda\omega\alpha - \text{sgn }\omega\alpha)}}{|\sqrt{N(\lambda)}|}. \tag{208}$$

In particular for $\alpha = 1$, we obtain from this

$$1 = \frac{C\left(\dfrac{-\gamma^2}{4\omega\lambda}\right)}{C\left(\dfrac{-\gamma^2}{4\omega}\right)} \frac{e^{(\pi i/4)S(\operatorname{sgn}\lambda\omega - \operatorname{sgn}\omega)}}{\left|\sqrt{N(\lambda)}\right|}. \tag{209}$$

Now as in the preceding proof, since 4λ and $\mathfrak{b}$ are relatively prime,

$$C\left(\frac{-\gamma^2\alpha}{4\omega\lambda}\right) = C\left(\frac{-\varkappa\mu\alpha}{4\lambda}\right)C\left(\frac{-4\lambda\varkappa\alpha}{\mu}\right) = \left(\frac{\alpha}{\mathfrak{b}}\right)C\left(\frac{-4\lambda\varkappa}{\mu}\right)C\left(\frac{-\varkappa\mu\alpha}{4\lambda}\right),$$

$$C\left(\frac{-\gamma^2\alpha}{4\omega}\right) = \left(\frac{\alpha}{\mathfrak{b}}\right)C\left(\frac{-4\varkappa}{\mu}\right)C\left(\frac{-\varkappa\mu\alpha}{4}\right).$$

Again in the special case $\alpha = 1$, if we divide we have

$$\frac{C\left(\dfrac{-\gamma^2}{4\omega\lambda}\right)}{C\left(\dfrac{-\gamma^2}{4\omega}\right)} = \frac{C\left(\dfrac{-4\lambda\varkappa}{\mu}\right)}{C\left(\dfrac{-4\varkappa}{\mu}\right)} \cdot \frac{C\left(\dfrac{-\varkappa\mu}{4\lambda}\right)}{C\left(\dfrac{-\varkappa\mu}{4}\right)}$$

$$\frac{C\left(\dfrac{-\gamma^2\alpha}{4\omega\lambda}\right)}{C\left(\dfrac{-\gamma^2\alpha}{4\omega}\right)} = \frac{C\left(\dfrac{-\gamma^2}{4\omega\lambda}\right)}{C\left(\dfrac{-\gamma^2}{4\omega}\right)} \frac{C\left(\dfrac{-\varkappa\mu\alpha}{4\lambda}\right)}{C\left(\dfrac{-\varkappa\mu}{4\lambda}\right)} \cdot \frac{C\left(\dfrac{-\varkappa\mu}{4}\right)}{C\left(\dfrac{-\varkappa\mu\alpha}{4}\right)} \tag{210}$$

where finally we can still replace $\varkappa\mu$ by ω. If we divide (210) by (209) and apply (208) we find

$$\left(\frac{\lambda}{\alpha}\right) = v(\alpha, \lambda) \cdot \frac{C\left(\dfrac{-\omega\alpha}{4\lambda}\right)}{C\left(\dfrac{-\omega}{4\lambda}\right)} \frac{C\left(\dfrac{-\omega}{4}\right)}{C\left(\dfrac{-\omega\alpha}{4}\right)}. \tag{211}$$

The Gauss sums with denominator $4\lambda = 4\mathfrak{l}\mathfrak{r}$ can again be reduced to those with denominator $4\mathfrak{l}$ and $\mathfrak{r}$ by (169). Then, if we choose auxiliary ideals $\mathfrak{m}, \mathfrak{n}$ which are odd and relatively prime to $\mathfrak{r}\alpha$ and for which

$$\lambda_1 = \mathfrak{l}\mathfrak{m}, \qquad \rho = \mathfrak{r}\mathfrak{n}, \qquad \sigma = \frac{\lambda_1\rho}{\lambda} = \mathfrak{m}\mathfrak{n},$$

we have

$$C\left(\frac{-\omega\alpha}{4\lambda}\right) = C\left(\frac{-\omega\sigma\alpha}{4\lambda_1\rho}\right) = C\left(\frac{-\omega\sigma\rho\alpha}{4\lambda_1}\right)C\left(\frac{-4\lambda_1\omega\sigma\alpha}{\rho}\right)$$

$$= \left(\frac{\alpha}{\mathfrak{r}}\right)C\left(\frac{-\omega\sigma\rho\alpha}{4\lambda_1}\right)C\left(\frac{-4\lambda_1\omega\sigma}{\rho}\right)$$

$$= \left(\frac{\alpha}{\mathfrak{r}}\right)C\left(\frac{-\omega\rho^2\alpha}{4\lambda}\right)C\left(\frac{-4\lambda_1\omega\sigma}{\rho}\right).$$

Finally if we set $\alpha = 1$ and substitute the results in (211), we obtain

$$\left(\frac{\lambda}{\alpha}\right)\left(\frac{\alpha}{\mathfrak{r}}\right) = v(\alpha, \lambda) \cdot \frac{C\left(\dfrac{-\omega\rho^2\alpha}{4\lambda}\right)}{C\left(\dfrac{-\omega\rho^2}{4\lambda}\right)} \cdot \frac{C\left(\dfrac{-\omega}{4}\right)}{C\left(\dfrac{-\omega\alpha}{4}\right)}.$$

Here ρ is an arbitrary odd number which is divisible by $\mathfrak{r}$. These last sums depend only on the behavior of α mod $4\mathfrak{l}$. In particular if we choose α to be a quadratic residue mod $4\mathfrak{l}$, then we obtain

Theorem 167. *If $\mathfrak{l}$ is an integral ideal without odd prime factors and λ is an integer with the decomposition $\lambda = \mathfrak{lr}$, where $\mathfrak{r}$ is an odd integral ideal, then*

$$\left(\frac{\lambda}{\alpha}\right)\left(\frac{\alpha}{\mathfrak{r}}\right) = (-1)^{\sum\limits_{p=1}^{r_1} ((\mathrm{sgn}\ \alpha^{(p)} - 1)/2)((\mathrm{sgn}\ \lambda^{(p)} - 1)/2)}$$

if the odd number α is a quadratic residue mod $4\mathfrak{l}$ and relatively prime to λ.

§60 Relative Quadratic Fields and Applications to the Theory of Quadratic Residues

We now consider the field $K = K(\sqrt{\mu}, k)$ which is generated, relative to k, by the square root of a number μ in k. The theorems in §39 with $l = 2$ hold for this field. It is useful to introduce a residue character which deviates somewhat from the quadratic residue symbol.

Definition. For an arbitrary prime ideal $\mathfrak{p}$ in k we set

$$Q(\mu, \mathfrak{p}) = \begin{cases} 1, & \text{if } \mathfrak{p} \text{ splits into two distinct} \\ & \text{factors in } K(\sqrt{\mu}, k). \\ -1, & \text{if } \mathfrak{p} \text{ remains irreducible in} \\ & K(\sqrt{\mu}, k). \\ 0, & \text{if } \mathfrak{p} \text{ is the square of a prime} \\ & \text{ideal in } K(\sqrt{\mu}, k). \end{cases}$$

By the results of §39, $Q(\mu, \mathfrak{p})$ is defined for all prime ideals if μ belongs to k but $\sqrt{\mu}$ does not. Moreover we have

$$Q(\mu, \mathfrak{p}) = \left(\tfrac{\mu}{\mathfrak{p}}\right) \qquad \text{if } \mathfrak{p} \text{ is odd and does not divide } \mu \tag{212}$$

$$Q(\mu\alpha^2, \mathfrak{p}) = Q(\mu, \mathfrak{p}) \quad \text{for each } \alpha \neq 0 \text{ in } k.$$

Moreover for arbitrary integral ideals $\mathfrak{a}\ (\neq 0)$ in k we set

$$Q(\mu, \mathfrak{a}) = Q(\mu, \mathfrak{p}_1)^{a_1} \cdot Q(\mu, \mathfrak{p}_2)^{a_2} \cdots Q(\mu, \mathfrak{p}_m)^{a_m} \tag{213}$$

if $\mathfrak{a}$ has the decomposition

$$\mathfrak{a} = \mathfrak{p}_1^{a_1} \cdots \mathfrak{p}_m^{a_m}$$

and for each square μ^2 in k

$$Q(\mu^2, \mathfrak{a}) = 1.$$

Thus we have for two integral ideals $\mathfrak{a}$, $\mathfrak{b}$ in k

$$Q(\mu, \mathfrak{a}\mathfrak{b}) = Q(\mu, \mathfrak{a})Q(\mu, \mathfrak{b}).$$

Finally for odd $\mathfrak{a}$, which are relatively prime to the integers μ and v, we have

$$Q(\mu v, \mathfrak{a}) = Q(\mu, \mathfrak{a})Q(v, \mathfrak{a}).$$

In the rational number field the introduction of this symbol would be superfluous, since there the number μ can always be assumed to be free of unnecessary square factors. But in other fields, where the class number is even, μ can have accessory square factors which can not be avoided.

With the help of the symbol Q, the zeta-function of K can be expressed through that of k and an additional series, as was shown in §49 for the quadratic fields. For if $\mathfrak{P}$ denotes a prime ideal of K then, in the notation of Theorem 108, the relative norm with respect to k is $N_k(\mathfrak{P}) = \mathfrak{p}$ or $\mathfrak{p}^2$ and $N(\mathfrak{P}) = n(\mathfrak{p})$ or $n(\mathfrak{p}^2)$, where $\mathfrak{p}$ is the prime ideal in k which is divisible by $\mathfrak{P}$. In the infinite product

$$\zeta_K(s) = \prod_{\mathfrak{P}} \frac{1}{1 - N(\mathfrak{P})^s}$$

we extract those factors which can be derived from all prime divisors $\mathfrak{P}$ of a fixed $\mathfrak{p}$. For these factors we then have

$$\prod_{\mathfrak{P}|\mathfrak{p}} (1 - N(\mathfrak{P})^{-s}) = (1 - n(\mathfrak{p})^{-s})(1 - Q(\mu, \mathfrak{p})n(\mathfrak{p})^{-s})$$

and hence

$$\zeta_K(s) = \zeta_k(s)Z(s)$$

$$Z(s) = \prod_{\mathfrak{p}} \frac{1}{1 - Q(\mu, \mathfrak{p})n(\mathfrak{p})^{-s}} = \sum_{\mathfrak{a}} \frac{Q(\mu, \mathfrak{a})}{n(\mathfrak{a})^s}.$$

By the formula for the class number in Theorem 123

$$\lim_{s \to 1} \frac{\zeta_K(s)}{\zeta_k(s)}$$

is equal to a finite nonzero value and thus we conclude:

Theorem 168. $\qquad\qquad \lim_{s \to 1} \log Z(s)$ *is finite.*

From this fact we obtain the analogue of Theorem 147:

Theorem 169. *Let $\mu_1, \mu_2, \ldots, \mu_m$ be integers in k such that a product of powers $\mu_1^{x_1} \cdots \mu_m^{x_m}$ is the square of a number in k only if all the exponents $x_1, \ldots, x_m$ are even. Let $c_1, c_2, \ldots, c_m$ be arbitrary values ± 1. Then there are infinitely many prime ideals $\mathfrak{p}$ in k which satisfy the m conditions*

$$\left(\frac{\mu_1}{\mathfrak{p}}\right) = c_1, \ldots, \left(\frac{\mu_m}{\mathfrak{p}}\right) = c_m.$$

For by the hypothesis, the square root of each of the $2^m - 1$ products of powers $\mu = \mu_1^{x_1} \cdots \mu_m^{x_m}$ ($x_i = 0$ or 1, not all $x_i = 0$) defines a relative quadratic field $K(\sqrt{\mu}, k)$. However, it now obviously follows as in §49, that for $s > 1$

$$\log \prod_{\mathfrak{p}} \left(1 - \frac{Q(\mu, \mathfrak{p})}{n(\mathfrak{p})^s}\right) = -\sum_{\mathfrak{p}} \frac{Q(\mu, \mathfrak{p})}{n(\mathfrak{p})^s} + \varphi(\mu, s),$$

where $\varphi(\mu, s)$ tends to a finite limit as $s \to 1$. Hence by Theorem 168 the first sum on the right also has this property. Consequently

$$L(s, \mu) = \sum_{\mathfrak{p}}' \left(\frac{\mu}{\mathfrak{p}}\right) \frac{1}{n(\mathfrak{p})^s}$$

also remains finite since, by (212), this sum differs from the former only in finitely many terms. The prime on the summation sign is to indicate that $\mathfrak{p}$ is only to run through the odd prime ideals which do not divide $\mu_1, \mu_2, \ldots, \mu_m$. On the other hand it again follows from the fact that $\zeta_k(s)$ becomes infinite as $s \to 1$ that

$$L(s, 1) = \sum_{\mathfrak{p}}' \frac{1}{n(\mathfrak{p})^s} \to \infty.$$

Consequently, the left-hand side of the equation

$$\sum_{x_1, \ldots, x_m = 0, 1} c_1^{x_1} \cdots c_m^{x_m} L(s, \mu_1^{x_1} \cdot \mu_2^{x_2} \cdots \mu_m^{x_m})$$

$$= \sum_{\mathfrak{p}}' \left(1 + c\left(\frac{\mu_1}{\mathfrak{p}}\right)\right) \cdots \left(1 + c_m\left(\frac{\mu_m}{\mathfrak{p}}\right)\right) \frac{1}{n(\mathfrak{p})^s}$$

becomes infinite as $s \to 1$ since only a single term becomes infinite. However, on the right-hand side only those terms whose $\mathfrak{p}$ satisfy the requirement of our assertion remain. Consequently there must be infinitely many $\mathfrak{p}$ of this type.

This existence theorem is the most important aid in the proof of the converse of Theorem 166 and Theorem 167, which we will now carry out.

§61 Number Groups, Ideal Groups, and Singular Primary Numbers

In subsequent investigations we are concerned with those factor groups of Abelian groups which are determined by the squares of elements. If $\mathfrak{G}$ is an Abelian group and $\mathfrak{U}_2$ is the subgroup of squares of all elements of $\mathfrak{G}$, we wish to designate each of the cosets which are defined by $\mathfrak{U}_2$ as a *complex* of elements of $\mathfrak{G}$. The factor group $\mathfrak{G}/\mathfrak{U}_2$ is the group of complexes by §9. The unit element in the factor group is the *principal complex*, that is, the system of elements of $\mathfrak{U}_2$. The square of each complex is the principal complex. If $\mathfrak{G}$ is a finite group, there are exactly 2^e different complexes where e is the basis number of $\mathfrak{G}$ belonging to 2. The number of independent complexes, that is, the number of independent elements of $\mathfrak{G}/\mathfrak{U}_2$, is then e.

We now introduce an important series of groups, complexes, and related constants:

1. The units of k form a group under composition by multiplication. The number of different *unit complexes* is 2^m, where $m = (n + r_1)/2$, since there are $r_1 + r_2 - 1 = m - 1$ fundamental units and in addition there is still a root of unity in k whose square root does not lie in k.

2. All the nonzero numbers of k form a group under composition by multiplication. Thus the system of all numbers $\alpha\xi^2$, where α is fixed and ξ runs through all numbers of k is a number complex. If we designate the r_1 values ± 1 given by sgn $\omega^{(1)}, \ldots,$ sgn $\omega^{(r_1)}$ as the sequence of signs of a number ω in k, then all numbers of the same number complex have the same sequence of signs. (For $r_1 = 0$ we understand the sequence of signs to be the number $+1$.) The group of all *totally positive number complexes* forms a subgroup of index 2^{r_1} in the group of all number complexes. For if $r_1 > 0$, there are numbers ω in k with an arbitrarily prescribed sequence of signs. To see this let θ be a generating number of k; then the r_1 expressions $a_0 + a_1\theta^{(i)} + \cdots + a_{r_1-1}\theta^{(i)r_1-1}(i = 1, \ldots, r_1)$ take on each system of real values for real a. Hence for rational a they take on each combination of signs.

3. In the group of ideal classes of k, there are exactly 2^e *different* class complexes, where e denotes the basis number belonging to 2 of the class group.

4. Those number complexes whose numbers are squares of ideals in k form a subgroup in the group of all number complexes. The order of this subgroup is 2^{m+e}. For by 3, there are e ideals $\mathfrak{a}_1, \ldots, \mathfrak{a}_e$ which define e independent class complexes and whose squares are principal ideals, say $\mathfrak{a}_i^2 = \alpha_i$ $(i = 1, 2, \ldots, e)$. The e numbers $\alpha_1, \ldots, \alpha_e$ define e independent number complexes. If ω is a number which is the square of an ideal $\mathfrak{c}$ in k, then $\mathfrak{c}$ is equivalent to a product of powers of the $\mathfrak{a}_1, \ldots, \mathfrak{a}_e$ and, after multiplication by a suitable unit, ω differs from a product of powers of $\alpha_1, \ldots, \alpha_e$ by a square factor. We call a number in k *singular*, if it is the square of an ideal in k. Thus there are $m + e$ *independent singular number*

complexes. They are represented by $\alpha_1, \ldots, \alpha_e$ and m units from the m independent complexes.

5. Let p denote the number of independent singular number complexes which consist of totally positive numbers. Accordingly there are 2^p *singular totally positive number* complexes. The 2^{m+e} singular number complexes thus indicate numbers with only 2^{m+e-p} different sequences of signs.

6. We regard two nonzero ideals $\mathfrak{a}$, $\mathfrak{b}$ in the same *strict ideal class* and all $\mathfrak{a}$ and $\mathfrak{b}$ *equivalent in the strict sense* if $\mathfrak{a}/\mathfrak{b}$ can be set equal to a totally positive number of the field. We again write $\mathfrak{a} \approx \mathfrak{b}$. The strict classes are again combined into an Abelian group, the strict class group. Those strict classes which contain a principal ideal in the broader sense form a subgroup of index h. The principal ideals obviously define at most 2^{r_1} distinct strict classes. Thus the strict class group has order at most $2^{r_1}h$. Let e_0 be the basis number belonging to 2 of this strict class group. We denote the *group of the strict ideal class* complexes by $\mathfrak{J}_0$. Its order is thus 2^{e_0}. By determining the order of $\mathfrak{J}_0$ in a second way we obtain the equation

$$e_0 = p + r_1 - m. \tag{214}$$

To see this we denote that subgroup of $\mathfrak{J}_0$ whose class complexes can be represented by principal ideals (in the broader sense) by $\mathfrak{H}$. Then by the general theorems on groups, the order of $\mathfrak{J}_0$ is equal to the order of the factor group $\mathfrak{J}_0/\mathfrak{H}$ multiplied by the order of $\mathfrak{H}$. Now the factor group $\mathfrak{J}_0/\mathfrak{H}$ has order 2^e. For if $\mathfrak{b}_1$, $\mathfrak{b}_2, \ldots, \mathfrak{b}_e$ are representatives of the e independent class complexes (in the broader sense), then the 2^e products of powers $\mathfrak{b} = \mathfrak{b}_1^{x_1} \cdots \mathfrak{b}_e^{x_e}$ ($x_i = 0$ or 1) define exactly 2^e distinct cosets in $\mathfrak{J}_0$ with respect to $\mathfrak{H}$. On the other hand to each ideal $\mathfrak{a}$ there exists a product of powers $\mathfrak{b}$ and a square of an ideal $\mathfrak{c}^2$ such that $\mathfrak{a} \sim \mathfrak{b}\mathfrak{c}^2$; hence $\mathfrak{a} = \alpha\mathfrak{b}\mathfrak{c}^2$ for a certain number α. The complex to which $\mathfrak{a}$ belongs thus differs from the complex to which $\mathfrak{b}$ belongs by the complex of α, that is, a complex from the group $\mathfrak{H}$. Hence the order of $\mathfrak{J}_0/\mathfrak{H}$ is equal to 2^e.

Now a principal ideal (γ) belongs to the unit element of $\mathfrak{J}_0$ if and only if (γ) is equivalent to the square of an ideal in the strict sense, that is, if γ is equal to a totally positive number multiplied by a singular number, that is, if and only if γ can be made totally positive by multiplication by a singular number. Of the 2^{r_1} possible sequences of signs for γ exactly 2^{m+e-p} are realized by singular numbers by 5, so that the principal ideals define exactly $2^{r_1-(m+e-p)}$ distinct strict ideal class complexes. Hence this is the order of $\mathfrak{H}$. Thus assertion (214) is proved.

7. Among the odd residue classes mod 4 there are exactly 2^n *distinct residue class complexes* mod 4. It follows from $\xi^2 \equiv 1 \pmod 4$ that $\xi \equiv 1 \pmod 2$, $\xi = 1 + 2\omega$ with ω an integer. Among these numbers there are $N(2) = 2^n$ incongruent ones mod 4.

8. We consider two numbers α and β to be in the same *strict residue class* mod $\mathfrak{a}$, if $\alpha \equiv \beta$ mod $\mathfrak{a}$ and α/β is totally positive. In each residue class mod $\mathfrak{a}$

moreover there are obviously numbers α whose r_1 conjugates have the same sign as the arbitrarily given integer ω since $\alpha + x|N(\mathfrak{a})|\omega$ belongs to the same residue class mod $\mathfrak{a}$ as α for each rational integer x and has the desired sign properties for all sufficiently large x. Thus each residue class mod $\mathfrak{a}$ splits into exactly 2^{r_1} strict residue classes mod $\mathfrak{a}$. In particular there are thus 2^{n+r_1} *distinct strict residue class complexes* mod 4.

9. Let $\mathfrak{l}$ be a prime factor of 2. Among the odd residue classes mod $4\mathfrak{l}$, there are 2^{n+1} *distinct residue class complexes* mod $4\mathfrak{l}$. It follows from $\xi^2 \equiv 1$ (mod $4\mathfrak{l}$) that $\xi = 1 + 2\omega$ with ω an integer and with ω satisfying the condition $\omega(\omega + 1) \equiv 0$ (mod $\mathfrak{l}$). Thus $\omega \equiv 0$ or 1 (mod $\mathfrak{l}$) and this yields exactly $2N(2) = 2^{n+1}$ incongruent numbers for ξ mod $4\mathfrak{l}$. In the corresponding fashion there are 2^{n+r_1+1} distinct strict residue class complexes mod $4\mathfrak{l}$.

10. The singular numbers which are at the same time primary numbers without being squares claim our main interest. Such numbers are called *singular primary numbers.* By Theorem 120 the singular primary numbers ω yield those fields $K(\sqrt{\omega}, k)$ which have relative discriminant 1 with respect to k. Suppose that there exist q independent complexes *of singular primary numbers.* Then by 4, $q \leq m + e$. The 2^{m+e} different singular number complexes thus define 2^{m+e-q} distinct residue class complexes mod 4, since precisely 2^q of these are primary, that is, they belong to the principal complex of residue classes mod 4.

11. Likewise let q_0 denote *the number of independent complexes of singular primary numbers which are totally positive.* The 2^{m+e} different singular number complexes thus define only 2^{m+e-q_0} distinct residue class complexes mod 4 in the strict sense, because each 2^{q_0} of the singular number complexes define the same strict residue class complex mod 4.

12. Finally we are led, by Theorem 166, to a new classification of all odd ideals modulo 4. Two integral odd ideals are considered to be in the same *"ideal class* mod 4" if there is a square ideal $\mathfrak{c}^2$ in k such that $\mathfrak{a} \sim \mathfrak{bc}^2$ and integers α, β can be chosen so that $\alpha\mathfrak{a} = \beta\mathfrak{bc}^2$ with $\alpha \equiv \beta \equiv 1$ (mod 4). The composition of these classes defined by multiplication of ideals determines the *"class group* mod 4"; let it be denoted by $\mathfrak{B}$.

To determine the order of $\mathfrak{B}$ we introduce the subgroup $\mathfrak{H}$ of those classes of $\mathfrak{B}$ which can be represented by odd integral principal ideals. The order of $\mathfrak{B}$ is then equal to the order of $\mathfrak{H}$ multiplied by the order of the factor group $\mathfrak{B}/\mathfrak{H}$. Now this factor group has order 2^e since if $\mathfrak{b}_1, \dots, \mathfrak{b}_e$ are odd representatives of the e independent ideal class complexes, then the 2^e products of power $\mathfrak{b}_1^{x_1} \cdots \mathfrak{b}_e^{x_e} = \mathfrak{b}$ ($x_i = 0$ or 1) define exactly 2^e distinct cosets in $\mathfrak{B}$ with respect to $\mathfrak{H}$. Furthermore for each odd ideal $\mathfrak{a}$ there exists one of these products $\mathfrak{b}$ and an odd ideal square $\mathfrak{c}^2$ such that $\mathfrak{a} \sim \mathfrak{bc}^2$. Thus the equation $\alpha\mathfrak{a} = \beta\mathfrak{bc}^2$ holds with odd numbers α, β. By multiplying by the same numerical factor on both sides, we can assume that $\alpha \equiv 1$ (mod 4). Consequently $\mathfrak{a}$ and $\beta\mathfrak{b}$ belong to the same ideal class mod 4. However $\beta\mathfrak{b}$ and $\mathfrak{b}$ differ only by an ideal in $\mathfrak{H}$ and hence each coset in $\mathfrak{B}$ is also represented by some $\mathfrak{b}$, that is, $\mathfrak{B}/\mathfrak{H}$ actually has order 2^e.

In order to make further progress in determining the order of $\mathfrak{H}$ we consider that in any case two odd integers γ_1, γ_2 define principal ideals (γ_1) and (γ_2) from the same ideal class mod 4, whenever γ_1 and γ_2 belong to the same residue class complex mod 4. The ideal class mod 4 to which the ideal (1) belongs consists of all odd ideals (γ) for which γ is congruent to a singular number mod 4. By 10 moreover, singular numbers define exactly 2^{m+e-q} distinct residue class complexes mod 4. Consequently among the 2^n residue class complexes mod 4 each 2^{m+e-q} belong to the same ideal class mod 4. Thus the order of $\mathfrak{H}$ is $2^{n-(m+e-q)}$. From this we obtain

the order of $\mathfrak{B}$ is equal to $2^{n-m+q} = 2^{m-r_1+q}$.

13. If $r_1 > 0$, then in the corresponding fashion we define the group $\mathfrak{B}_0$ of *strict ideal classes* mod 4. We consider two odd ideals $\mathfrak{a}$ and $\mathfrak{b}$ to be in the same strict ideal class mod 4, if there is a square of an ideal $\mathfrak{c}^2$ such that $\mathfrak{a} \sim \mathfrak{b}\mathfrak{c}^2$ and the numbers α and β can be chosen so that $\alpha\mathfrak{a} = \beta\mathfrak{b}\mathfrak{c}^2, \alpha \equiv \beta \equiv 1$ (mod 4) and moreover α and β are totally positive.

The order of $\mathfrak{B}_0$ is determined in a manner similar to that in which the order of $\mathfrak{B}$ is determined. If $\mathfrak{H}_0$ is the subgroup of $\mathfrak{B}_0$ which is represented by odd principal ideals, then the order of $\mathfrak{B}_0/\mathfrak{H}_0$ is again 2^e. However, by 11, the order of $\mathfrak{H}_0$ is found to be $2^{n+r_1-(m+e-q_0)}$, since among the 2^{n+r_1} strict residue class complexes mod 4, each 2^{m+e-q_0} differ by a singular number complex.

Hence

the order of $\mathfrak{B}_0$ is equal to $2^{n+r_1-m+q_0} = 2^{m+q_0}$.

§62 The Existence of the Singular Primary Numbers and Supplementary Theorems for the Reciprocity Law

Now we determine q and q_0 by a very simple enumeration method.

Lemma (a). *We have $q_0 \leq e$ and $q \leq e_0$.*

Suppose that there are q_0 independent totally positive singular primary numbers $\omega_1, \omega_2, \ldots, \omega_{q_0}$ and let us consider the q_0 functions

$$\chi_i(\mathfrak{a}) = Q(\omega_i, \mathfrak{a}) \qquad i = 1, \ldots, q_0,$$

of the odd ideal $\mathfrak{a}$. These depend only on the ideal class complex to which $\mathfrak{a}$ belongs. For if $\mathfrak{a} \sim \mathfrak{b}\mathfrak{c}^2$ holds with odd $\mathfrak{a}$, $\mathfrak{b}$, $\mathfrak{c}$ and if the odd numbers α and β are chosen so that $\alpha\mathfrak{a} = \beta\mathfrak{b}\mathfrak{c}^2$, then if we assume the ω_i relatively prime to $\alpha\mathfrak{a}$, we obtain

$$\chi_i(\alpha\mathfrak{a}) = \chi_i(\beta\mathfrak{b}\mathfrak{c}^2) = \left(\frac{\omega_i}{\alpha\mathfrak{a}}\right) = \left(\frac{\omega_i}{\beta\mathfrak{b}\mathfrak{c}^2}\right) = \left(\frac{\omega_i}{\beta\mathfrak{b}}\right).$$

However, by the reciprocity law, we have

$$\left(\frac{\omega_i}{\gamma}\right) = \left(\frac{\gamma}{\omega_i}\right),$$

for each integer γ which is relatively prime to $2\omega_i$ since ω_i is primary and totally positive. The last symbol, moreover, is $+1$ because ω_i is singular. Hence it actually follows that

$$\chi_i(\mathfrak{a}) = \left(\frac{\omega_i}{\mathfrak{a}}\right) = \left(\frac{\omega_i}{\mathfrak{b}}\right) = \chi_i(\mathfrak{b}) \quad \text{if } \mathfrak{a} \sim \mathfrak{b}\mathfrak{c}^2.$$

Furthermore, since $\chi_1(\mathfrak{a}_1\mathfrak{a}_2) = \chi(\mathfrak{a}_1) \cdot \chi(\mathfrak{a}_2)$, the q_0 functions $\chi_i(\mathfrak{a})$ are group characters of the group of ideal class complexes, by §10. By Theorem 169 they are also independent characters. On the other hand, by Theorem 33 the group of ideal class complexes has exactly e independent characters since this group has order 2^e; hence $q_0 \leq e$.

When we get to the bottom of the concept of strict equivalence we prove the relation $q \leq e_0$ in analogous fashion.

Lemma (b). *Let $\varepsilon_1, \ldots, \varepsilon_{m+e}$ be $m + e$ independent singular numbers. Then the $m + e$ functions of the odd ideal $\mathfrak{a}$*

$$Q(\varepsilon_i, \mathfrak{a}) \qquad (i = 1, 2, \ldots, m + e)$$

form a system of independent group characters of the group $\mathfrak{B}_0$.

It again follows from Theorem 165 that these functions are group characters of $\mathfrak{B}_0$. Theorem 169 shows that they are independent.

By the general theorems on groups of §10 we thus have

$$m + e \leq m + q_0,$$

since by 13 the order of $\mathfrak{B}_0$ is $m + q_0$. Hence $q_0 \geq e$ and consequently, by Lemma (a), we have $q_0 = e$. With this lemma the following two theorems are proved.

Theorem 170. *There are exactly e independent singular primary numbers, say $\omega_1, \ldots, \omega_e$, which are totally positive. Here e is the basis number belonging to 2 of the group of broader ideal classes of the field. The e characters $Q(\omega_i, \mathfrak{a})$ form the complete system of characters of the group of class complexes.*

Theorem 171. *In order that an odd ideal $\mathfrak{a}$ can be made into a totally positive and primary number of the field by multiplication by the square of an ideal, it is necessary and sufficient that the conditions*

$$Q(\varepsilon, \mathfrak{a}) = +1$$

are satisfied for every singular number ε.

If we consider the group $\mathfrak{B}$ instead of $\mathfrak{B}_0$, it follows in analogous fashion:

Lemma (c). *Let $\varepsilon_1, \ldots, \varepsilon_p$ be the $p = e_0 + m - r_1$ independent totally positive singular numbers. Then the p functions $Q(\varepsilon_i, \mathfrak{a})$ $(i = 1, \ldots, p)$ form a system of independent group characters of the group $\mathfrak{B}$ for odd $\mathfrak{a}$.*

Since $\mathfrak{B}$ has order $2^{m - r_1 + q}$, it again follows from this that

$$m - r_1 + q \geq p = m - r_1 + e_0, \qquad e_0 \leq q.$$

Hence by lemma (a) $e_0 = q$, and thus $\mathfrak{B}$ has order 2^p. With this we have proved:

Theorem 172. *There are exactly e_0 independent singular primary numbers, say $\omega_1, \ldots, \omega_{e_0}$. Here e_0 is the basis number belonging to 2 of the group of strict ideal classes of the field. The e_0 characters $Q(\omega_i, \mathfrak{a})$ form the complete system of characters of the group of the strict class complexes for odd $\mathfrak{a}$.*

Theorem 173. *In order that an odd ideal $\mathfrak{a}$ can be made into a primary number of the field by multiplication by a square of an ideal, it is necessary and sufficient that the conditions*

$$Q(\varepsilon, \mathfrak{a}) = +1$$

are satisfied for each totally positive singular number ε.

One usually calls Theorems 171 and 173 the *first supplementary theorem*.

In similar fashion we obtain the converse of Theorem 167 which concerns the residue character modulo numbers which are not odd. We call an odd integer α *hyperprimary modulo* $\mathfrak{l}$, where $\mathfrak{l}$ denotes a prime factor of 2, if $\alpha \equiv \xi^2 \pmod{4\mathfrak{l}}$ can be satisfied by a number ξ in k. Thus the hyperprimary numbers modulo $\mathfrak{l}$ define the principal complex of residue classes mod $4\mathfrak{l}$. By Number 9 of the preceding section there are 2^{n+1} distinct complexes mod $4\mathfrak{l}$ but only 2^n distinct complexes mod 4. Hence each complex mod 4 contains exactly two distinct complexes mod $4\mathfrak{l}$. Hence the primary numbers define exactly two distinct residue class complexes mod $4\mathfrak{l}$. Let these be denoted by R_1 and R_2, where we choose R_1 as the principal complex mod $4\mathfrak{l}$.

Theorem 174. *If the prime ideal $\mathfrak{l}$ which divides 2 belongs to the principal class complex in the strict sense, then all e_0 independent singular primary numbers are also hyperprimary modulo $\mathfrak{l}$. On the other hand, in the other case, only $e_0 - 1$ independent singular primary numbers are also hyperprimary modulo $\mathfrak{l}$.*

Proof: Let $\mathfrak{c}$ be an odd ideal chosen so that $\mathfrak{l}\mathfrak{c}^2 = \lambda$ is a totally positive number, which is possible in the first of the cases stated in Theorem 174. Then for each odd number α, which we assume at first to be relatively prime to $\mathfrak{l}\mathfrak{c}$, we have by Theorem 167

$$\left(\frac{\lambda}{\alpha}\right) = \left(\frac{\lambda}{\alpha}\right)\left(\frac{\alpha}{\mathfrak{c}^2}\right) = +1,$$

provided α belongs to the complex R_1. If we now just consider the functions $\left(\frac{\lambda}{\alpha}\right) = Q(\lambda, \alpha)$ for primary numbers α, then we have $Q(\lambda, \alpha_1) = Q(\lambda, \alpha_2)$ if α_1 and α_2 belong to the same complex R_1 or R_2. Moreover, $Q(\lambda, \alpha_1\alpha_2) = Q(\lambda, \alpha_1)Q(\lambda, \alpha_2)$, so that $Q(\lambda, \alpha)$ is a group character of the group of order two which is formed from the elements R_1, R_2 where $R_2^2 = R_1$. Nevertheless this character is not the principal character; for by Theorem 169 there are infinitely many prime ideals $\mathfrak{p}$ for which $\left(\frac{\lambda}{\mathfrak{p}}\right) = -1$ while the characters $Q(\varepsilon, \mathfrak{p})$ are equal to $+1$ for each of the p independent totally positive squares ε of ideals. Then by Theorem 173, $\mathfrak{p}$ can be made into a primary number by multiplication by a suitable $\mathfrak{m}^2$, say $\alpha = \mathfrak{p}\mathfrak{m}^2$. Then $Q(\lambda, \alpha) = \left(\frac{\lambda}{\mathfrak{p}}\right) = -1$. Consequently $Q(\lambda, \alpha)$ is the uniquely determined group character of the group (R_1, R_2) which is not the principal character; hence it is $= 1$ if and only if the primary number α belongs to R_1, that is, if α is also hyperprimary modulo $\mathfrak{l}$. Now for each singular primary number ω we have $Q(\lambda, \omega) = +1$, thus all odd singular primary numbers modulo $\mathfrak{l}$ are also hyperprimary modulo $\mathfrak{l}$.

Secondly, if $\mathfrak{l}$ does not belong to the principal class complex in the strict sense, then let us choose an odd ideal $\mathfrak{r}$ such that $\lambda = \mathfrak{l}\mathfrak{r}$ is a totally positive number. Since $\mathfrak{r}$ also does not belong to the strict principal class complex, there are, by Theorem 172, among the e_0 singular primary numbers exactly $e_0 - 1$ independent numbers, say $\omega_2, \ldots, \omega_{e_0}$, such that $Q(\omega_i, \mathfrak{r}) = +1$ for $i = 2, 3, \ldots, e_0$, and one number ω_1, independent of these numbers, for which $Q(\omega_1, \mathfrak{r}) = -1$. This ω_1 is then surely not hyperprimary modulo $\mathfrak{l}$ for otherwise

$$\left(\frac{\omega_1}{\mathfrak{r}}\right) = \left(\frac{\lambda}{\omega_1}\right)\left(\frac{\omega_1}{\mathfrak{r}}\right) = +1$$

would hold, by Theorem 167, while the product is equal to -1 by the definition of ω_1. Hence ω_1 belongs to the complex R_2 mod $4\mathfrak{l}$. Therefore every primary number belongs to the complex ω_1 or ω_1^2 mod $4\mathfrak{l}$. If, however, the odd numbers α and β belong to the same complex mod $4\mathfrak{l}$, then, if we set $\chi(a) = \left(\frac{\lambda}{\alpha}\right)\left(\frac{\alpha}{\mathfrak{r}}\right)$, we have

$$\chi(\alpha) \cdot \chi(\beta) = \chi(\alpha\beta) = 1$$

because $\alpha\beta$ is hyperprimary mod $\mathfrak{l}$, that is,

$$\chi(\alpha) = \chi(\beta).$$

Consequently, none of the numbers $\omega_2, \ldots, \omega_{e_0}$ can belong to the complex R_2 represented by ω_1, since then $\chi(\omega_2)$ would be $= -1$, while $\chi(\omega_2)$ is equal to 1 by the definition of ω_2. Consequently, $\omega_2, \ldots, \omega_{e_0}$ are hyperprimary modulo $\mathfrak{l}$ and ω_1 is not; with this Theorem 174 is proved.

Theorem 175. *Let $\lambda = \mathfrak{l}\mathfrak{r}$ be a totally positive number, $\mathfrak{r}$ an odd ideal, and let $\mathfrak{l}$ be a prime factor of 2. In order that the primary integer α, which is relatively* *prime to λ, be hyperprimary, it is necessary and sufficient that*

$$\chi(\alpha) = \left(\frac{\lambda}{\alpha}\right)\left(\frac{\alpha}{\mathfrak{r}}\right) = +1.$$

Theorem 167 asserts that the condition is necessary. The proof of the preceding theorem shows in the following way that it is sufficient. First suppose that $\mathfrak{l}$ is equivalent in the strict sense to the square of an ideal. Then we can find integers β, ρ, λ such that

$$\lambda\beta^2 = \lambda_0\rho, \qquad \lambda_0 = \mathfrak{l}\mathfrak{r}_1^2$$

$$\rho = \mathfrak{r}\mathfrak{r}_1^2, \qquad \lambda_0, \rho \text{ totally positive,}$$

where β is odd and relatively prime to $\alpha\mathfrak{r}$. Then

$$\chi(\alpha) = \left(\frac{\lambda\beta^2}{\alpha}\right)\left(\frac{\alpha}{\mathfrak{r}}\right) = \left(\frac{\lambda_0\rho}{\alpha}\right)\left(\frac{\alpha}{\mathfrak{r}}\right) = \left(\frac{\lambda_0}{\alpha}\right)\left(\frac{\alpha}{\rho}\right)\left(\frac{\alpha}{\mathfrak{r}}\right) = \left(\frac{\lambda_0}{\alpha}\right),$$

and as shown above, $\left(\frac{\lambda_0}{\alpha}\right) = +1$ is the necessary and sufficient condition that the primary number α is also hyperprimary.

However, if $\mathfrak{l}$ is not in a principal class complex, then there is indeed a singular primary number ω_1, for which $\left(\frac{\omega_1}{\mathfrak{r}}\right) = -1$; and 1, ω_1 represent simultaneously the two distinct residue class complexes mod $4\mathfrak{l}$ which arise from primary numbers. If α and ω_1^a $(a = 0$ or $1)$ belong to the same complex mod $4\mathfrak{l}$, then by Theorem 166 $\chi(\alpha) = \chi(\omega_1^a) = (-1)^a$. Thus $\chi(\alpha) = +1$, if α is hyperprimary mod $\mathfrak{l}$; otherwise $\chi(\alpha) = -1$.

Theorem 175 is called the *second supplementary theorem*.

§63 A Property of Field Differents and the Hilbert Class Field of Relative Degree 2

In conclusion we wish to make two applications of the reciprocity law. The first deals with the ideal class to which the different $\mathfrak{d}$ of the field belongs.

Theorem 176. *The different $\mathfrak{d}$ of the field k is always equivalent to the square of an ideal in k.*

If we choose an integer ω in k, which is divisible by $\mathfrak{d}$ with the decomposition

$$\omega = \mathfrak{a}\mathfrak{d}, \qquad \mathfrak{a} \text{ odd,}$$

then, by Theorem 170, we need only show for the proof of our theorem that for each singular totally positive primary number ε, such that $(\varepsilon, \mathfrak{a}) = 1$, the residue symbol is $\left(\frac{\varepsilon}{\mathfrak{a}}\right) = +1$.

To prove this we go back to Formula (199) for Gauss sums and use Theorem 156 which determines the value of a sum that belongs to a square denominator. By (169) we decompose the sum $C(\frac{\varepsilon}{4\omega})$ belonging to the denominator $4\mathfrak{a}$, where $(\varepsilon, \mathfrak{a}) = 1$, into a sum with denominator 4 and a sum with denominator $\mathfrak{a}$, by introducing an odd auxiliary ideal $\mathfrak{c}$ such that

$$\mathfrak{a}\mathfrak{c} = \text{a number } \alpha, \qquad \gamma = \frac{\alpha}{\omega} = \frac{\mathfrak{c}}{\mathfrak{b}}.$$

Then by (169)

$$C\left(\frac{\varepsilon}{4\omega}\right) = C\left(\frac{\varepsilon\gamma}{4\alpha}\right) = C\left(\frac{4\varepsilon\gamma}{\alpha}\right) C\left(\frac{\alpha\varepsilon\gamma}{4}\right),$$

and if ε is primary, the right-hand side is

$$= \left(\frac{\varepsilon}{\mathfrak{a}}\right) C\left(\frac{\gamma}{\alpha}\right) C\left(\frac{\alpha\gamma}{4}\right).$$

In particular, it follows for $\varepsilon = 1$ that

$$C\left(\frac{1}{4\omega}\right) = C\left(\frac{\gamma}{\alpha}\right) C\left(\frac{\alpha\gamma}{4}\right)$$

and consequently

$$\left(\frac{\varepsilon}{\mathfrak{a}}\right) = \frac{C\left(\dfrac{\varepsilon}{4\omega}\right)}{C\left(\dfrac{1}{4\omega}\right)}. \tag{215}$$

We now apply reciprocity formula (199) to the last sums, by which these sums transform into sums with denominator ε, which can be determined directly by Theorem 156.

We obtain

$$\frac{C\left(\dfrac{\varepsilon}{4\omega}\right)}{|\sqrt{N(4\mathfrak{a})}|} = \left|\sqrt{N\left(\frac{2}{\varepsilon}\right)}\right| e^{(\pi i/4)S(\text{sgn } \omega\varepsilon)} C\left(-\frac{\gamma^2\omega}{\varepsilon}\right).$$

Likewise

$$\frac{C\left(\dfrac{1}{4\omega}\right)}{|\sqrt{N(4\mathfrak{a})}|} = |\sqrt{N(2)}| e^{(\pi i/4)S(\text{sgn } \omega)}.$$

Thus it follows from (215) that

$$\left(\frac{\varepsilon}{\mathfrak{a}}\right) = e^{(\pi i/4)S(\text{sgn } \omega\varepsilon - \text{sgn } \omega)} \frac{C\left(\dfrac{-\gamma^2\omega}{\varepsilon}\right)}{|\sqrt{N(\varepsilon)}|}$$

is valid for each primary number ε relatively prime to $\mathfrak{a}$. If we now assume that ε is also a singular number then, by Theorem 156, we obtain the value $|\sqrt{N(\varepsilon)}|$ for the sum $C(-\gamma^2\omega/\varepsilon)$. Consequently

$$\left(\frac{\varepsilon}{\mathfrak{a}}\right) = e^{(\pi i/4)S(\text{sgn } \omega\varepsilon - \text{sgn } \omega)} \quad \text{if } \omega = \mathfrak{a}\mathfrak{d}, \mathfrak{a} \text{ odd,}$$

and ε is a singular primary number, $(\varepsilon, \mathfrak{a}) = 1$.

Finally if, in addition, ε is totally positive, it follows that $\left(\frac{\varepsilon}{\mathfrak{a}}\right) = +1$ and, by Theorem 170, that $\mathfrak{a}$ as well as the different $\mathfrak{d}$ belongs to the principal class complex.

Since the differents of relative fields compose according to Theorem 111, it also follows from what has just been proved:

The relative different $\mathfrak{D}_k$ of a field K with respect to a subfield k is always equivalent to the square of an ideal in k.

Moreover, since the relative norm of $\mathfrak{D}_k$ is equal to the relative discriminant of K with respect to k, we see that the relative norm is also equivalent to a square in K. Thus we have shown

Theorem 177. If the ideal $\mathfrak{d}_k$ in k is the relative discriminant of a field with respect to k, then $\mathfrak{d}_k$ is equivalent to a square in k.

As a second application of the reciprocity theorem we wish to investigate the Hilbert class fields of k of relative degree 2. Following Hilbert we call a field unramified with respect to k if its relative discriminant is equal to 1. The unramified fields which are obtained by adjoining to k the square root of a number in k can then be specified, for, by Theorem 120, these fields arise by adjoining the square root of a singular primary number in k. However, the number of distinct complexes of singular primary numbers in k is equal to $2^{e_0} - 1$ by Theorem 172 (the square numbers are not to be considered as singular primary numbers).

Hence we have

Theorem 178. Relative to k there are exactly $2^{e_0} - 1$ distinct unramified fields of relative degree 2.

Accordingly, these fields are related to the ideal classes of k. If the class number, in the strict sense, of k is odd, then there is no unramified field of relative degree 2 at all. The connection with the ideal classes shows up even more clearly in the formulation of the decomposition theorem.

Theorem 179. Let ω be a singular primary number. Then there is a subgroup $\mathfrak{G}(\omega)$ of order $h_0/2$ in the group of the h_0 ideal classes in the strict sense such that a prime ideal $\mathfrak{p}$ splits in the field $K(\sqrt{\omega}, k)$ if and only if $\mathfrak{p}$ belongs to $\mathfrak{G}(\omega)$.

The set of odd ideals $\mathfrak{r}$, for which $Q(\omega, \mathfrak{r}) = +1$, determines a subgroup of order 2^{e_0-1} in the group of class complexes in the strict sense, by Theorem 172. Since each class complex consists of $h_0/2^{e_0}$ classes in the strict sense, the odd ideals $\mathfrak{r}$ with $Q(\omega, \mathfrak{r}) = +1$ are identical with the odd ideals which lie in the $h_0/2$ strict classes of this group $\mathfrak{G}(\omega)$.

Moreover, this also holds for the prime ideals $\mathfrak{l}$ which divide 2 since by Theorem 119 we have $Q(\omega, \mathfrak{l}) = +1$ for the splitting symbol defined in §60, if the odd number ω is congruent to the square of a number in k mod $\mathfrak{l}^{2c+1}$, where $\mathfrak{l}^c$ is the highest power of $\mathfrak{l}$ dividing 2. In the other case $Q(\omega, \mathfrak{l}) = -1$ for odd ω. Now, however, ω is primary and $\mathfrak{l}^{2c+1}$ and $4/\mathfrak{l}^{2c}$ are relatively prime; hence $Q(\omega, \mathfrak{l})$ is $= +1$ if and only if ω is a quadratic residue mod $4\mathfrak{l}$. However, by Theorem 175 only the ideal class to which $\mathfrak{l}$ belongs actually satisfies this condition. For if $\lambda = \mathfrak{l}\mathfrak{r}$ is totally positive and $\mathfrak{r}$ is odd, then ω is hyperprimary relative to $\mathfrak{l}$ if and only if $\left(\frac{\omega}{\mathfrak{r}}\right) = +1$.

Because of this close relation to the ideal classes, the fields $K(\sqrt{\omega}, k)$ are called the *class fields* of k.

In the manner in which we have laid the foundations for the theory of relatively quadratic fields, the reciprocity law appears as the first result; the existence of class fields appears as a consequence of this law. In the classical development of *Hilbert* and *Furtwängler* (also in the investigation of residues of higher powers) the train of thought runs in the reverse direction. First the existence of class fields is proved by another method which, by the way, is very complicated. Their connection with ideal classes is then discussed, and from this the reciprocity law is then derived. For this the so-called *Eisenstein* reciprocity law is an indispensible aid. One proceeds in this way in all cases which are concerned with fields of relative degree higher than 2. No transcendental functions have yet been discovered which, like the theta-functions of our theory, yield a reciprocity relation between the sums which occur for higher power residues in place of the Gauss sums. A new and very fruitful contribution which is related to that of Hilbert has been made by *Takagi*[3] who also has succeeded in gaining a complete overview of all relative fields of k, which are "relatively Abelian," that is, which have the same relation to k as cyclotomic fields do to $k(1)$.

[3] Uber eine Theorie des relativ-Abelschen Zahlkörpers, *Journal of the College of Science, Imperial University of Tokyo*, Vol. XLI (1920).

Chronological Table

Euclid (about 300 B.C.)
Diophantus (about 300 A.D.)
Fermat (1601–1665)
Euler (1707–1783)
Lagrange (1736–1813)
Legendre (1752–1833)
Fourier (1768–1830)
Gauss (1777–1855)
Cauchy (1789–1857)
Abel (1802–1829)
Jacobi (1804–1851)
Dirichlet (1805–1859)
Liouville (1809–1882)

Kummer (1810–1893)
Galois (1811–1832)
Hermite (1822–1901)
Eisenstein (1823–1852)
Kronecker (1823–1891)
Riemann (1826–1866)
Dedekind (1831–1916)
Bachmann (1837–1920)
Gordan (1837–1912)
H. Weber (1842–1913)
G. Cantor (1845–1918)
Hurwitz (1859–1919)
Minkowski (1864–1909)

References

The reader will find further expositions of the theory presented in this book in the following books:

P. Bachmann, *Allgemeine Arithmetik der Zahlkörper* (= *Zahlentheorie*, Vol. V). Leipzig 1905.
―――― *Die analytische Zahlentheorie* (= *Zahlentheorie*, Vol. II). Leipzig 1894.
―――― *Grundlehren der neueren Zahlentheorie*, 2nd edition. Berlin-Leipzig 1921.
―――― *Die Lehre von der Kreisteilung und ihre Beziehungen zur Zahlentheorie*. Leipzig 1872.
P. G. Lejeune-Dirichlet, *Vorlesungen über Zahlentheorie*, Herausgegeben und mit Zusätzen versehen von R. Dedekind, 4th edition. Braunschweig 1894.
R. Fueter, *Synthetische Zahlentheorie*. Leipzig 1917.
―――― Die Klassenkörper der komplexen Multiplikation und ihr Einfluss auf die Entwickelung der Zahlentheorie. (Bericht mit ausführlichem Literaturverzeichnis.) *Jahresbericht der Deutschen Mathematiker-Vereinigung*, Vol. 20 (1911).
K. Hensel, *Zahlentheorie*. Leipzig 1913.
D. Hilbert, Bericht über die Theorie der algebraischen Zahlkörper. *Jahresbericht der Deutschen Mathematiker-Vereinigung*, Vol. 4 (1897). Here one also finds references to the older literature.
L. Kronecker, *Vorlesungen über Zahlentheorie*, Herausgegeben von K. Hensel. Vol. 1. Leipzig 1913.
E. Landau, *Handbuch der Lehre von der Verteilung der Primzahlen*, Vols. 1, 2. Leipzig 1910.
―――― *Einführung in die elementare und analytische Theorie der algebraischen Zahlen und der Ideale*. Leipzig 1918.
H. Minkowski, *Diophantische Approximationen. Eine Einführung in die Zahlentheorie*. Leipzig 1907.
H. Weber, *Lehrbuch der Algebra*, Vols. 1–3, 2nd edition. Braunschweig 1899–1908. (Vol. 3 is entitled *Elliptische Funktionen und algebraische Zahlen*.)

Graduate Texts in Mathematics

Soft and hard cover editions are available for each volume up to Vol. 14, hard cover only from Vol. 15.